Astrophysics and Space Science Library

Astrophysics and Space Science Library

More information about this series at http://www.springer.com/series/5664

Loris Magnani • Steven N. Shore

A Dirty Window

Diffuse and Translucent Molecular Gas in the Interstellar Medium

Springer

Loris Magnani
Department of Physics and Astronomy
The University of Georgia
Athens
Georgia, USA

Steven N. Shore
Dipartimento di Fisica "E. Fermi"
Università di Pisa
Pisa, Italy

ISSN 0067-0057 ISSN 2214-7985 (electronic)
Astrophysics and Space Science Library
ISBN 978-3-662-54348-1 ISBN 978-3-662-54350-4 (eBook)
DOI 10.1007/978-3-662-54350-4

Library of Congress Control Number: 2017937359

Cover illustration: Optical image of the barred spiral galaxy NGC 7497 shining through the dust of the Galactic diffuse molecular cloud MBM 54. Credit: Adam Block/Mount Lemmon SkyCenter/University of Arizona.

Printed on acid-free paper

This Springer imprint is published by Springer Nature
The registered company is Springer-Verlag GmbH Germany
The registered company address is: Heidelberger Platz 3, 14197 Berlin, Germany

First, to Butler Burton who had the audacity to extend the invitation to write this book and the faith it would someday appear.

To the memory of Francesco Palla

a Gabriella, la sefirot che ha mutato le tenebre in luce.

And to Sandy, finally.

Preface

> *...and so there ain't nothing more to write about, and I am rotten glad of it, because if I'd a knowed what a trouble it was to make a book I wouldn't a tackled it, and ain't a-going to no more...*
>
> Mark Twain in *Huckleberry Finn*

Suspected but undetected only a bit more than a century ago, the matter between the stars in our Galaxy and others is a vast medium of extremes. Hendrik van de Hulst once described the interstellar medium as a place where an atom could sit for a million years and still act spontaneously. It is fuel for the birth of stars and the disposal site of their nucleosynthesis, where lengths are measured in parsecs or kiloparsecs yet whose energetics are governed on the scale of solar systems. Its kinetics are nearly collisionless yet its dynamical state is the result of large-scale ordering as if a fluid. Its density varies by five to six orders of magnitude in the diffuse phase and even more in the densest clouds. Its velocities range from diffusion to hypersonic, and its temperatures range from a few K to 10^7 K, from ices to plasmas.

During the last half-century, the interstellar medium has been intensively studied and its complexity increasingly appreciated. The launching of the *Planck* and *Herschel* orbiting observatories has lent a new, multi-wavelength approach to studies of the interstellar medium, and the astounding databases that are now available from radio waves to gamma-rays will keep astrophysicists busy for decades to come.

Most of the recent work on the interstellar medium has focused on star and planetary formation. Necessarily, these studies have concentrated on the denser parts of molecular clouds and cores where these objects form. However, both of us have spent a good portion of our careers thinking about the more diffuse molecular and atomic gas, where star formation plays no role. In light of this, we decided to share some of our thoughts and views on this somewhat neglected component of the Galaxy.

Our goal in this book is to present a broad view of the diffuse interstellar medium, with emphasis on diffuse and translucent molecular gas. The book is not intended to be a textbook or a lengthy review; rather, it's our point of view of the diffuse interstellar medium as our experiences led us to see it. For this reason, for example, we have chosen to not discuss the physics of shocks in the diffuse medium. Although they play an important role in the dynamics and energetics of the ISM, we concentrate on the neutral and molecular gas, for which they are less important. This book is intended for graduate students who are entering the field and postdocs who are switching fields. It is meant to be accessible to someone who does not necessarily have the astrophysical background. However, some prior knowledge of radiative transfer and the theory of turbulence (Chaps. 2 and 11) is required as our treatments of these topics presuppose some basic familiarity with them.

The seeds for this book were planted nearly 20 years ago from conversations between one of us (LM) and Dap Hartmann and Eugene de Geus. The original plan was to produce a book on high-latitude molecular clouds. For a number of reasons, that work was stillborn. The idea was resurrected about 5 years ago when the two of us decided, with encouragement from Butler Burton, to turn some of our research on high-latitude translucent clouds into a more complete treatment than was possible in a review article. Given the work we have done over the last 15 years (much of which with our colleague, Ted LaRosa), we decided to use the high-latitude molecular clouds as illustrations of many of the concepts we wished to discuss. For most of these objects, gravity does not dominate; instead, they show flows and are perhaps the best objects to study the original turbulence that gave rise to them, undisturbed by star formation and its effects.

We have learned so much from our colleagues and students over the years. To thank all of them would be impossible. We limit ourselves to mentioning those who interacted with us particularly on the topics of this book: Mike A'Hearn, Gianni Aiello, Joao Alves, Tom Bania, Giuseppe Bertin, Claude Bertout, John Black, Leo Blitz, Jan Brand, J.-P. Caillault, Bruce Elmegreen, Nye Evans, Edith Falgarone, Steve Federman, Daniele Galli, Riccardo Giovanelli, Dap Hartmann, Thomas Hartquist, Mike Hauser, Carl Heiles, Mark Heyer, Anthony Jones, Namir Kassim, Elizabeth Lada, Alex Lazarian, Mordecai-Mark Mac Low, Lee Mundy, Joe Onello, Jan Palous, Francesco Pegoraro, John Raymond, Tim Robishaw, Göran Sandell, John Scalo, Bob Silverberg, Mattia Sormani, Steve Stahler, Snezana Stanimirovic, Pat Thaddeus, Barry Turner, Enrique Vasquez-Semadeni, Fran Verter, Malcolm Walmsley, Jan Wouterloot, and, our dear friend and colleague, Ted LaRosa.

We thank our students who worked on the interstellar medium: Ivan de Gennaro Aquino, Matteo Cantiello, Alessandro Cilla, Francesco Costagliola, Fabio del Sordo, Lapo Fanciullo, Marco Monaci, Jeremy Gordon, Meredith McCarthy, Samantha Blair, Ray Chastain, Elizabeth Wennerstrom, David Cotten, Emmanuel Donate, Thom Hearty, and Allison Smith.

We thank Jan Brand, Tara Cotten, Renata Cumbee, Tom Dame, Lapo Fanciullo, Jeffrey Gritton, Jinhee Lee, Tim Robishaw, Robin Shelton, Allison Smith, Kyle Walker, and Ziwei Zhang for a critical reading with comments and suggestions for several chapters of the manuscript.

A special thanks to the late Lars E.B. Johansson, Magnus Thomasson, Henrik Olofsson and the staff of the Onsala Observatory, and Lucy Ziurys and the staff of the Arizona Radio Observatory for their generosity in allocating telescope time and for help with observations. We also acknowledge support from the EU-funded RadioNet Transnational Access program, and from the NASA Guest Observer programs for IUE and HST.

We have made use of data from the Sloan Digital Sky Survey: Funding for the SDSS and SDSS-II has been provided by the Alfred P. Sloan Foundation, the Participating Institutions, the National Science Foundation, the US Department of Energy, the National Aeronautics and Space Administration, the Japanese Monbukagakusho, the Max Planck Society, and the Higher Education Funding Council for England. The SDSS website is http://www.sdss.org/. The SDSS is managed by the Astrophysical Research Consortium for the Participating Institutions. The Participating Institutions are the American Museum of Natural History, Astrophysical Institute Potsdam, the University of Basel, the University of Cambridge, Case Western Reserve University, the University of Chicago, Drexel University, Fermilab, the Institute for Advanced Study, the Japan Participation Group, Johns Hopkins University, the Joint Institute for Nuclear Astrophysics, the Kavli Institute for Particle Astrophysics and Cosmology, the Korean Scientist Group, the Chinese Academy of Sciences (LAMOST), Los Alamos National Laboratory, the Max-Planck-Institute for Astronomy (MPIA), the Max-Planck-Institute for Astrophysics (MPA), New Mexico State University, Ohio State University, the University of Pittsburgh, the University of Portsmouth, Princeton University, the US Naval Observatory, and the University of Washington.

Some of the data presented in this work were obtained from the Mikulski Archive for Space Telescopes (MAST). STScI is operated by the Association of Universities for Research in Astronomy, Inc., under NASA contract NAS5-26555. Support for MAST for non-HST data is provided by the NASA Office of Space Science via grant NNX09AF08G and by other grants and contracts.

This publication makes use of data products from the Wide-Field Infrared Survey Explorer, which is a joint project of the University of California, Los Angeles, and the Jet Propulsion Laboratory/California Institute of Technology, funded by the National Aeronautics and Space Administration. We also made use of NASA's Astrophysics Data System.

SkyView has been developed with generous support from the NASA AISR and ADP programs (P.I. Thomas A. McGlynn) under the auspices of the High Energy Astrophysics Science Archive Research Center (HEASARC) at the NASA/GSFC Astrophysics Science Division.

And let's not forget our best friends past and present, the dogs: Lupo, Barkum, Thanatos, Laika, Astro, Buddy, Ella, Jerry, Kismet, Beau, Tony, Cody, Quillo, and Ricky.

Athens, GA, USA
Pisa, Italy
September 2016

Loris Magnani
Steven N. Shore

Contents

1 A Quick Look at the Diffuse Interstellar Medium 1
1.1 Introduction 1
1.2 Overview of the ISM and Its Role in Spiral Galaxies 2
1.2.1 A Remark on Physical Processes in the ISM 4
1.3 How Does the ISM Manifest Itself in a Galaxy Like the Milky Way? 5
1.4 Why Does It Break Up into Phases? 9
1.5 The Phases of the ISM: Observational Signatures 10
1.5.1 The Hot Ionized Medium 11
1.5.2 The Warm Ionized Medium 12
1.5.3 The Warm Neutral Medium 13
1.5.4 The Cold Neutral Medium 14
1.6 Molecular Gas: Why Some of It Is a Phase and Some Isn't 16
1.7 The Transition from Atomic to Molecular Gas 18
1.7.1 Photodissociation Regions: PDRs 18
1.7.2 Diffuse vs. Dark Clouds 20
1.8 The Role of Dust 21
1.9 Cosmic Rays 23
1.10 The Concept of a "Molecular Cloud" 24
References 25

2 Radiative Transfer Considerations 29
2.1 Introduction 29
2.2 The Transfer of Radiation Through the Interstellar Medium 30
2.2.1 Statistical Balance 32
2.2.2 Radiative Processes in the Rayleigh-Jeans Limit 34
2.2.3 The Line Profile Function 34
2.2.4 Obtaining Column Densities from Absorption Lines: Equivalent Width and Curve of Growth Methods 39
2.2.5 Obtaining Column Densities from Emission Lines 42

2.2.6 An Example: HI—The 21 Centimeter Line 43
2.2.7 Line Pumping by the Cosmic Background Radiation: The Optical CN Transitions.................... 47
2.2.8 Collisional Excitation in a Nearly Collisionless Medium 48
2.2.9 Dynamical and Nonlocal Complications to Radiative Transfer 51
2.2.10 Elemental Abundances and Depletion in the Diffuse Medium 57
2.2.11 H_2 and Fine Structure Transitions 60
2.2.12 Diffuse Interstellar Features in Emission and Absorption: Radiative Excitations and Fluorescence 62
2.3 Radio Continuum Measurements of the Diffuse Medium.......... 64
References.. 66

3 The Diffuse ISM from the Ground: Chemistry and Tracers 69
3.1 Introduction... 69
3.2 The First Detections of Interstellar Molecules...................... 70
3.3 Astrochemistry 72
3.3.1 General Considerations.................................... 73
3.3.2 The Oxygen Network....................................... 77
3.3.3 The Carbon Network 78
3.3.4 The Nitrogen Network 79
3.3.5 Cosmic Rays as Agents of Astrochemical Processing 81
3.4 Molecular Tracers of the Diffuse ISM in the Radio Regime 83
3.4.1 The CO Rotational Transitions............................ 84
3.4.2 The OH 18 cm Lines 85
3.4.3 The CH 3.3 GHz Lines 91
3.4.4 The H_2CO 4.8 GHz Line.................................. 92
3.4.5 The HCO^+(1-0) Transition 95
3.4.6 C_3H_2, C_2H, and C_3H^+ 96
3.4.7 HF, a Tracer of Molecular Hydrogen 97
3.5 Optical Manifestations of Diffuse Molecular Gas and Dust 97
3.5.1 CH, CH^+, CN 97
3.5.2 C_2.. 99
3.5.3 The Diffuse Interstellar Bands 99
3.5.4 Red Luminescence... 100
3.6 The Molecular Medium in a Cosmological Line of Sight: PKS 1830-211 102
References.. 103

4 Observing the Diffuse ISM: Making Sense of the Radio Observations 107
4.1 Introduction 107
4.2 Radio Observations: Practical Considerations 108
4.2.1 The Radiometer Equation 108
4.2.2 Frequency Switching vs. Position Switching 111
4.2.3 On-Off Radio Measurements 114
4.2.4 Polarization Considerations 115
4.2.5 Sidelobes and Stray Radiation 118
4.2.6 Converting from Antenna Temperature to Beam or Radiation Temperature 119
4.2.7 Interference 121
4.2.8 Mapping Techniques 123
4.3 Analysis of Radio Spectroscopic Data 124
4.3.1 Virial Masses from Spectral Maps 124
4.3.2 A Brief Mention of Statistical Tools 125
References 128

5 Observing the Diffuse ISM: The Space Missions 131
5.1 Introduction 131
5.2 The Radio Regime 132
5.2.1 The *Planck* Observatory 132
5.2.2 WMAP 137
5.3 Probing the Infrared Sky 139
5.3.1 IRAS 140
5.3.2 COBE and ISO 141
5.3.3 Spitzer, Herschel, and WISE 143
5.4 Spectral Tracers of Diffuse Molecular Gas in the Far-Infrared 145
5.5 Probing the Ultraviolet Sky 146
5.5.1 IUE and HST 147
5.5.2 FUSE 148
5.6 Probing the X-Ray Sky 149
5.7 Probing the Gamma-Ray Sky 150
5.7.1 How Gamma Rays Trace Molecular Gas 151
References 152

6 Observing in the Dark: The Dust-Gas Connection 155
6.1 Introduction 155
6.2 Extinction and Color Excess 156
6.2.1 A Bit of Grain Optics 159
6.3 The Correlation Between Gas and Dust 162
6.4 Diffuse Reflection Nebulae 162
6.5 Diffuse Galactic Ultraviolet Emission 163
6.6 Diffuse Galactic Infrared Emission 164
6.6.1 Light Echoes, Halos, and Reflection Nebulae 165
6.6.2 Polarization 168

6.6.3 Anomalous Microwave Emission and Spinning Grains 169
6.6.4 Grain Size Distributions 171
6.7 Tracing Molecular Gas Using Dust Maps 172
6.7.1 The Schlegel, Finkbeiner, and Davis (SFD) Dust Maps 173
6.8 The E(B-V) Molecular Threshold 174
6.9 Ices and Their Signatures 177
6.9.1 Other Sources of Information: Ices in the Solar System and Laboratory 178
6.10 Gas to Dust Ratio from Cosmic Ray-Produced γ-Rays 178
6.10.1 Extinction from Hydrogen Lyman Continuum Absorption Measurements 180
6.10.2 Time Variability of the Diffuse Gas 180
References 181

7 Surveying the Molecular Milky Way 185
7.1 Introduction 185
7.2 The CO Surveys Along the Galactic Plane 186
7.2.1 Early Results 186
7.2.2 Recent Surveys 188
7.3 Giant Molecular Clouds 190
7.4 The Smaller Molecular Clouds 192
7.4.1 Dark Clouds 193
7.4.2 Translucent Molecular Clouds 196
7.4.3 Diffuse Molecular Clouds 199
7.5 What Is a Molecular Cloud? 201
References 201

8 The Relationship Between CO and H_2 205
8.1 Introduction 205
8.2 The CO-H_2 Conversion Factor 206
8.2.1 X_{CO} from Extinction 207
8.2.2 X_{CO} Using ^{13}CO(1-0) 208
8.2.3 X_{CO} from Virial Equilibrium Considerations 208
8.2.4 X_{CO} from Gamma-Rays 210
8.2.5 X_{CO} from Far-Infrared Emission 211
8.2.6 X_{CO} Using the CH 3335 MHz Line 212
8.3 Do the Galactic CO Surveys Trace Most of the Molecular Gas? 214
8.4 On the Question of Dark Gas 217
8.4.1 Sensitive CO(1-0) Observations 219
8.4.2 OH as the Best Tracer of Dark Gas? 221
References 223

9 Surveys for High-Latitude Molecular Clouds 227
9.1 Introduction 227
9.2 The Initial Searches 228
9.3 The MBM Survey 231
9.4 The IRAS Cirrus: High Latitude Cloud Connection 235
9.5 Post-MBM CO Surveys at High Latitudes 237
9.6 The Georgia-Harvard CfA High Latitude Survey 241
9.7 High-Latitude Molecular Gas via Infrared Techniques 246
References 247

10 Distances 249
10.1 Introduction 249
10.2 Distances to the Clouds: Galactic Rotation Curve 250
10.3 Distances to the Clouds: Statistical 250
10.3.1 A Practical Example: High-Latitude Molecular Clouds 253
10.4 Distances to the Clouds: Direct Measurements 255
10.4.1 The Method of Wolf Diagrams 255
10.4.2 Distances from Spectroscopy 258
10.4.3 The Method of Photometry 259
10.5 Asymmetries in the Distribution of High-Latitude Molecular Clouds 261
10.6 Molecular Clouds in the Halo? 263
10.7 The Southern Extension of the Taurus Dark Clouds 264
References 265

11 Dynamical Considerations: Instabilities and Turbulence 267
11.1 Introduction 267
11.2 Virial Theorem and Stability 268
11.3 Reprise: What Is a Cloud? 271
11.3.1 Isothermal Cloud Models 271
11.3.2 Gravothermal Instability 273
11.4 Thermal Instability 275
11.5 Turbulence 277
11.5.1 Introduction 277
11.5.2 Dynamical Separations 279
11.5.3 The Source Scale and the Cascade 280
11.5.4 Statistical Methods for Dynamical Analyses 285
11.5.5 Empirical Line Width-Size Relations 292
11.5.6 Some Open Questions 293
References 293

12 Coda 295
12.1 What Is a Molecular Cloud? 295
Reference 296

A Regulatory Institutions for Radio Frequency Interference (Chap. 4) 297

B Radio Band Designations (Chap. 4) 299

C The Kramers-Kronig Relations and the Relation Between C_{abs} and C_{sca} Through Reciprocity (Chap. 6) 301

Index 303

Acronyms

ACF	Autocorrelation Function
AU	Astronomical Unit
CF	Correlation Function
CGRO	*Compton* Gamma-Ray Observatory
CMB	Cosmic Microwave Background
CMM	Cold Molecular Medium
CNM	Cold Neutral Medium
COBE	Cosmic Background Explorer
COS	Cosmic Origins Spectrograph
COSPAR	Committee on Space Research
CR	Cosmic Rays
DIB	Diffuse Interstellar Bands
DIG	Diffuse Ionized Gas
DIRBE	Diffuse Infrared Background Experiment
DIRF	Diffuse Interstellar Radiation Field
DM	Dispersion Measure
DSS	Digitized Sky Survey
EGRET	Energetic Gamma-Ray Experiment Telescope—CGRO
ELDM	Extended Low-Density Medium
EM	Emission Measure
ERE	Extended Red Emission
ESA	European Space Agency
ESO	European Southern Observatory
FAST	Five Hundred Meter Aperture Spherical Telescope
FCC	Federal Communications Commission
FCRAO	Five College Radio Astronomy Observatory
FUSE	Far Ultraviolet Spectroscopic Explorer
FUV	Far Ultraviolet
FWHM	Full Width Half Maximum
GALEX	Galaxy Evolution Explorer
GBT	Green Bank Telescope

GC	Galactic Center
GHRS	Goddard High Resolution Spectrograph
GMC	Giant Molecular Cloud
GRO	Gamma-Ray Observatory
HIM	Hot Ionized Medium
HRI	High Resolution Imager - ROSAT
HST	Hubble Space Telescope
IAU	International Astronomical Union
IP	Ionization Potential
IRAC	Infrared Array Camera - *Spitzer*
IRAS	Infrared Astronomy Satellite
IUE	International Ultraviolet Explorer
ISM	Interstellar Medium
ISRF	Interstellar Radiation Field
KAO	Kuiper Airborne Observatory
KKR	Kramers-Kronig relations
LAB	Leiden-Argentina-Bonn HI survey
LAT	Large Area Telescope—*Fermi*
LCP	Left Circular Polarization
LSR	Local Standard of Rest
LVG	Large Velocity Gradient
MAST	Mikulski Archive for Space Telescope
MBM	Magnani, Blitz, and Mundy—when followed by a number, a cloud from catalog in paper by Magnani, Blitz, and Mundy (1985) ApJ, 295, 402
MCMC	Markov chain Monte Carlo
MWO	Millimeter Wave Observatory
NASA	National Aeronautics and Space Administration
NIST	National Institute of Standards and Technology
NGH	North Galactic Hemisphere
NRAO	National Radio Astronomy Observatory
NSF	National Science Foundation
OTF	On-the-Fly Mapping Technique
PAHs	Polycyclic Aromatic Hydrocarbons
PDF	Probability Distribution Function
PDRs	Photo-dissociation Regions—sometimes, Photon-Dominated Regions
POSS	Palomar Observatory Sky Survey
PSF	Point Spread Function
PSPC	Position Sensitive Proportional Counter—ROSAT
RCP	Right Circular Polarization
RFI	Radio-Frequency Interference
RGS	Reflection Grating Spectrometer—ROSAT
RM	Rotation Measure
RRLs	Radio Recombination Lines
SDSS	Sloan Digital Sky Survey

SGH	South Galactic Hemisphere
SKA	Square Kilometer Array
STIS	Space Telescope Imaging Spectrograph
UNESCO	United Nations Educational, Scientific and Cultural Organization
URSI	Union Radio-Scientifique Internationale
VLBI	Very Long Baseline Interferometry
VT	Virial Theorem
WHAM	Wisconsin H-Alpha Mapper
WIM	Warm Ionized Medium
WISE	Wide-Field Infrared Survey Explorer
WMAP	Wilkinson Microwave Anisotropy Probe
WNM	Warm Neutral Medium
VUV	Vacuum Ultraviolet

Chapter 1
A Quick Look at the Diffuse Interstellar Medium

It is clear to everyone that astronomy at all events compels the soul to look upwards, and draws it from the things of this world to the other.

—Plato

Abstract To introduce our subject, we provide an overview of the interstellar medium (ISM) in our Galaxy and its relation to the stellar component. We then discuss the phase structure of the ISM, with a special emphasis on why the hydrogen takes the different forms it does. We then outline the observational signatures of the Hot Ionized, Warm Ionized, Warm Neutral, and Cold Neutral media, followed by a brief introduction to the molecular component of the Galaxy. A description of Photo-Dissociation Regions is presented with emphasis on the transition region from atomic to diffuse molecular gas. The role of dust and cosmic rays is also briefly described. Finally, we discuss the concept of a molecular cloud, a central theme of this book.

1.1 Introduction

The interstellar medium or ISM is literally "the stuff between the stars". Although stars dominate the baryonic component of a galaxy like the Milky Way, the gas and dust play a key role in their life cycle and in the evolution of galaxies. The ISM in spiral galaxies comprises primarily hydrogen and helium gas with a dust component that has less mass than the gas by a factor of 100, and a cosmic ray component that has an energy density similar to that of starlight.

In this book, we focus primarily on the lower density molecular component of the ISM, i.e., those regions where the volume density, n, is less than or about 10^3 particles cm^{-3}. In particular, we will examine the low-density, low-extinction molecular gas contained in the so-called diffuse and translucent molecular clouds and in the envelopes or interclump regions of Giant Molecular Clouds (GMCs). In this regime "molecular clouds" are often not very molecular, and so we will also discuss their atomic component. We will refer to these objects as the molecular part of the diffuse ISM and mostly ignore the ionized and very hot gas, even though these components are clearly diffuse. The molecular portion of the diffuse ISM does not play a direct role in the star formation process, but it provides the canvas against

L. Magnani, S.N. Shore, *A Dirty Window*, Astrophysics and Space Science Library 442, DOI 10.1007/978-3-662-54350-4_1

which the proto-star forming cores are projected or embedded. Moreover, it does play a key role in the formation of dark molecular clouds and GMCs, the sites of all star formation in the Galaxy.[1]

To begin this exploration of the diffuse ISM, we will take a quick look at the ISM as a whole using the unifying concept of the Star-Gas Cycle. We then briefly discuss how the *atomic* gas component (including both neutral and ionized species) distinguishes itself by breaking itself up into distinct "phases" which have specific combinations of density and temperature. The basic reasons why this happens are discussed and then the observational signatures of the various phases are examined. We then take a quick look at the molecular component and try to put it in the context of the atomic phases by examining the role self-gravity and radiative self-shielding play in determining what form is taken by the molecular gas. Another unifying concept, the Photo-Dissociation Region or PDR (sometimes also known as a Photon-Dominated Region) is covered in some detail. The dust component is addressed next and then cosmic rays which, though very important in their own right, mostly play an indirect role in the diffuse ISM. We end this introductory chapter with a re-examination of the concept of a "molecular cloud".

1.2 Overview of the ISM and Its Role in Spiral Galaxies

The gas in the ISM assumes several distinct, long-lived configurations of volume density ranging from 10^{-3} cm^{-3} to 10^{7} cm $^{-3}$ and temperature ranging from 10 K to 10^{7} K. With this wide range of temperatures and densities, the structures formed are very different in distribution, occupied galactic volume, morphology, relevant physical processes, and interplay with the stellar component. These density and temperature configurations or phases are descriptively called the Hot Ionized Medium (HIM), the Warm Ionized Medium (WIM), the Warm Neutral Medium (WNM), and the Cold Neutral Medium (CNM) (Kulkarni and Heiles 1987; Shull 1987). The latter component is the most dense with volume densities in the range $1 - 10^{2}$ cm^{-3} and temperatures of order 10^{2} K. The Cold Neutral Medium (CNM) is atomic, but an even denser and colder component of the ISM is mostly molecular and is considered separately because, unlike the atomic phases, self-gravity is a key feature in its dynamics. Dense concentrations of cold atomic and molecular gas are readily detected via centimeter- and millimeter-wave spectral techniques and are whimsically called clouds. It is from these predominantly molecular entities that stars form.

To determine the relative importance of the various gaseous components of the Galaxy, it is useful to inventory them in a typical location such as the Solar neighborhood (i.e., away from unique regions such as the Galactic Center or the

[1]We will follow the standard astronomy convention that, when referring to the Milky Way, the word *Galaxy* be capitalized.

outermost regions of the Galaxy). If the mass surface density is defined to be the total mass of material contained in a column whose footprint is a square parsec perpendicular to the Galactic plane, i.e., along the z-direction, then the stellar mass surface density, Σ_*, of the Galaxy at the Solar Circle ($\sim$ 8 kpc from the center) is about 50 $M_\odot$ pc^{-2}, Σ_{HI},[2] is 5 $M_\odot$ pc^{-2}, and Σ_{H_2} is 1.3 $M_\odot$ pc^{-2} (Gould 1996; Liszt 1983; Dame et al. 1987). These values are typical of the solar vicinity; it is important to realize that the mass surface densities change significantly from the Galactic Center to the Outer Galaxy. For instance, in the inner 200 pc of the Galaxy, Σ_{H_2} increases to several hundred $M_\odot$ pc^{-2} (Morris and Serabyn 1996), while at distances greater than 18 kpc from the center Σ_{H_2} is likely to be less than 0.01 $M_\odot$ pc^{-2} (Digel et al. 1994).

In spiral galaxies the ISM can make up to 20% of baryonic mass and there is a constant interplay between the stars and the gas and dust. The stellar component dominates the baryonic mass, contributing 10^{11} $M_\odot$ in galaxies like the Milky Way. The mass in gas is lower by an order of magnitude from the stellar mass and the dust is down two orders of magnitude from the gas.

The interplay between the stars and the ISM is often referred to as the "star-gas cycle" or the baryonic cycle and is schematically depicted in Fig. 1.1. All stars, regardless of mass, form in molecular clouds. As the stars evolve they return gas to the ISM in the form of stellar winds. At the end of their lives, high-mass stars explode in supernova explosions, leaving behind supernova remnants. Low-mass stars return a significant amount of material in gentler fashion in the form of planetary nebulae. The dynamics of these mass exchanges are important because they involve supersonic processes which drive shocks and produce large cavities in the ISM. Eventually, the hot gas from the stellar winds, supernova remnants, and planetary nebulae cools, forming cold, neutral hydrogen clouds and a warmer, distributed, neutral hydrogen layer. A fraction of the CNM becomes molecular perhaps under the mediation of gravity, turbulence, and/or magnetic fields. Once molecular clouds form, the denser portions form stars and the process continues. Although the star-gas cycle is a superb organizing concept for understanding the interplay of stars and the ISM, it does not do justice to some important components of the ISM. Magnetic fields play an important role in structuring the ISM but are present only implicitly in the diagram. Similarly, cosmic rays, another important constituent of a galaxy, are accelerated by supernova explosions, but do not appear in the diagram. Finally, the role of the hot ionized gas, though present in the cycle, is probably not given its due as far as the global structure of a galaxy and its interactions with the exo-galactic environment are concerned. Despite these shortcomings, the star-gas cycle is critical to understanding the evolution of galaxies.

[2]HI is the traditional astronomical way to refer to neutral hydrogen; ionized hydrogen is not typically referenced to as H^+, but, rather, HII. This convention continues with other neutral and ionized atomic species.

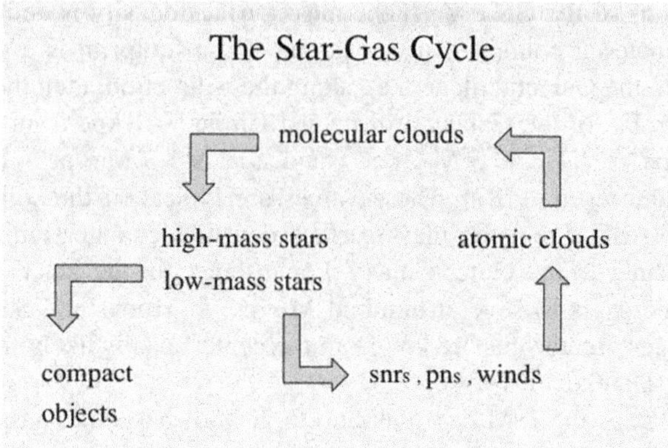

Fig. 1.1 The star-gas cycle in a galaxy. SNRs are supernova remnants; PNs are planetary nebula. These objects recycle stellar material into the ISM by way of their ejecta. "Winds" represent the recycled material produced by stellar winds during the normal lifetime of stars

1.2.1 A Remark on Physical Processes in the ISM

The galactic component that we will examine in this book is not as far from human experience as sometimes supposed. We actually come rather close to its densest parts when examining the ionosphere and exosphere of the Earth. Aside from the role of the planet's gravity in maintaining mechanical balance on large scale, the densities (from a few to $10^6\,\mathrm{cm}^{-3}$), the highly (but not completely) ionized state of the gas, and the presence of high energy particles in the form of precipitating electrons and protons from the magnetosphere, all resemble conditions we will encounter in the ISM. Further, the thermal state of the gas is governed by an external irradiating and ionizing source, at large distance (1 AU in this case is considerable compared to $1\,R_\oplus$). The incident solar ultraviolet and X-ray photons are heating agents along with cosmic ray ionization, while recombination and collisional de-excitations cool the gas. It is optically thin to visible light, completely opaque to the Lyman continuum,[3] and completely transparent to radio waves above a few tens of MHz.[4] In short, this terrestrial gas is a close match to the diffuse ISM; they share many of the properties that make the ISM especially difficult to treat.

One of the most important distinctions between, say, a stellar atmosphere and these rarified environments is the role of collisions. When we use the word "temperature" to discuss the properties of the gas, note that it has many different meanings and should never be automatically taken to mean kinetic energy of a Maxwellian distribution of particles. We will return to this point in the following chapters.

[3]The continuous range of wavelengths less than the Lyman limit (912 Å).

[4]In the millimeter-wave regime there are transparent windows, but also very opaque ones.

1.3 How Does the ISM Manifest Itself in a Galaxy Like the Milky Way?

Even a quick look at the Milky Way on a dark, moonless night reveals dark patches in the stellar distribution such as the Aquila Rift or the Coalsack. In the nineteenth and early twentieth centuries there was spirited debate concerning why certain regions of the sky were devoid of stars. The two principal points of view consisted of postulating the existence of some obscuring medium between us and the background stars or that the dark patches were just "holes" in the heavens, that is, regions where the stellar density dropped precipitously.[5] The answer turned out to be obscuration by dust and the ISM became an established part of the Galaxy. After the early work by Barnard and others, the availability of the plates of the Palomar Observatory Sky Survey (POSS) allowed Beverly Lynds (1962) to produce a catalog of so-called "dark clouds" which were later established to be molecular clouds. In the interim, the widespread distribution of cold atomic gas was established by the detection of the 21 cm spin-flip transition of neutral hydrogen—dubbed HI by astronomers (Ewen and Purcell 1951; Muller and Oort 1951). Absorption-emission observations (Radhakrishnan et al. 1972) implied that a portion of the 21 cm emission arose from a warmer, less dense, atomic component of the ISM and the theoretical idea of "phases"—regions of HI at different temperatures and densities, but roughly in pressure equilibrium—gained observational confirmation. The theory behind the phases is discussed in the next section but, besides dust and the gas, radio spectroscopic observations from the late 1960s and early 1970s showed a widespread molecular component along the Galactic plane. The key development in this area was the detection of the $^{12}C^{16}O$ ground state rotational transition ($J = 1 \rightarrow 0$) in 1970 at 115.271 GHz (Wilson et al. 1970). We will refer to this transition as just CO(1-0) throughout the rest of the book unless clarity dictates otherwise. Carbon monoxide is the second most abundant molecule in the ISM after H_2 and the CO(1-0) line almost immediately became the workhorse tracer of molecular gas in the Milky Way. Surveys of this transition along the Galactic plane established the broad properties of the molecular emission of the Galaxy (Stark 1979; Burton and Gordon 1978; Dame 1983; Sanders et al. 1984). By the early 1980s, ground based observations in the optical and radio regimes had established the existence of a pervasive ISM in the Milky Way consisting of well-mixed regions of gas and dust, with a widespread atomic component that was both warm and cold, and a denser molecular component which was localized in entities named "molecular clouds". This latter component coincided with regions of star formation and the paradigm that stars came exclusively from collapsing clumps in molecular clouds was established.

[5] The history of this debate is nicely told in Gerrit Verschuur's book *The Invisible Universe Revealed* (Verschuur 1989) and in Lyman Spitzer Jr's *Searching Between the Stars* (Spitzer 1982).

The multiwavelength extension of the electromagnetic spectrum during the 1980s and 1990s and improved instrumentation enabled new forms of the ISM to be identified. Sensitive Hα observations with a Fabry-Perot spectrometer allowed the detection of a widespread warm hydrogen component that was ionized (Reynolds 1983). This differed from the well-known HII regions such as the Orion Nebula because the medium was at much lower densities and was distributed throughout the sky. Interestingly, its existence had long been suspected by pulsar astronomers who had attributed pulsar dispersion measures to a significant ionized component of the ISM.

A major breakthrough in directly imaging the dust component of the ISM came from the Infrared Astronomy Satellite (IRAS) launched in 1983 (Neugebauer et al. 1984). IRAS imaged the whole sky at 12, 25, 60, and 100 μm and revealed the thermal continuum emission from the dust component. Figure 1.2 shows the 100 μm emission over the whole sky. In addition to the extensive dust distribution along the Galactic plane, a wispy, low-level emission was detected at high Galactic latitudes ($|b| \geq 20°$) and dubbed the IRAS "cirrus" (Low et al. 1984).

It is not an exaggeration to say that the IRAS images revolutionized our understanding of the ISM because even low levels of dust were now visible and dust-gas comparisons could be made with much greater precision. While IRAS traced infrared emission in the mid- and far-infrared, the Two Micron All Sky Survey (2MASS) project imaged the sky in the near-infrared (1.24, 1.63, and 2.19 μm) from the ground and could trace dust *absorbing* background starlight (see Fig. 1.3).

Ultraviolet satellites (*Copernicus*, IUE, HST, FUSE, and GALEX) have revealed widespread spectral lines of OVI. X-ray satellites detected emission from more highly ionized species and the diffuse x-ray background, so the hottest component of the ISM could be traced. Although found primarily in the halo of the Galaxy,

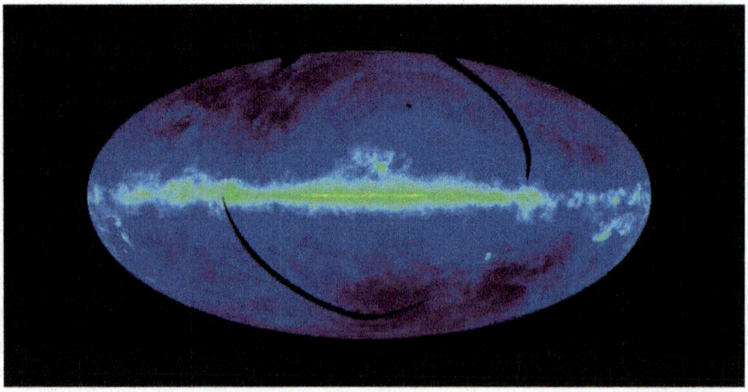

Fig. 1.2 IRAS 100 μm emission from the whole sky at 2′ resolution in an Aitoff projection in Galactic coordinates centered on the Galactic Center. The emission along the Galactic plane is primarily from thermal emission by dust particles in the ISM. The units are in MJy/ster and range from -1.68×10^{-8} to 1.43×10^{4} on a log scale. The *black gaps* in map represent regions where no data were taken. The image was made with the Skyview Virtual Observatory

Fig. 1.3 Three color (J, H, and K) 2MASS map of the whole sky in an Aitoff projection centered on the Galactic Center. The Galactic plane is clearly noticeable but, unlike Fig. 1.2, the infrared emission in the near infrared (1.24, 1.63. and 2.19 μm) is primarily from stellar photospheres so that the Galactic Bulge is evident. The dust component is present in this data in *absorption* as dark regions rather than emission as in the IRAS 100 μm data. The Large and Small Magellanic Clouds are prominent in the lower right-hand side of the image. Atlas image mosaic obtained as part of the Two Micron All Sky Survey (2MASS), a joint project of the University of Massachusetts and the Infrared Processing and Analysis Center/California Institute of Technology, funded by NASA and the NSF

large cavities produced by supernovae seemed to permeate the ISM in the disk and the question of the morphology of the ISM vis-a-vis the cold, warm, and hot hydrogen components began to be debated with inconclusive results (see review by Cox 2005). The Galactic ISM can be seen even in gamma rays. Figure 1.4 shows an EGRET whole-sky image at energies > 100 MeV. The plane of the Galaxy is easily visible with the emission produced as cosmic rays collide with hydrogen nucleons and produce intermediate particles which decay and produce gamma-rays.

Although designed to study the Cosmic Microwave Background radiation (CMB), the WMAP satellite mapped the whole sky at 22, 30, 40, 60, and 90 GHz. Radio continuum emission from the Galaxy is clearly noticeable at all wavelengths (e.g., see Fig. 1.5) and has to be removed to allow study of the CMB.

More recently, the Spitzer Space Telescope and the Herschel Space Observatory have again revolutionized our view of the infrared sky. The *Planck* satellite, following on the heels of COBE and WMAP is giving new insights into the molecular distribution of the ISM in addition to its primary mission of studying the CMB.

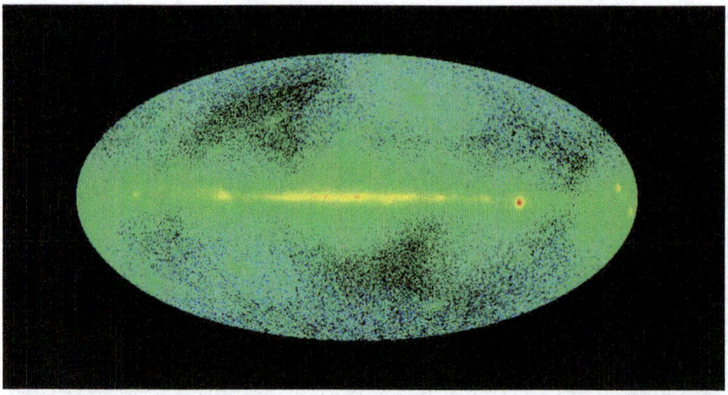

Fig. 1.4 EGRET gamma-ray emission at > 100 MeV from the whole sky in an Aitoff projection. The Galactic plane is clearly noticeable; this emission arises from cosmic rays colliding with hydrogen nucleons. Units are in counts/s/cm^2/steradian and depicted on a log scale from 0 to 0.00422. The image was made with the Skyview Virtual Observatory

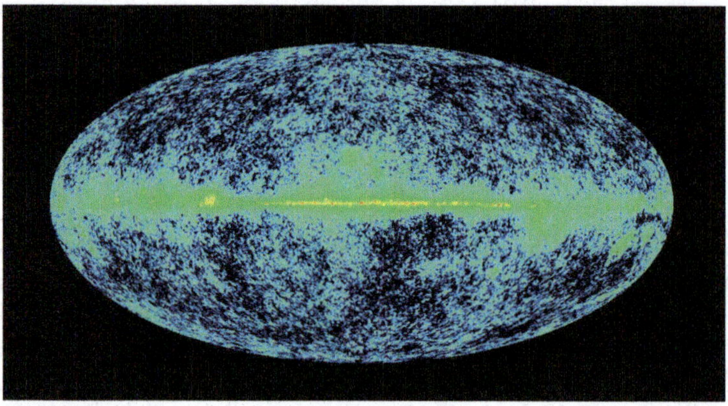

Fig. 1.5 WMAP image of the whole sky in an Aitoff projection at 0.88° resolution at Q band (centered at 40 GHz—see Appendix B). Non-thermal emission from the Galactic plane is evident and the background fluctuations away from the plane are from the CBR. The units of the image are in milliKelvin and depicted on a log scale from −0.378 to 90.578. The image was made with the Skyview Virtual Observatory

With all this data available, the ISM can be studied in unprecedented detail across the electromagnetic spectrum. We have seen that there is a molecular, an atomic neutral, and an ionized component with the latter two taking many different temperature and density configurations known as phases. A basic question immediately arises: Why does the atomic and ionized component of the ISM break up into different phases? We turn to that question in the next section.

1.4 Why Does It Break Up into Phases?

As an initial approximation, the various distinct phases of the ISM are thought to be in thermal and pressure equilibrium and are produced when heating and cooling mechanisms in a region equilibrate at a particular temperature. A generalized energy loss function, $\mathscr{L}$ (*cgs* units: erg s^{-1} cm^{-3}), is defined as the difference between the gas cooling function, Λ, and the gas heating rate, Γ. These functions depend on the composition and physical conditions of the medium (n and T). If the density dependence of the heating and cooling functions is included explicitly, $\mathscr{L}$ can be written as

$$\mathscr{L} = n^2 \Lambda - n\Gamma \quad . \tag{1.1}$$

For values of the loss function greater than 0 the gas is cooling down, for values less than 0 the gas is heating up, and for $\mathscr{L} = 0$, the locus of the resulting points in the n-T plane defines thermal equilibrium.

In the low-density limit we can write the condition for thermal equilibrium as

$$\Gamma/(nkT) = \Lambda/(kT) \quad . \tag{1.2}$$

If heating is produced by cosmic rays and starlight, then at constant pressure Γ/P is constant and can have specific values of n and T for a given P. For typical values of Γ and P, there are four equilibrium points along the curve $\mathscr{L} = 0$ (see, e.g., Shull 1987, p. 225). Two of these are stable points and two are unstable (i.e., small perturbations in n and T will drive the gas parameters away from the original values). The criteria for determining stable vs. unstable equilibrium points is determined by examining stability criteria such as those derived by Field (1965) or Balbus (1986). A plot of Λ vs. T is known as the cooling curve (see, e.g., Sutherland and Dopita 1993) and the various stability criteria predict instability for temperature regimes where the cooling curve is steeply rising. The cooling curve as a function of T can be calculated if the abundance of the various elements in a region of the ISM is known.

The low-density ISM is heated primarily by stellar photons ejecting electrons from grains while the cooling occurs via radiation losses. For regions with densities around 1–10 cm^{-3} excitation of far-infrared fine-structure levels of ionized carbon (CII) dominates the cooling and produces a long-lived, stable phase with equilibrium temperatures $\sim$ 100 K, hence the name Cold Neutral Medium (CNM).[6] The Warm Neutral Medium (WNM) exists at lower densities (10^{-1} cm^{-3}) and the cooling from neutral oxygen and the neutral hydrogen Lyman series produces theoretical equilibrium temperatures close to 10,000 K. However, Kulkarni and Heiles (1987) and Heiles and Troland (2003) find that about half of the WNM gas has temperatures

[6] Although C^+ is present, this gas is overwhelmingly neutral.

of only a few thousand K. Gas at temperatures in this range is thermally unstable and so the simple physics described above does not seem to tell the whole story. Nevertheless, the assumption of stable phases in the ISM provides some insight into what is occurring there. The earliest models of the ISM involved the CNM and WNM and were known as "two-phase" models (Field et al. 1969). Cox and Smith (1974) proposed that supernova explosions radically change this picture because the remnants cool slowly, overlap, and produce a network of hot tunnels in the ISM. A few years later, McKee and Ostriker (1977) developed a supernova remnant-dominated model for the ISM consisting of three phases, the CNM, the WNM, and a hot (10^6 K), low-density (10^{-3} cm^{-3}) phase, the Hot Ionized Medium (HIM). Because of the long cooling times, the HIM dominates the volume of the galaxy in this model even though most of the gas mass is concentrated in the CNM. The HIM is therefore a background medium in which the other phases are embedded. A key component of the theory was the idea that mass is exchanged between the hot and cold phases of the ISM through the evaporation of cold gas when it comes into contact with the HIM, and the formation of dense, cold shells as supernova remnants expanded and cooled. McKee and Ostriker also predicted the existence of a low-density widespread ionized component which become known as the Warm Ionized Medium (WIM). When confronted with observations these theoretical ideas yielded some puzzling failures (see review by Cox 2005). Four decades after the seminal work by Field and Habing, the detailed nature of the ISM is still controversial as far as the morphology, mass, surface density with Galactic location, and interplay of the phases.

1.5 The Phases of the ISM: Observational Signatures

Given the vast range in temperature and density of the gas in the ISM it is not surprising that very different techniques are needed to probe the various phases. As a start, the temperature of the particular phase dictates in which region of the electromagnetic spectrum the gas is best observed. Thus, the hottest phase (the HIM) is detected in x-rays and by UV radiation. The coldest phase (the CNM and the molecular component) are best observed in the radio and infrared portions of the spectrum. The Warm Neutral Medium (WNM) presents some interesting problems as far as its direct detection, but the main probe is the HI 21 cm line. Finally, the Warm Ionized Medium (WIM) is observed by direct observations of the Hα line. We examine the techniques for observing and studying each of the phases in detail below.

1.5.1 The Hot Ionized Medium

Spitzer (1956) predicted the possibility that a hot phase of the ISM should be present surrounding the disk of the Galaxy. He reasoned that a hot, low-density ambient medium is required to maintain the neutral, high-velocity, HI clouds that had been detected far from the disk in what we now call the "halo" of the Galaxy. Once formed, this phase should be long-lived because its low density makes cooling difficult. It took nearly two decades before techniques became available to detect this gas. The advent of UV and x-ray detector technology on board satellites orbiting above the Earth's atmosphere finally revealed the presence of a pervasive, low-density ($n \sim 10^{-3}\,\mathrm{cm}^{-3}$), high-temperature (T $\sim$ 0.3–2 x 10^6 K) medium surrounding the disk (Bowyer et al. 1968; Jenkins and Meloy 1974; York 1974). Gas at these temperatures is buoyant in the Galactic gravitational field and has a thermal scale height of several kpc. It is likely produced by supernova explosions and stellar winds from early-type stars heating and accelerating cold or warm HI by shocks. The gas is thought to exist in large, long-lived cavities (because of the relatively long cooling times) distributed throughout the disk (McKee and Ostriker 1977). Some of these cavities vent into the lower halo and the buoyant gas escapes the disk (see review by Cox 2005). Eventually, the gas cools and likely falls back onto the disk as in the "Galactic fountain" models of Shapiro and Field (1976) and Bregman (1980), but the constant activity within the disk ensures that the hot phase always blankets the disk. The exact morphology and distribution of the HIM in the disk is poorly known although its filling factor is thought to be as high as 0.5. The hot cavities and bubbles certainly exist (we live in one known as the Local Bubble), and many different cartoon models for the distribution of this gas have been proposed and are more or less plausible (see Savage 1987; Cox 2005; Shelton 2009). It is worth noting that additional sources of hot gas are required to account for the observed x-ray brightness, particularly at 0.7–0.8 keV. The currently considered source is infall from the intergalactic medium (Shelton 2006).

Ultraviolet satellites such as *Copernicus*, IUE, and FUSE provided the tools for direct detection of the HIM. Jenkins and Meloy (1974) and York (1974) used *Copernicus* to observe absorption lines of OVI at λ1032 and λ1038 Å along many lines of sight. In subsequent years, lines from other ions like NV, CIV, OVII, and OVIII established the existence of a hot diffuse component with temperature ranging from 10^5 to 10^7 K. The discovery of this component is what led McKee and Ostriker (1977) to propose their famous Three-Phase Model for the ISM. The development of x-ray satellites and detectors allowed the HIM to be routinely detected and studied from continuum and emission line observations. The resemblance of the physical conditions of the HIM to those of the outermost solar atmosphere lead to its alternate name, coronal halo gas.

1.5.2 The Warm Ionized Medium

Early-type (massive) stars power the HIM by mechanical injection of energy via stellar winds. In addition, their ionizing radiation produces another phase of the ISM: the Warm Ionized Medium or WIM. As the name implies, this phase has a significantly lower temperature ($\sim 10^4$ K) and is produced as the Lyman continuum luminosity from these stars propagates throughout the disk and into the halo producing extensive regions of ionized hydrogen. The WIM differs from HII regions in that it is more extended and is comprised of lower density gas ($\sim 0.1\,\mathrm{cm}^{-3}$ vs. $10^2\,\mathrm{cm}^{-3}$). Moreover, the SII/Hα ratio in the WIM is several times higher than in HII regions. This is characteristic of gas that is photoionized by a very weak radiation field from distant stars (Mathis 1986). Globally, 90% of the Galactic ionized hydrogen mass is contained in the WIM and not in HII regions and the overall mass in the WIM rivals that in molecular form (a few times $10^9\,\mathrm{M}_\odot$). Locally, the mass surface density is 2–3 $\mathrm{M}_\odot\,\mathrm{pc}^{-2}$. Heiles et al. (1996) proposed that the WIM is composed of "chimneys" or "worms" (filamentary structures extending from the disk into the lower halo) that facilitate the transport of ionizing photons beyond the Galactic plane. As discussed above, the morphology of the ISM is not precisely known, but the large, low-density cavities somehow permit the ionizing radiation from these stars to reach regions that are hundreds of pc away from the stars, so that the scale height of the WIM is 1–2 kpc, less than the HIM, but significantly more than the neutral phases described below (see Reynolds and Ogden 1979; Norman and Ikeuchi 1989; Reynolds 1989; Gaensler et al. 2008). The WIM is also sometimes referred to as the Extended Low-Density Medium or ELDM (Mezger 1978).

The WIM was first postulated by Hoyle and Ellis (1963) as they analyzed the Galactic radio continuum spectrum from 1–100 GHz at high Galactic latitudes. In 1970, Gottesman and Gordon used radio recombination lines (RRLs) to detect an ionized medium not associated with HII regions. Pulsar dispersion measures (DM $= \int n_e ds$ where n_e is the electron density and ds is the differential path length calibrated for distance) showed the existence of a distributed ionized component (Lyne et al. 1985; Lazio and Cordes 2002). If the distribution of pulsar dispersion measures and the ensuing estimates of n_e are assumed to have an exponential distribution, then the scale height of this component of the ISM is estimated to be $\sim$1 kpc. In addition, observations of the Hα recombination line can be used to estimate the emission measure (EM $= \int n_e n_{\mathrm{HII}} dV = \int n_e^2 dV$, where n_{HII} is the density of ionized gas) which depends on the temperature and density of ionized gas. The early radio work on emission measure also stemmed from observations of bremsstrahlung emission and absorption. More recently, RRLs have become powerful probes of the ionized gas in the ISM, including the WIM. RRLs do not suffer from interstellar extinction (see Sect. 1.8) and they are optically thin allowing for easier interpretation of the line profiles (see Sect. 2.1). Most importantly, they provide velocity information which can help locate the distance of the emitting region along the line of sight. RRLs are also excellent probes of the denser HII regions. In the infrared, fine structure lines of [NII] (at 122 and 205 μm) indicate that there is widespread ionized hydrogen, because nitrogen has an ionization potential of 14.532 eV and so where nitrogen is ionized, hydrogen must also be ionized. Other tracers of the WIM are the forbidden lines [SII] λ6716 and [NII] λ6584 (Madsen et al. 2006).

However, the most successful way to observe the WIM is to use high throughput Fabry-Perot spectroscopy to detect faint optical interstellar emission lines, particularly the Hα recombination line. This work was pioneered by Ron Reynolds at Wisconsin in the mid-1980s, so much so that the WIM is sometimes referred to as the "Reynolds Layer". A dedicated Fabry-Perot on Kitt Peak, whimsically called WHAM (Wisconsin Hα Mapper), mapped the Hα emission throughout large regions of the Galaxy establishing temperatures (6000–10,000 K within 2–3 kpc of the midplane), densities (0.03–0.08 $\mathrm{cm^{-3}}$) and distribution of the phase (filling factor 0.2–0.4 within 2–3 kpc of the midplane - Reynolds 1991; Haffner et al. 2003; Hill et al. 2008). This phase of the ISM has also been detected in other galaxies, where it is more commonly known as the Diffuse Ionized Gas or DIG.

1.5.3 The Warm Neutral Medium

The neutral atomic gas is traced primarily by the 21 cm spin-flip transition of atomic hydrogen (1420.406 MHz). Line profiles of the 21 cm line show narrow features (FWHM $< 10\,\mathrm{km\ s^{-1}}$) thought to be fairly cold and dense ($\sim$100 K; 10 $\mathrm{cm^{-3}}$) concentrated in filamentary or planar structures (discussed in the next sections), and broader features thought to represent a pervasive, warm (of order 10^3 K), lower density ($\sim$0.1 $\mathrm{cm^{-3}}$) phase known as the Warm Neutral Medium (WNM).[7] Heiles and Troland (2003) derive a volume filling factor for the WNM of ≈ 0.5 with a minimum of 0.48 of the gas in the thermally unstable 500–5000 K region. Wolfire et al. (2003) find that the WNM has a thermal velocity dispersion of $\sim$8 $\mathrm{km\ s^{-1}}$ implying a scale height of less than 200 pc.[8] The heating of this medium is primarily photoelectric with the electrons ejected from dust grains or polycyclic aromatic hydrocarbons (PAHs), although cosmic rays also contribute directly to the heating. Cooling proceeds primarily by photon losses by visible transitions of oxygen and the Lyman-α line. The WNM was first detected observationally by emission/absorption observations of the 21 cm line (Radhakrishnan et al. 1972). By comparing single-dish or interferometric observations of HI absorption towards extragalactic sources with single-dish observations of HI emission along adjacent lines of sight, these authors noted that the absorption lines were narrow, while

[7]In their classic 2003 paper, Heiles and Troland (2003) refer to the WNM as "... the key to the Universe...".

[8]The relationship between velocity dispersion of an ensemble of Galactic objects and their scale height is discussed in Sect. 10.3.

the adjacent emission lines had both narrow and broad components. The lack of broad absorption counterparts indicated that the broad emission had low opacity and, thus, was not very dense. Historically, the opacity was deemed to be inversely proportional to the excitation temperature of the 21 cm line (often referred to as the "spin temperature"—see, e.g., Kulkarni and Heiles 1987). Thus, the broad features are warm (and, conversely, the narrow features are relatively cold). In this manner, Radhakrishnan et al. (1972), Dickey et al. (1978) and Payne et al. (1983) conclusively showed the existence of the WNM as a pervasive phase of the ISM, likely the background "bath" in which the colder HI filaments and clouds are embedded. Although the conclusion is correct, the manner in which it was obtained is suspect as it now appears that the opacity-excitation temperature relationship (known as the $\tau_o - T_s$ relation) has little physical meaning (see Heiles and Troland 2003, for details).

The WNM can also be deduced directly from Gaussian decomposition of HI emission spectra. Mebold (1972) analyzed about 1300 HI emission spectra separating the profiles into narrow and wide Gaussian components. If the intrinsic velocity dispersion of the wide component (8.8 km s^{-1}) is interpreted as a kinetic temperature, an upper limit for the warm neutral medium of 9600 K is obtained. Other decomposition studies (e.g., Mebold et al. 1982; Verschuur and Schmeltz 1989; Verschuur and Magnani 1994; Alexander 2006; Haud and Kalberla 2007) broadly confirmed the original conclusions proposed by Mebold. Some of the later studies also found evidence of a very broad component (velocity dispersion > 25 km s^{-1}) whose width cannot be identified with a kinetic temperature and whose origin is unknown and may even be an instrumental artifact (e.g., see discussion of stray radiation; Sect. 4.2.5).

1.5.4 The Cold Neutral Medium

The 21 cm line emission that showed the presence of the WNM also provides clear evidence of a colder, denser phase that is not as broadly distributed. This is the CNM with temperatures of less than or around 100 K and densities of a few particles cm^{-3} up to tens of particles cm^{-3} (see Heiles and Troland 2003). Wolfire et al. (2003) find characteristic thermal velocities for this component ~ 1.5 km s^{-1}. The main heating sources for this component are photoelectric heating from dust grains and cosmic rays (as for the WNM). Perhaps surprisingly, the primary coolant is the ${}^2P_{3/2}{-}{}^2P_{1/2}$ transition of singly-ionized carbon at 158 μm. This is so for two main reasons: (1) The ionization potential is 11 eV. Therefore it is a substantial presence even if the hydrogen is predominantly neutral. (2) With $h/k = 91$ K, after the hyperfine states of hydrogen this fine-structure state of CI is the lowest-lying energy state of all gas-phase CNM constituents. Thus even very low energy electrons can collisionally excite this optically thin transition.

The CNM is distributed not so much in atomic "clouds" (though that was an early interpretation) as in filaments and sheets (Verschuur 1991; Heiles and Troland

2003). A significant fraction of this component is also associated with the envelopes of molecular clouds (e.g., Elmegreen and Elmegreen 1987). An excellent review of this phase of the ISM was given by Dickey and Lockman (1990), and we will not repeat their exposition. However, because molecular clouds are always associated with atomic gas, the relationship between the CNM and the molecular gas holds the key to understanding how the latter forms.

Although the CNM is the easiest phase of the ISM to observe (because of the intensity of the 21 cm line and the relatively low technology required to detect it), its morphology is still controversial, with various authors interpreting atomic clouds as filaments seen along their length while others claim that the filaments are sheets seen along the side. The association of a portion of this phase with the molecular gas makes it easier to discern its morphology as the molecular gas tends to be much more clumped and confined both spatially and in velocity. A clear problem in untangling the distribution of the CNM is angular resolution (see Fig. 1.6). Even with the 100-m Green Bank radiotelescope (GBT) and its 10′ resolution at 21 cm, the relationship between the HI associated with the high-latitude cloud MBM 40 is not at all clear. At this resolution, the atomic hydrogen appears to be distributed in a cocoon enveloping the molecular core. However, with the factor of 3 greater resolution permitted by the 305-m Arecibo radiotelescope, the cocoon appears to be more like a partial atomic ring with molecular gas "filling in" the ring or perhaps forming a helix shape (Verschuur 1974; Shore et al. 2003). Given the difficulty in understanding the intricacies of the relationship between the cold atomic and molecular components, it is clear that observations at the highest *spatial* resolution are needed to understand the relationship between the two phases. Thus, the Square Kilometer Array (SKA) will likely be the key future telescope.

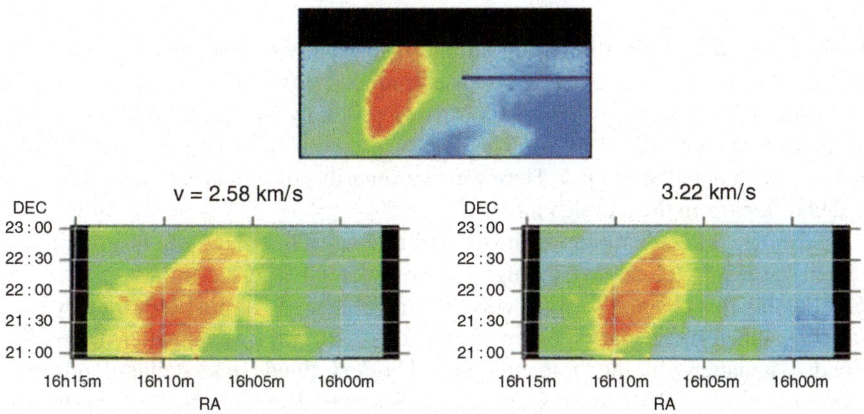

Fig. 1.6 The importance of angular resolution in observing the 21 cm HI lines. The top image is the 21 cm line integrated in velocity from 2–5 km s^{-1} from a map of the high-latitude cloud, MBM 40, made with the 100 m GBT (10′ resolution). The lower two figures are velocity slices 0.32 km s^{-1} in width and centered at 2.6 and 3.2 km s^{-1} made with the 305-m Arecibo radiotelescope (resolution 4′)

At high Galactic latitudes, the optical depth of the 21 cm transition is small, so that the column density of HI is directly obtained from the observations. The relevant formula is

$$N(\text{HI}) = 1.823 \times 10^{18} \int T_B \text{dv} \quad cm^{-2} \tag{1.3}$$

where T_B is the brightness temperature (see section 4.2.6) of the 21 cm line and dv is the differential velocity over the line. However, the lines of sight through the Galactic disk can be several hundred parsecs, and untangling the radial distribution of the HI is often impossible. In particular, associating a fraction of the HI gas along a line of sight with a small molecular cloud is very difficult. Sometimes, a distinct HI spectral feature at the molecular cloud velocity is clearly seen [e.g., HI spectra from the Draco region taken by Goerigk et al. (1983) and Gir et al. (1994)], but, more often, there is no narrow spectral feature and the HI line extends over tens of km s^{-1} . Thus, deciding which fraction of the velocity range to associate with the molecular gas (typically with linewidths of less than a few km s^{-1}) is based only on plausibility arguments. Although some high-latitude clouds have been specifically identified by looking for isolated HI concentrations (e.g., Heiles et al. 1988), there is no obvious correlation between N(HI) and the presence of a high-latitude molecular cloud. In this respect, dust emission is a better signpost of the possible presence of molecular gas in a given region (see below).

1.6 Molecular Gas: Why Some of It Is a Phase and Some Isn't

Thus far, we have been discussing atomic hydrogen, whether neutral or ionized. In the Galactic disk there exists a dense, clumped, molecular component which gives rise to stars and was historically traced primarily by extensive observations of the CO(1-0) line. We will discuss the global distribution and properties of this component in detail in Chap. 7. Here we are primarily concerned with how it relates to and differs from the atomic phases.

The molecular gas is found in the densest regions of the CNM and forms entities called "molecular clouds". Although we do not want to discuss these objects in detail in this first chapter, for now we can think of them as coming in primarily two varieties: Giant molecular clouds with masses $\gtrsim 10^4\, M_\odot$, and smaller molecular clouds. The latter—for the moment—are either *dark* clouds (interstellar clouds that contain significant obscuring dust and are primarily molecular in content), or *diffuse* clouds (where the dust extinction is low and atomic hydrogen is at least as abundant as the molecular gas). We will expand on these categories in Chap. 7.

Unlike atomic structures, the majority of molecular clouds are gravitationally bound. Thus, Elmegreen (1993) proposed categorizing interstellar clouds of all types on the basis of their gravitational state: diffuse, self-gravitating, and unbound. In this schema, diffuse clouds are bound structures with insufficient mass to be held together by gravity. Instead, they are confined by external pressure. Their formation is likely due to shocks and thermal instabilities in which pressure is the important dynamical factor. In contrast, self-gravitating clouds contain enough mass to be gravitationally bound. Such clouds can form in regions of the interstellar medium where turbulence dissipates and dense gas becomes quiescent enough for gravity to be the most important structuring agent. Although these types of clouds are almost always molecular, there are some primarily atomic examples as well (see below). The third group, unbound clouds, are not confined by either pressure or gravity and are thus dispersing on about their sound-crossing time. They may form in the same manner as diffuse clouds, but with the high pressure dissipated. To distinguish between diffuse and self-gravitating clouds, Elmegreen introduced the dimensionless parameter $P(G\sigma_m^2)^{-1}$ to effectively measure the importance of pressure versus gravity, Here, P is the external pressure, G is the gravitational constant, and σ_m is the average mass surface density of a cloud. If the parameter is significantly greater than 1, then gravity is relatively unimportant and the dynamics are dictated by external flows and pressure gradients. In contrast, if $P(G\sigma_m^2)^{-1} \ll 1$, the cloud is self-gravitating and its internal dynamics are governed by internal pressures and density gradients. In the marginal case where $P(G\sigma_m^2)^{-1} \sim 1$, the dynamical state of the cloud depends on which of the two terms begins to dominate. If the mass term dominates, the cloud tends to become self-gravitating on the free-fall timescale ($\sim 10^4$ years). If P begins to dominate, the mass surface density quickly drops on the sound crossing timescale. Thus, the transition to one case or the other is expected to be relatively rapid compared to the lifetime of the cloud (10^6–10^7 years).

Interstellar clouds can be further divided into two sub-populations depending on their atomic/molecular fraction. For H_2 formation, dust is required to serve as a nucleation site for the molecule as gas phase reactions do not proceed rapidly enough under ISM conditions (see, e.g., Duley and Williams 1984). Once H_2 has formed, it must be shielded from the background diffuse UV radiation field, ϕ, as the molecule is readily photodissociated by the absorbed UV flux which is proportional to $\phi N^{1/3}$. In terms of the volume density of hydrogen nucleons, n, the cloud size, D, the grain volume density, n_g, the grain cross-sectional area, σ_g, and the thermal velocity of the hydrogen, v_t, the hydrogen-grain collision rate is proportional to $n n_g \sigma_g v_t$. If we define the grain surface area per atom as $\mathscr{Z}$, then this rate is proportional to $n^2 D\mathscr{Z} \sim nN\mathscr{Z}$. Molecular formation occurs when the path integral through the cloud of the hydrogen-grain collision rate is greater than the absorbed UV flux. The shielding function can then be written (following Elmegreen 1993) as

$$S = (n/60\,cm^{-3})(N/5\times 10^{20} cm^{-2})^{2/3}(\mathscr{Z}/\mathscr{Z}_o)(\phi/\phi_o)^{-1} \tag{1.4}$$

where $\mathscr{Z}_o$ and ϕ_o are the values of the respective parameters in the solar neighborhood. Similarly, the dimensionless quantity, $P(G\sigma_m^2)^{-1}$, can be parameterized as

$$P(G\sigma_m^2)^{-1} = 15(n/60\,cm^{-3})(c_s/1\,km\,s^{-1})^2(N/5\times10^{20}\,cm^{-2})^{-2} \tag{1.5}$$

where c_s is the speed of sound in the cloud.

With these two quantity, clouds can be categorized depending on the relative values of S and $P(G\sigma_m^2)^{-1}$. There are thus four cases: (1) If S is small and $P(G\sigma_m^2)^{-1}$ is large, the objects with these characteristics have low column densities of atomic gas and so the cloud is predominantly atomic and diffuse as far as the volume density of gas is concerned. This situation represents a typical CNM entity. (2) If S is large and $P(G\sigma_m^2)^{-1}$ is small, the column density of gas is now larger, so self-shielding allows the gas to transition from primarily atomic to molecular. Self-gravity of the cloud now dominates the dynamics and this object is a typical molecular cloud of the type seen in the large-scale CO surveys (e.g., Dame et al. 2001). Most CNM filaments or sheets and molecular clouds fall into these two categories. However, when n is small, but N is large, i.e., when a large mass of gas is spread out over a large volume, then we have case (3), both S and $P(G\sigma_m^2)^{-1}$ are small. This would represent a self-gravitating cloud that is mostly atomic. An example of this type of structure could be the virialized, giant HI clouds observed in our Galaxy and in external galaxies (Elmegreen and Elmegreen 1987; Skillman 1987). Finally, case (4) is when both S and $P(G\sigma_m^2)^{-1}$ are large so the objects are molecular, but self-gravity is not very important. Most of the diffuse molecular clouds which are the subject of this book fall in this category. Because this component is not self-gravitating, it has to interact with the atomic phases we have discussed and may constitute an unrecognized fifth phase of the ISM (because like the other four phases, gravity can be ignored); a cold, molecular medium (CMM).

1.7 The Transition from Atomic to Molecular Gas

Molecular clouds form from the neutral atomic hydrogen ISM. These have a wide range of size, densities, and molecular fraction that lead to very different types of structures which we will discuss in more detail in Chap. 7. Here some general considerations on the physics of the transition from atomic to molecular gas are presented that will be useful for better understanding the role these objects play in the ISM.

1.7.1 Photodissociation Regions: PDRs

Molecular formation in the diffuse ISM is the competition between gas-phase formation mechanisms and photodissociation. Because H_2 molecules self-shield by line-driven instead of continuum dissociation, they form a "sacrificial" stratum of molecules. A sufficient atomic gas column illuminated by UV radiation on only one side rapidly turns molecular once it is sufficiently optically thick in the photo-dissociating transition (some of the dissociating photons are also soaked up by the dust present in the region). This self-shielding and the ensuing precipitous rise in $N(H_2)$ typically occurs at hydrogen column densities of $\sim 10^{19}\,cm^{-2}$. Thus, the diffuse ISM can assume a layered structure; from the source of UV radiation inwards, a series of zones consisting of atomic gas, a transition region, and molecular gas quickly develops (on timescales of 10^3 yr—see, e.g., Millar 2000). These layered regions are called Photo-dissociation Regions or PDRs (less commonly, a PDR is sometimes referred to as a Photon-Dominated Region) and their physical and chemical properties have been extensively discussed (Hollenbach and Tielens 1997, 1999, and references therein) . When metals are included in the PDR models, a more complex layer structure develops with the outermost region (closest to the photodissociating source) consisting of HI and C^+, an inner region where the carbon begins transitioning to C^0, the beginnings of a molecular region with first-generation molecules such as OH and CH, but with the carbon primarily locked up in C^0 and C^+, a transition to a CO-dominated region, and, finally, a region where more complex molecules appear (see, e.g., Wolfire 2010). Note that the ionization potential of C^0 is less than hydrogen.

The precise layer structure of PDRs requires numerical modeling but the basic parameter that governs the structure is the ratio of the photodissociation rate, which is proportional to $G_o n(H_{total})$, where G_o is the standard intensity of the interstellar radiation field and $n(\mathrm{H}_{total})$ is the volume density of hydrogen nucleons (ionized, atomic, molecular), and the binary collisional rate for the creation of the various species in the PDR. These latter processes scale as n^2 and so the parameter can be written $G_o/n(H_{total})$. The diffuse ISM as a whole can be thought of as a PDR since it contains H, C^0, and C^+, and so can the outer regions of molecular clouds. Astronomers don't normally think in this way because they usually study one particular layer of the PDR, ignoring the others. For example, the early studies of molecular clouds generally ignored the transition region to the atomic ISM, and focused almost exclusively on the densest regions where star formation was taking place.

The physical processes needed for calculations of PDR structure include the extinction of the interstellar radiation field (ISRF) by dust grains, and the micro-physics of heating and cooling of the gas. The most important heating mechanisms in the various regions include photoelectric heating by grain ionization, H_2 ejection by grains, and H_2 excitation by UV photons. In the most opaque regions of the PDR (the dense molecular cores), the UV field is quenched and the heating is dominated by cosmic rays and x-rays. The cooling mechanisms include the radiative emission from the various species with CII dominating the cooling of the outer layers, giving way to CI, giving way to CO as the depth into the PDR increases.

Complicating matters further is the chemistry which depends sensitively on variations in the density and ISRF. We briefly discuss some aspects of astrochemistry relevant to the diffuse ISM in Chap. 3.

1.7.2 *Diffuse vs. Dark Clouds*

Applying the PDR concept for molecular clouds, van Dishoeck and Black (1988) studied the photodissociation of CO in detail and arrived at a new way of categorizing molecular gas. Historically, the smaller molecular clouds had been divided into diffuse and dark molecular clouds on the basis of their opacity to background starlight. Before the 1990s, diffuse clouds had been the province of optical and near-UV astronomers, who detected them by observing narrow, optical, spectral lines seen against the background stellar continuum of an early-type star. It is worth remembering that the first detections of interstellar molecules were accomplished by this technique as CH, CH^+, and CN transitions were detected in the late 1930s (Dunham and Adams 1937a,b; McKellar 1940). This technique precluded any mapping of the molecular structures around these special lines of sight so the discovery of actual molecular "clouds" (as opposed to molecule bearing lines of sight) had to wait until radio spectroscopy was developed after World War II. Dark clouds, cataloged by Barnard (1927) and, later, Lynds (1962) were known to contain molecules as early as 1963, when the OH lines were detected in emission by Weinreb et al. By 1970, the CO(1-0) line was detected (Wilson et al. 1970) and the picture of a dark cloud as a *molecular* cloud was already firmly established. Within a decade, the CO(1-0) transition had become the line of choice for detecting molecular gas in the Galaxy.

The relevant observational differences between diffuse and dark clouds included: the optical opacity, with $A_V < 1$ for the diffuse clouds and greater than several for the dark clouds; N(H_2), with diffuse clouds prior to the 1990s believed to have N(H_2) $< 10^{20}$ cm^{-2} and dark clouds $> 10^{21}$ cm^{-2}; and the CO/H_2 ratio, with diffuse clouds showing ratios $< 10^{-6}$ and dark clouds $\sim 10^{-4}$. The bulk of the carbon in diffuse molecular clouds is in the form of C^0 and C^+, while in dark clouds the bulk is tied up in CO. In this early scheme, dictated primarily by the inability of millimeter-wave technology of the 1970s and early 1980s to detect molecular emission from the traditional diffuse clouds, an oversimplified idea was widespread: diffuse clouds are primarily atomic and dark clouds are primarily molecular. That this was insufficient should have been clear even then, since the work by Spitzer et al. (1973) and Savage et al. (1977) using the *Copernicus* UV data had established that the transition to a mostly molecular environment has a threshold of total hydrogen column density $\approx 5 \times 10^{19}$ cm^{-2} and E(B-V) $\approx$ 0.08 mag rather than 10^{20} cm^{-2} and 0.3 mag. These threshold objects were difficult to detect in CO until the mid-1980s so radio astronomers tended to ignore diffuse molecular clouds.[9]

As technology allowed the detection of CO emission from regions with $A_V \sim 1$ mag, Van Dishoeck and Black introduced a new category: the translucent cloud. These were differentiated by having visual extinctions between 1 and 5 magnitudes (in this scheme, diffuse clouds were defined to have $A_V < 1$ mag and

[9] A few of the people who tried to bridge the gap in this heroic age were Bob Dickman, Steve Federman, Göran Sandell, Robert Willson, and Ken Lang.

dark clouds > 5 mag) and column densities $\sim 10^{20}$ cm^{-2}. More importantly, CO self-shields from photodissociation and is additionally shielded by H_2 and dust in the clouds. At values of N(H_{total}) $\approx 1 \times 10^{20}$ cm^{-2}, the CO/H_2 abundance rises precipitously, from 10^{-6} to 10^{-4} even as the column density increases by less than an order of magnitude. In this sense, the translucent clouds are a transitional regime where N(CO) rises rapidly enough to allow detection of the lowest CO rotational transitions even with the technology of the 1980s. The exact column density range where the transition occurred depends on the radiation field, but, for most regions distant from O and B stars, this occurs at N(H_2) values of a few times 10^{20} cm^{-2}. This scheme was certainly an improvement in categorizing small molecular clouds but still did not address the issue that significant, molecular-dominated entities could exist at A_V levels as low as 0.3 mag.

Most of the high-latitude clouds are thought to be translucent (but see remarks below) and it may be that a significant portion of the molecular gas in a galaxy (residing in the interclump medium of GMCs) is as well according to the definition of van Dishoek and Black (see Polk et al. 1988; Wright et al. 1992; Chiar et al. 1994). Moreover, some authors have dubbed all translucent gas a PDR because it represents the region where carbon transitions from primarily C^0 and C^+ to CO.

van Dishoeck and Black (1988) also considered the chemistry in diffuse gas and realized that in translucent regimes, the chemistry is also in transition from primarily photoprocesses (which dominate diffuse cloud chemistry), to collisional processes that dominate dark clouds. In translucent clouds, both types of processes are important.

This simple, and attractive, classification scheme for molecular clouds has however recently been challenged by Liszt et al. (2010). They find very strong CO lines in what can only be defined as diffuse molecular gas according to the definition of van Dishoeck and Black (1988). Perhaps the radiation field is weaker than normal in these objects so that the transition from C^0 and C^+ dominated to CO-dominated regions occurs at lower extinction, or perhaps other factors are involved. We will discuss this issue more deeply in Chaps. 7 and 8.

1.8 The Role of Dust

An excellent rule of thumb about the ISM is: Where there is gas, there is dust. Although its total Galactic mass is factor of 100 lower than the gaseous component, the dust is more than a trace constituent. With the advent of infrared astronomy, it is easier to detect than the gas and the resolution of the SDSS and 2MASS surveys, and observations from the Spitzer Space Telescope and the Herschel Space Observatory provide higher spatial resolution than spectral line maps from single-dish radio telescopes. In many respects, infrared observations of the dust constituent of molecular clouds have supplanted the spectral line maps of clouds so popular in the 1970s and 1980s.

Fig. 1.7 The dark cloud Barnard 68. The quantity of dust in the cloud is great enough that it obscures all background stars. The proximity of the cloud to the Earth means that foreground stars are not present so that the region looks completely dark. The image was obtained with the 8.2-m VLT ANTU telescope and the multimode FORS1 instrument in March 1999. Credit: ESO

The presence of dust in the ISM was first demonstrated by Trumpler (1930a,b,c). Even a perfunctory glance at a good photograph of the Milky Way will show regions with distinctly reduced numbers of stars (e.g., Barnard 1927). Nineteenth century astronomers dubbed these regions "dark nebulae" and Fig. 1.7 shows a prominent example. The pre-discovery can be assigned to John Herschel's first glimpse of the Coalsack.

With the advent of infrared astronomy in the 1980s, emission from interstellar dust in the ISM was detected and studied (see Chap. 6). It soon became clear that the basic spectral properties of the dust on large scales could be explained by a large grain component (sizes ranging from 0.01 μm to 0.3 μm), a very small grain component (sizes less than 10 nm), and a component of very large molecules (PAHs). The large grains are in radiative equilibrium with the radiation field in the ISM and typically have temperatures of 17 K (a bit less in the denser, molecular regions of the ISM where the radiation field is attenuated). The other two components experience significant temperature fluctuations after photon absorption and emit over a wide range in the infrared below 60 μm.

Because the gas and dust are well-mixed, and since it is now relatively easy to image the entire sky in the infrared, dust emission is often used to trace the gas. Dust provides the nucleation site for molecular hydrogen formation (Duley and Williams 1984) so its role in producing the molecular component of the ISM must be addressed and we will do so in Chap. 6.

1.9 Cosmic Rays

Filling the diffuse medium, there is a population of superthermal charged particles whose energies require distinctive acceleration and propagation mechanisms, called cosmic rays (hereafter CR).[10] The cosmic ray particles have several distinctive, universal properties. Their composition is mainly protons, electrons, and helium nuclei, all stripped bare. A heavier population of accelerated, completely charged nuclei reaches energies of tens to hundreds of MeV. In the standard parlance, *primary* cosmic rays are those originating at cosmic accelerators, for instance supernova explosions, where synchrotron emission by electrons evinces a process of energization above GeV (see Sect. 3.3.5 for further discussion). The mean CR energy density is inferred from synchrotron measurements although, strictly speaking, this applies only to the electrons. The sources are distributed throughout the Galaxy (Strong et al. 2007). These are mainly supernova remnants and pulsars, for the energy range below a few GeV, and confinement to the Galaxy results for energies less than TeV by diffusion of the particles in the turbulent magnetic field of the diffuse medium. Since this population is not thermalized, there is no reason to expect complete charge neutrality everywhere. Whatever its spatial distribution, the mean (or *rms*) magnetic energy density is assumed to be in rough equipartition with the relativistic particles. The synchrotron luminosity in some volume is the product of u_{rel}, the energy density in CR electrons, and B^2, the magnetic energy density. Thus, for an observed nonthermal surface brightness, minimizing the sum of the two components with respect to the magnetic field, the equipartition argument gives a minimum magnetic pressure and, by extension, the CR density. This is about 1 eV cm^{-3} in the Galactic disk corresponding to a mean particle density of about 10^{-6} cm^{-3} at about 1 MeV, and similar values are obtained everywhere but in the Galactic center.

The energy spectrum of the photons is important as their number since the cross sections especially for ionization are strongly energy dependent. Below 10^{15} eV and above $\sim$ 1 GeV, the CR spectrum is a single power law, $N(E) \propto E^{-2.7}$ for both electrons and protons. But at the lowest energies, below several hundred MeV, their energy is strongly modulated by the heliosphere. The Voyager interplanetary probes have now passed out of the heliosphere, beyond the immediate zone of modulation by the solar wind, and have measured the local CR energy distribution. For the $N(E) \propto E^{-\alpha}$ relation, they yielded an exponent of 1.45 ± 0.15 below about 100 MeV and 3.15 ± 0.05 above that. The lower energy range is immediately relevant since this is the first *direct* determination of the energy spectrum in the interval that governs the chemistry. The contrast with near Earth and outer heliosphere measurements is striking. The flux for He nuclei is $N(E) \approx 1.8(E/GeV)^{-2.7}$

[10]First discovered by electrostatic measurements of the upper atmosphere (the ionosphere) in balloon ascents a century ago by Hess (1912) they were recognized as particles and, in that sense, penetration of new regimes by probes has been characteristic of studies of the lower energy range of such particles.

nucleons cm^{-2} s^{-1} GeV^{-1} with approximately 80% of the primary CR nuclei being protons and about 1% electrons.

As we will discuss in later chapters, CR interactions with the diffuse gas are extremely important for governing the thermal and ionization state of the medium and its chemical makeup. They also provide a unique probe, through secondary γ-ray production, of the dust to gas ratio.

1.10 The Concept of a "Molecular Cloud"

GMCs were identified early during the Galactic plane surveys and were thought to have very well-defined edges (Blitz 1979). Thus, calling a localized concentration of molecular gas a "cloud" was a consequence of the observational methods. Average densities of GMCs are of order 10^2 cm^{-3} while the average density of the ISM is roughly 1 cm^{-3}, so the idea of molecular clouds plowing through the ISM like cannonballs through air was quickly established. This mindset was reinforced by virial analyses of the stability of clouds which implied that clouds were long-lived, stable structures (see Sect. 4.3.1). Some estimates of the ages and lifetimes of clouds from the early 1980s were as high as $> 10^8$ years (e.g., Scoville and Hersh 1979), and the idea of massive, long-lived entities that are distinct from their atomic surroundings took hold.

Of course, there were disturbing issues with this picture even then. GMCs were surrounded by large HI envelopes which had clearly played key roles in their formation (Elmegreen and Elmegreen 1987). Moreover, applying the PDR concept to these structures argued against a truly sharp boundary between molecular and non-molecular gas. An edge as traced by CO(1-0) does not require the absence of molecular hydrogen. The idea of gravitationally bound clouds also implied a too-high star formation rate for the Galaxy, so some type of supporting mechanism was needed to prevent the bulk of the gas in a GMC from collapsing and forming stars. It was thought a gravitational instability could be prevented by magnetic fields or turbulence. While a magnetically controlled cloud seemed dynamically reasonable, a turbulent molecular cloud posed a whole series of problems that "cannonballs" just do not address. Finally, a dissenting opinion on the lifetimes of GMCs was presented by Blitz and Shu (1980) who proposed 10^7 years as a more reasonable age estimate.

Still unresolved is how low the molecular content of a cold hydrogen structure must be for the object to be called a *molecular* cloud. The cores of GMCs and dark clouds are easy; inside their HI envelopes the hydrogen nucleons are virtually all in the form of H_2. There is some cold HI that is well-mixed with the H_2 but the ratio of this atomic component to the molecular gas is $\sim 10^{-3}$-10^{-4} (Goldsmith and Li 2005). These objects are clearly molecular clouds in all senses. The situation is more complicated for less well-shielded regions with lower hydrogen column densities, such as the envelope or interclump medium of a GMC or in the diffuse and translucent clouds. It is here that the PDR models show that the HI content rises as one moves away from the opaque regions. With a visual extinction of 1 mag, it is very common to have sites where the H_2/H_{total} ratio is less than 0.5. Yet the objects show plenty of CO(1-0) emission and are considered "molecular" clouds even though there is more hydrogen nucleons in atomic form than locked in H_2. For the moment, we will call something a molecular cloud if it has a well defined boundary region where the CO(1-0) line can be detected in emission by a single-dish mm-wave radiotelescope with less than fifteen minutes of integration time given current (2017) mm-wave technology. For all practical purposes, these are regions with $N(H_2) >$ several times 10^{19} cm^{-2} and E(B-V) $\gtrsim$ 0.1 mag. Even excluding GMC cores and dark molecular clouds, this observational definition encompasses a vast zoo of objects; from those that are mostly atomic, diffuse and dominated by turbulence to those that are mostly molecular, gravitationally bound, and forming stars. Understanding the theoretical and observational differences between these groups is one of our principal themes.

Despite the caveats listed above, the idea of a molecular "cloud" as a distinct, stable, long-lived entity was the dominant paradigm throughout the 1980s and early 1990s. It wasn't until numerical simulations studies of cloud structure revealed serious problem with molecular clouds supported by magnetic fields that the issue of what constitutes a mostly molecular interstellar entity was re-examined. We will discuss this development throughout the remainder of the book.

References

Alexander, M. 2006, M.S. Thesis, University of Georgia
Balbus, S. 1986, ApJ, 303, L79
Barnard, E.E. 1927, *Photographic Atlas of Selected Regions of the Milky Way*, eds. E.B. Frost and M.R. Calvert (Washington: Carnegie Institution of Washington)
Blitz, L. 1979, PhD Thesis, Columbia University
Blitz, L. and Shu, F. 1980, ApJ, 238, 148
Bowyer, C.S., Field, G.B., and Mack, J.E. 1968, Nature, 217, 32
Bregman, J.N. 1980, ApJ, 236, 577
Burton, W.B. and Gordon, M.A. 1978, A&A, 63, 7
Chiar, J.E., Kutner, M.L., Verter, F., and Leous, J. 1994, ApJ, 431, 658
Cox, D.P. and Smith, B.W. 1974, ApJ, 189, 105
Cox, D.P. 2005, ARAA, 43, 337
Dame, T.M. 1983, Ph.D. Thesis, Columbia University
Dame, T.M. *et al.* 1987, ApJ, 322, 706
Dame, T.M., Hartmann, D., and Thaddeus, P. 2001, ApJ, 547, 792
Dickey, J.M., Terzian, Y., and Salpeter, E.E. 1978, ApJS, 36, 77
Dickey, J.M. and Lockman, F.J. 1990, ARAA, 28, 215
Digel, S., DeGeus, E., and Thaddeus, P. 1994, ApJ, 422, 92
Duley, W.W. and Williams, D.A. 1984, *Interstellar Chemistery*, (London: Academic Press)
Dunham, T.,Jr. and Adams, W.S. 1937a, PAAS, 9, 5
Dunham, T.,Jr. and Adams, W.S. 1937a, PASP, 49, 26
Elmegreen, B.G. and Elmegreen, D.M. 1987, ApJ, 320, 182

Elmegreen, B.G. 1993, in Protostars and Planets III, eds. E.H. Levy, and J.I. Lunine, (Arizona, Arizona U. Press), 97
Ewen, H.I. and Purcell, E.M. 1951, Nature, 168, 356
Field, G.B. 1965, ApJ, 142, 531
Field, G.B. Goldsmith, D.W. and Habing, H.J. 1969, ApJ 155, L149
Gaensler, B.M., Madsen, G.J., Chatterjee, S., and Mao, S.A. 2008, PASP, 25, 184
Gir, B.-Y., Blitz, L., and Magnani, L. 1994, ApJ, 434, 162
Goerigk, W., Mebold, U., Reif, K., Kalberla, P.M.W., and Velden, L. 1983, A&A, 120, 63
Goldsmith, P.F. and Li. D. 2005 ApJ, 622, 938
Gottesman, S.T. and Gordon, M.A. 1970, ApJ, 162, L93
Gould, A. 1996, in *Proceedings of the* 169th *Symposium of the IAU*, eds. L. Blitz and P. Teuben, (Dordrecht: Kluwer), 651
Haffner, L.M., et al. 2003, ApJS, 149, 405
Haud, U. and Kalberla, P.M.W. 2007, A&A, 466, 555
Heiles, C., Reach, W.T., and Koo, B.-C. 1988, ApJ, 332, 313
Heiles, C., Reach, W.T., and Koo, B.-C. 1996, ApJ, 466, 191
Heiles, C. and Troland, T.H. 2003, ApJ, 586, 1067
Hess, V.F. 1912, Phys. Zeit., 13, 1084
Hill, A.S. et al. 2008, ApJ, 686, 363
Hollenbach, D.J. and Tielens, A.G.G. 1997, ARAA, 35, 179
Hollenbach, D.J. and Tielens, A.G.G. 1999, RMP, 71, 173
Hoyle, F. and Ellis, G.R.A. 1963, Aust.J.Phys, 16, 1
Jenkins, E.B. and Meloy, D.A. 1974, ApJ, 193, 121
Kulkarni, S.R. and Heiles, C. 1987, in *Interstellar Processes*, eds. D.J. Hollenbach and H.A. Thronson, Jr., (Dordrecht: Reidel), 87
Lazio, J. and Cordes, J. 2002, http://arxiv.org/abs/astro-ph/0207156v3
Liszt, H.S. 1983, ApJ, 275, 163
Liszt, H.S., Pety, J., and Lucas, R. 2010, A&A, 518, 45
Low, F.J. et al. 1984, ApJ, 278, 19
Lynds, B.T. 1962, ApJS, 7, 1
Lyne, A.G., Manchester, R.N., and Taylor, J.H. 1985, MNRAS, 213, 613
Madsen, G.J., Reynolds, R.J., and Haffner, L.M. 2006, ApJ, 652, 401
Mathis, J.S. 1986, ApJ, 301, 423
McKee, C.F. and Ostriker, J.P. 1977, ApJ, 218, 148
McKellar, A. 1940, PASP, 52, 187
Mebold, U. 1972, AAp, 19, 13
Mebold, U., Winnberg, A., Kalberla, P.M.W., and Goss, W.M. 1982, AAp, 115, 223
Mezger, P.G. 1978, A&A, 70, 565
Millar, T.J. 2000, in *Atomic & Molecular Data and Their Applications: ICAMDATA Second International Conference*, eds. K A Berrington and K L Bell (AIP Conference Proceedings, 543), pp. 81–91
Morris, M. and Serabyn, E. 1996, ARAA, 34, 645
Muller, C.A and Oort, J.H. 1951, Nature, 168, 357
Neugebauer, G. et al. 1984, ApJ, 278, L1
Norman, C.A. and Ikeuchi, S. 1989, ApJ, 345, 372
Payne, H.E., Salpeter, E.E., and Terzian, Y. 1983, ApJ, 272, 540
Polk, K.S., Knapp, G.R., Stark, A.A., and WIlson, R.W. 1988, ApJ, 332, 432
Radhakrishnan, V., Murray, J.D., Lockhart, P., and Whittle, R.P.J. 1972, ApJS, 24, 15
Reynolds, R.J. and Ogden, P.M. 1979, ApJ, 229, 942
Reynolds, R.J. 1983, ApJ, 268, 698
Reynolds, R.J. 1989, ApJ, 339, L29
Reynolds, R.J. 1991, ApJ, 372, L17
Sanders, D.B., Solomon, P.M., and Scoville, N.Z. 1984, ApJ, 276, 182
Savage, B.D., Bohlin, R.C., Drake, J.F., and Budich, W. 1977, ApJ, 216, 291

Savage, B.D. 1987, in *Interstellar Processes*, eds. D.J. Hollenbach and H.A. Thronson, Jr., (Dordrecht: Reidel), 123
Scoville, N.Z. and Hersh, K. 1979, ApJ, 229, 578
Shapiro, P.R. and Field, G.B. 1976, ApJ, 205, 762
Shelton, R.L. 2006, ApJ, 638, 206
Shelton, R.L. 2009, Space Sci Rev, 143, 231
Shore, S.N., Magnani, L., LaRosa, T.N., and McCarthy, M.N. 2003, ApJ, 593, 413
Shull, J.M. 1987, in *Interstellar Processes*, eds. D.J. Hollenbach and H.A. Thronson, Jr., (Dordrecht: Reidel), 225
Skillman, E.D. 1987, NASCP, 2466, 263
Spitzer, L., .Jr. 1956, ApJ, 124, 20
Spitzer, L., Jr. et al. 1973, ApJ, 181, 116
Spitzer, L., Jr. 1982, *Searching Between the Stars*, Yale University Press
Stark, A.A. 1979, Ph.D. Thesis, Princeton University
Strong, A.W., Moskalenko, I.V., and Ptuskin, V.S. 2007, Ann. Rev. Nucl. Part. Sci., 57, 285
Sutherland, R.S. and Dopita, M.A. 1993, ApJS, 88, 253
Trumpler, R.J. 1930*a*, Lick Obs. Bulletin, 420 (Berkeley: U. California Press), 154
Trumpler, R.J. 1930*b*, PASP, 42, 214
Trumpler, R.J. 1930*c*, PASP, 42, 267
van Dishoeck, E.F. and Black, J.H. 1988, ApJ, 334, 771
Verschuur, G.L. 1974, ApJS, 27, 283
Verschuur, G.L. 1989, *Interstellar Matters*, (New York: Springer)
Verschuur, G.L. and Schmeltz, J.T. 1989, AJ, 98, 267
Verschuur, G.L. 1991, ApSS, 185, 305
Verschuur, G.L. and Magnani, L. 1994, AJ, 107, 287
Weinreb, S., Barret, A.M., Meeks, M.L., and Henry, J.C. 1963, Nature, 200, 829
Wilson, R.W., Jefferts, K.B., and Penzias, A.A. 1970, ApJ, 161, 43
Wolfire, M.G., McKee, C.F., Hollenbach, D., and Tielens, A.G.G.M. 2003, ApJ, 587, 278
Wolfire, M.G. 2010, APSS, 10.1007/s10509-010-0566-9
York, D.G. 1974, ApJ, 194, 77
Wright, E.L. et al. 1992, ApJ, 336, L13

Chapter 2
Radiative Transfer Considerations

Know your lines and don't bump into the furniture.

—Spencer Tracy

Abstract In this chapter we describe some of the elements of radiative transfer that are necessary to understand the radiation coming from the diffuse interstellar medium. We concentrate on line processes. After a discussion of absorption and emission lines, we use the 21 cm line of hydrogen as an illustrative example. We also briefly discuss the CN transitions that are pumped by the Cosmic Microwave Background radiation. Collisional excitation, complications to radiative transfer from dynamical and nonlocal effects, and elemental abundances and depletions are also discussed. Molecular hydrogen and the Diffuse Interstellar Bands (DIBs) are used as examples of these effects. Finally, radio continuum processes are examined because they provide a way of measuring the properties of interstellar plasmas.

2.1 Introduction

Virtually everything we know about the interstellar medium derives from observing its effects on the passage of light.[1] Thus, it is important to understand how radiation is absorbed and emitted by gas and dust in the ISM as it propagates from its source to us. This comes under the broader subject of radiative transfer and it is covered in detail by many fine texts.[2] We assume in this section that the reader is familiar with the basics of radiative transfer and will limit ourselves to the more relevant and subtle aspects related to the diffuse ISM. We start with the radiative transfer equation as a general framework and concentrate on line transfer. After a discussion of absorption and emission lines, we use the 21 cm line of hydrogen as an illustrative example. We also briefly discuss the CN transitions that are pumped by the Cosmic Microwave Background (CMB) radiation. Collisional excitation, complications to radiative transfer from dynamical and nonlocal effects, and elemental abundances

[1] A part of the information also comes from studying cosmic rays but these are more indirect probes of the ISM.

[2] For instance, Chandrasekhar (1950), Cowley (1970), Athay (1972), Rybicki and Lightman (1979), Osterbrock and Ferland (2006), Hubeny and Mihalas (2014).

L. Magnani, S.N. Shore, *A Dirty Window*, Astrophysics and Space Science Library 442, DOI 10.1007/978-3-662-54350-4_2

and depletions are also discussed. Molecular hydrogen and the Diffuse Interstellar Bands (DIBs) are used as examples of some of these concepts. Finally, radio continuum processes are examined because they provide a way of measuring the properties of interstellar plasmas. The role of continuum processes in general for studying dust and the far infrared emission of the diffuse medium is postponed to Chap. 6.[3]

2.2 The Transfer of Radiation Through the Interstellar Medium

Radiative transfer theory describes how a beam of photons in a direction $\hat{k}$ of monochromatic intensity I_ν changes as it propagates through a medium. The beam can be extinguished by absorption (with coefficient κ_ν) and scattering (coefficient σ_ν) out of the line of sight, but photons can also be created by emission (with a coefficient j_ν) or injected by scattering *into* the line of sight. The intensity, having the units of a surface brightness (*cgs* units: ergs s^{-1} cm^{-2} ster^{-1} Hz^{-1} or per unit wavelength in appropriate units, or Jy sterad^{-1}) is independent of distance and only changes along a line of sight by interactions. Calling the general monochromatic extinction coefficient κ_ν and the emission $j_\nu\rho$ (*cgs* units: ergs s^{-1} cm^{-3} ster^{-1} Hz^{-1}), the simplest form of the equation of transfer is:

$$dI_\nu = -(\kappa_\nu I_\nu - j_\nu)\rho ds. \tag{2.1}$$

The *intrinsic* emissivity is independent of I_ν but any scattering terms depend on the *mean intensity*

$$J_\nu = \frac{1}{4\pi}\int I_\nu d\Omega \tag{2.2}$$

where Ω is the solid angle. The scattering includes the angular distribution of the photons coming from some direction $\mathbf{k}'$ and scattered into the direction $\mathbf{k}$ through the coefficient α:

$$\text{(scattering term)} = \frac{1}{4\pi}\int \alpha(\hat{k}',\hat{k})\sigma_\nu I(\hat{k}')d\Omega'. \tag{2.3}$$

When diffuse radiation is important, scattered light can change level populations depending on the background spectrum. This is important for the far infrared, in the

[3]A word of warning to the reader. In this chapter, because of the broad range of applications found in line formation theory to atoms and molecules, we will treat them on the same footing. Sometimes, we will switch between them in a single discussion but we hope the context will make clear what we are referring to.

densest clouds, and certainly in stellar atmospheres and even HII regions, but it can be safely neglected in the diffuse ISM. Scattered light is especially significant when treating dust but also a primary mechanism for making dark lines. The radiative energy density is related to the mean intensity by

$$\epsilon = \frac{1}{4\pi c} \int\int I_\nu d\Omega d\nu. \tag{2.4}$$

Since the *point spread function* (*psf*) of any telescope is finite, the source is always observationally convolved with the beam. However, intensity is not measured if the source is smaller than the beam. For these what is routinely measured is actually the monochromatic *flux* (also called the flux density), the intensity integrated over the solid angle of the *psf* or its decomposition of flux density per beam in the direction relative to the surface normal:

$$F_\nu = \int_{beam} I_\nu(\hat{k})\hat{\mathbf{k}}' \cdot \hat{\mathbf{n}} d\Omega \;/ \int d\Omega \tag{2.5}$$

where $\mathbf{n}$ or $\hat{n}$ is the surface normal relative to the line of sight and the integral is over the *psf* or beam of the telescope.

A photon can disappear either altogether or simply deviate from the direction of the observer. The former is a true absorption event and, as we will see, couples the radiation directly to the gas. The latter conserves photon number if it is coherent, or redistributes the energy among several emergent photons. The distinction is important. In a sufficiently dense medium, the incident photons are destroyed by collisions and the energy is redistributed within the "thermal pool" of the ambient particles that are not the absorbers. In such case, the excitation of the ions is governed by the energy distribution of the colliding particles. Instead, if the collisions are relatively rare, as in the diffuse ISM, the absorption event ends in a radiative de-excitation, the decay of the excited state, and the populations are determined by the spectrum of the incident photons. The same is true of the emergent spectrum. The ratio of the emissivity to absorption is called the *source function*

$$S_\nu = \frac{j_\nu}{\kappa_\nu} \tag{2.6}$$

which, for thermal equilibrium, is the Planck function, $B_\nu(T)$, at temperature T.

The optical depth of the medium is a measure of the extinction and is defined[4]

$$d\tau_\nu = -(\kappa_\nu + \sigma_\nu)\rho ds. \tag{2.7}$$

[4]The choice is to use the absorption coefficient per unit mass, which requires the mass density ρ, or per absorber, so one uses the number density n. Only consistency is needed, the two are formally the same.

In the absence of internal sources, the solution to equation 2.1 for an incident intensity $I_\nu(0)$ is:

$$I_\nu = I_\nu(0)e^{-\tau_\nu}. \tag{2.8}$$

This result differs from the usual presentation in discussion of stellar atmospheres since we are treating a single line of sight toward a point source rather than specifying a geometry for the intervening medium and treating diffuse radiation.

Although simple, equation 2.8 is actually profound. Re-interpreted, it gives that the probability (ignoring emission by the medium) of a photon at any frequency penetrating a scaled distance τ_ν in a medium in any direction. The photon's probability of emerging from a medium if its source function is uniform is a Poisson process (individual photons with an exponential distribution in penetration)

$$I_\nu = S_\nu(1 - e^{-\tau_\nu}). \tag{2.9}$$

The optical depth is the measure for a gas of its transparency to photons. and should *not* be confused with a probability of transmission. More importantly, it is an integrated quantity along the line of sight and cannot distinguish individual contributors. In the same way that independent probabilities multiply, the optical depths simply add at a fixed frequency. Therefore, if we have individual clouds along a line of sight, each contributes separately depending on the frequency.

2.2.1 *Statistical Balance*

We can now derive the emission and absorption coefficients. The level populations are determined by the competition between radiative and collisional rates. These form the link between the phenomenological measures and the microphysical properties. Since in this chapter we will usually deal with resonance transitions, we will use a two level system as our paradigm. The two levels are $E_i = 0$ and $E_j > E_i$. The governing equation, neglecting ionizations, is given by local statistical equilibrium, linking all radiative and collisional processes. The *steady state* rate equation for this two-level system links the level populations n_i and n_j to the ambient radiation J_ν and electron density n_e:

$$(n_jB_{ji}\Phi_{\nu,em} - n_iB_{ij}\Phi_{\nu,abs})J_\nu + (n_jC_{ji} - n_iC_{ij})n_e + (n_jA_{ji}\Phi_{\nu,em}) = 0 \tag{2.10}$$

where we have grouped together the stimulated absorption and emission terms that depend on the arrival rate for photons of frequency $\nu = (E_j - E_i)/h$ in the first parentheses, the collisional terms in the second, and the spontaneous decay from the upper level in the third. The rates for the radiative processes are the Einstein

A and B transition probabilities, and $\Phi_{\nu,abs}$ and $\Phi_{\nu,em}$ are the profile functions for absorption and emission, respectively. The collision rates are C_{ij} and C_{ji} where

$$C_{ij} = \int_{\Delta E}^{\infty} \sigma_{ij}(E)\, vf(E)dE. \tag{2.11}$$

In this equation $\sigma_{ij}(E)$ is the energy dependent collision cross section, v is the relative velocity, and $f(E)$ is the energy distribution of ambient electrons (not necessarily a Maxwellian).[5] The downward rate has no threshold, so $\Delta E = 0$, while for the excitation rate the lower limit on the integral is $\Delta E = E_j$. If collisions are sufficiently frequent that they balance separately from radiative rates, the level populations are fixed by the kinetic temperature of the electrons, T_{kin}. The radiation is then in thermal equilibrium with matter, this condition is called *local thermodynamic equilibrium* (LTE) because it is valid on the microscopic scale set by the collisional mean free path and $T = T_{kin}$. The level populations are governed by collisions separately from the radiative processes in what is called *detailed balance*. If the emission and absorption profiles are the same, an assumption called *complete redistribution*, then the ratio of the radiative terms determines the energy distribution of the radiation

$$J_\nu = A_{ji}n_j/[n_iB_{ij} - n_jB_{ji}]. \tag{2.12}$$

This is the source function we defined earlier, the ratio of the emission to absorption coefficients. For LTE, J_ν is the Planck function

$$B_\nu(T) = \frac{2h\nu^3}{c^2}\frac{1}{e^{h\nu/kT} - 1} \tag{2.13}$$

and level populations follow a Boltzmann distribution

$$\frac{n_j}{n_i} = \frac{g_j}{g_i}e^{-h\nu/kT} \tag{2.14}$$

where the statistical weights of the two levels are (g_i, g_j). Substituting equation 2.14 into 2.12 provides the *definition* of the Einstein transition probabilities:

$$A_{ji} = \frac{2h\nu^3}{c^2}B_{ji}, \qquad g_iB_{ij} = g_jB_{ji}. \tag{2.15}$$

The emission is then given by

$$j_\nu = h\nu A_{ji}\Phi_\nu n_j \tag{2.16}$$

[5]Note that the perturbers can also be H atoms and H_2 molecules, or even He atoms. The electrons will be the most important, mainly because of their mass, and they have much larger mean velocities than the protons. But the same formalism applies to molecules and atoms.

and the absorption coefficient is

$$n_i \kappa_\nu = n_i B_{ij} \Phi_\nu (1 - e^{-h\nu/kT}); \tag{2.17}$$

where the exponential accounts for the stimulated emission rate.

2.2.2 *Radiative Processes in the Rayleigh-Jeans Limit*

We are going to deal extensively with molecules and long wavelength observations so we will often be in the regime $h\nu_0 \ll kT$. This is the called the Rayleigh-Jeans limit for the Planck function implying LTE. The solution for the transfer equation then becomes

$$T_B = (T_{ex} - T_{bg})(1 - e^{-\tau_\nu}). \tag{2.18}$$

Here, T_B is the brightness temperature of the emitting gas, defined as $T_B = I_\nu c^2/2k\nu^2$—see Sect. 4.2.6), T_{ex} is the excitation temperature of the medium, and T_{bg} is the brightness temperature of a background beyond the medium. Properly speaking, T_B is really ΔT_B, the difference in a radio measurement towards a source and off the source. We will discuss *on/off observations* as these types of measurements are called by radio astronomers in Sect. 4.2.2. When $\tau \gg 1$, equation 2.18 becomes $\Delta T_B = T_{ex} - T_{bg}$ and observations of T_B provide information on the excitation temperature of the medium. In the optically thin case, when $\tau \ll 1$, equation 2.18 reduces to

$$\Delta T_B = (T_{ex} - T_{bg})\tau_\nu. \tag{2.19}$$

Most molecular transitions in the diffuse ISM are optically thin so that this equation may be used together with an expression for the absorption coefficient, κ_ν to derive the column density of the ionic species in question.[6] We show how this is done in Sect. 2.2.5 .

2.2.3 *The Line Profile Function*

The line emission and absorption coefficients require knowing the intrinsic profile for the transition. This is a Lorentzian function that has the same form as a damped

[6]Ironically, the principal transition for tracing molecular gas in the ISM, the CO(1-0) line at 115 GHz or 2.6 mm, is often optically thick even in the diffuse ISM.

harmonic oscillator.[7] For the classical emission mechanism, a radiation damped harmonic oscillator (e.g. Jackson 1999), the classical damping rate is

$$\Gamma_{rad} = \frac{2}{3}\frac{e^2\nu_0^2}{4\pi^2 m_e c^3}. \tag{2.20}$$

This, however, is modified by the quantum mechanical selection rules depending on the multipolarity of the transition. This Γ_{rad} now includes the *oscillator strength*, f_{ij},

$$\frac{\pi e^2}{mc} f_{ij} = \frac{h\nu}{4\pi} B_{ij}. \tag{2.21}$$

The generalized Γ_{rad} value depends on the transition probabilities (the equivalent of a Q-value for an oscillator)—the lifetimes of the levels i and j together. However, most of the observable absorption lines from the diffuse ISM are from the ground state so the broadening is due to the excited state so that $\Gamma_{rad} = \Gamma_{ij} = A_{ji}/4\pi$ and the absorption coefficient is

$$\kappa_\nu = \frac{\pi e^2}{mc} f_{ij} \frac{\Gamma_{ij}}{4\pi} \frac{1}{\Delta\nu^2 + (\Gamma_{ij}/4\pi)^2}. \tag{2.22}$$

The line optical depth is proportional to $g_i f_{ij}$.

Besides the intrinsic profile, each atom also has a random thermal motion that Doppler shifts the frequency of any absorbed or emitted photon. Hence, although the rest frequency is precisely defined, the ensemble of particles has a three dimensional Gaussian velocity distribution relative to the photons. In an absorbing medium, the effect is only along the line of sight so the chance that a photon will be seen at any frequency $\Delta\nu$ relative to ν_0 is

$$G(\Delta\nu) = (2\pi\Delta\nu_D^2)^{-1/2} e^{-(\Delta\nu/\Delta\nu_D)^2} \tag{2.23}$$

where

$$\Delta\nu_D = \frac{\nu_0}{c}\left(\frac{2kT}{Am_p}\right)^{1/2} \tag{2.24}$$

is the *thermal* Doppler width of the line for a species of atomic mass A. For the diffuse ISM it would be around $0.1(T/10\ K)^{1/2}A^{-1/2}$ km s^{-1}. However, observed line widths are often much larger than this, up to several km s^{-1}, and independent of the atomic mass. So nonthermal motions are clearly important. If such dynamical broadening is due only to uncorrelated motions, with a typical random speed σ, the

[7] See, e.g., Griem (1964, 1997), Spitzer (1978), Rybicki and Lightman (1979), Draine (2011), and Hubeny and Mihalas (2014).

thermal and dynamical contributions add in quadrature, $\Delta\nu_D^2 = \Delta\nu_{D,th}^2 + (\nu_0\sigma/c)^2$. Regardless of the microphysical intrinsic profile, the random motions broaden the line at each $\Delta\nu$ and the total profile is the convolution $P(\nu) = G \star \Phi$. For an intrinsic Lorentzian profile, the convolution function is the so-called Voigt profile

$$H(v,a) = \frac{a}{\pi^{1/2}} \int_{-\infty}^{\infty} \frac{e^{-x^2} dx}{(v-x)^2 + a^2} \tag{2.25}$$

where $a = \Gamma/(4\pi\Delta\nu_D)^{1/2}$ is the dimensionless line width and the integration variable (whether in wavelength or frequency) is normalized to the Doppler width. If a is very small (for a Gaussian much broader than the Lorentzian for any reason), the Voigt profile is like a convolution of a Gaussian and a δ-function so the resulting profile is also Gaussian. For each species, the full width half maximum is given by $2.35\nu_0\sigma/c$ and if T is constant along the line of sight, the FWHM of lines of different atoms depends only on the atomic weight. If, on the other hand, the thermal broadening is negligible compared to the intrinsic line width, the profile will be Lorentzian. For such a profile, the FWHM is defined but the dispersion (second moment) is not and the wings will be much broader than for a Gaussian. For large $\Delta\nu$, the profile varies as $\Delta\nu^{-2}$ (see Fig. 2.1).

Only an uncorrelated random velocity field will produce a Gaussian profile. This is sometimes referred to as *macroturbulence*, although the term is a misnomer. The hypothesis is that whatever the magnitude of the velocities, they are distributed in a sort of fully developed (homogeneous, isotropic) chaos. So being independent of the thermal motions, this macroscale contributes another Gaussian with a dispersion σ that depends on the larger scale random motions and *not* on the thermal width, σ_T.

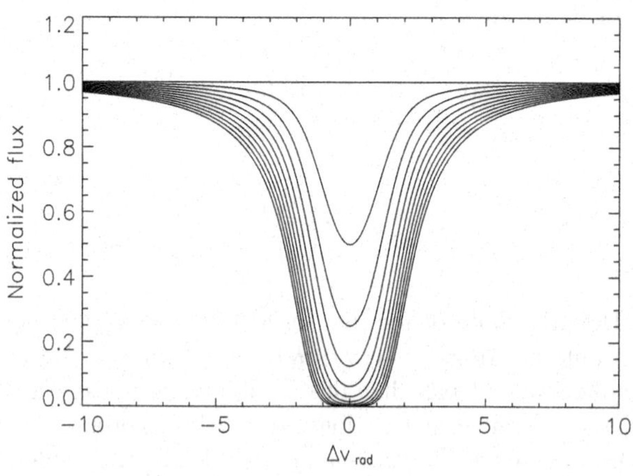

Fig. 2.1 Variation in the absorption (in normalized flux units) as a function of radial velocity (in scaled units) for a series of Voigt profiles with equivalent widths in scaled units from 0 to 1000 in steps of 10% starting from zero (so 0,100,200,. . . ,1000). Note the increase in the wings once the profile saturates at line center

Since the convolution of two Gaussian profiles is another Gaussian with the width being given by $(\sigma^2 + \sigma_T^2)^{1/2}$, the relative contribution of the Lorentzian wings to the line profile is reduced. Macroscopic motions may not be drawn from a normal distribution but they shift the frequency of the transition from ν_0 along the line of sight. This is where the ISM is very different from a stellar or planetary atmosphere. Since the medium is *not* in hydrostatic equilibrium, there are virtually no constraints on the velocity field. Not only can the motions be supersonic but the line of sight can pass through regions that have nothing mechanically to do with each other. We will return to this in more detail when discussing observations of turbulence in Chap. 11, but for now we simply ask what macroscopic motions do to the line formation.

In denser plasmas, collisions significantly reduce the lifetime of excited states and broaden the levels. In general, this is not important under interstellar conditions. Yet collisions cannot be completely ignored. In the absence of a strong radiation field, excitation can still occur if the energy differences are of the same order or less than the thermal energy. For instance, consider the CO(1-0) transition, the lowest rotational line at 115 GHz for the $^{12}C^{16}O$ isotopic composition. For cold environments, in the absence of a background source of radiation, the excitation is governed by collisions followed by radiative de-excitation. Therefore the strength of the line depends on the collision rate, hence on the thermal pool of the gas. While this is not really in LTE, the excitation temperature, hence the brightness temperature, is close to the kinetic temperature of the colliding H_2 molecules that are principally responsible for the excitation in the molecular medium. Every emitted photon, corresponds to a collision no matter how low density the medium may be so the luminosity depends on the CO/H_2 ratio. This is the origin of the so-called "CO to H_2 conversion factor", conventionally designated as X_{CO} (this conversion factor is discussed extensively in Chap. 8). The masses of the emitting region are thus determined by a proxy measure based on the line that can be *observed* while the H_2 is virtually invisible in wavelengths longer than the far ultraviolet (FUV) where the electronic state absorptions are detectable (in contrast to the atomic case, aside from the UV there is no strong background against which the lines can absorb and the transition probabilities are relatively low).

The lack of collisional de-excitation greatly simplifies the treatment of *absorption* line formation. Furthermore, for atomic resonance line absorption, collisional excitation plays no role, so the only temperature dependence of the column density comes from the ionization fraction.[8] For the transfer equation, large scale velocity fields (i.e. shear flows or ordered motions that have large velocity gradients), turbulence, and thermal Dopper broadening all combine to desaturate a line profile. The overall transparency of the gas at any frequency increases and, for the curve of growth (see the next section), this extends the range of column density over which the absorbed fraction of light along the line of sight is linearly proportional to the optical depth at line center and $\tau \sim N$. The normalization is over a broader profile

[8]The only complication arises when treating fine structure lines, such as [CII] at 158 μm, because collisions affect the population ratios for the atomic levels.

so the column density required to produce the same equivalent width increases. This also delays the onset of the saturated portion of the curve of growth and the column density at which the Lorentzian wings contribute to the total absorption.

It should be emphasized that *any* macroscopic motions, provided that they exceed the thermal broadening, will produce a similar effect. The difference between different mechanisms is in the detailed change in the profile. For instance, an ordered magnetic field also desaturates the profile since it decreases the optical depth at line center by redistributing the absorption over the Zeeman components. The profile is not, however, Gaussian so the change in the curve of growth with column density will not be the same as desaturation of a purely random field. This is true even though each component in the split line is thermally broadened (and also macroscopically affected). The one important difference is that since the Zeeman effect is coherent for an ordered field there is no coupling between the states and the profile can be distinguished by polarization. This is important for Zeeman measurements, for instance using the 21 cm line, since the magnetic fields are usually very weak in the ISM. For *integrated, unpolarized* measurements, it is possible that the random motions dominate and the profile looks Gaussian.

The effect of a magnetic field on line profiles is rather simply treated for interstellar lines of sight since the individual lines are formed under collisionless conditions. Observationally, however, measuring the field strengths is very challenging. The splitting of optical lines is far too small for any measurements of the splitting since the fields are of order μG. but since the separation of components varies as λ^2, radio observations can measure the fields. Direct measurements of Zeeman splitting in integrated profiles is exceedingly difficult since a field of microGauss, typical of the diffuse phase, produces a negligible splitting relative to the dynamical and thermal broadening of the profiles (of order 1.4 Hz μG^{-1}; e.g., Fig. 4.5). However, using circular polarization, the degree of polarization measured from the difference in intensity between two circular polarization states, is directly proportional to the field strength (see the recent review article by Crutcher 2012). These measurements are unaffected by self-absorption or other optical depth effects so the only requirement is sufficient signal to noise ratio to obtain the field. Successful measurements of the Zeeman splitting of low-lying CN transitions ($\mathrm{N} = 2-1$, $\mathrm{N} = 1-0$) have been achieved in dense clouds (e.g. Crutcher 2012, 2014). The competition between collisional coupling and radiative transitions in the presence of velocity shears (large scale flows) can enhance the polarization by changes in the optical depths of the Zeeman components but this has not been seen, yet, in low density clouds (Goldreich and Kylafis 1981). It is another case in which the escape probability formalism with flows makes the treatment of the radiative level couplings simpler.

2.2.4 Obtaining Column Densities from Absorption Lines: Equivalent Width and Curve of Growth Methods

The discovery of the diffuse interstellar medium was a *spectroscopic* accident. Observations of the radial velocity variations of several unresolved binary stars, especially δ Ori, led Hartmann in 1904 to identify stationary absorption lines with the presence of unbound gas interposed between the stellar system and the Earth. The centroid velocities of these features was neither that of the center of mass of the binary nor directly related to any of the neighboring stars. They were also specific, two absorption lines that were normally not seen in the spectrum of the hot stars being studied. The two main optical atomic absorption features are the doublets of NaI 5889, 5895 Å and CaII 3933, 3968 Å. Importantly, both are resonance lines of their respective ions and both are doublets. The NaI D lines, $2p^6 3s^2 S_{1/2} - 2p^6 3p^2 P^o_{3/2,1/2}$ share a common lower state and CaII H and K, similarly, have almost identical behavior, $3p^6 4s^2 S_{1/2} - 3p^6 4p^2 P^o_{3/2,1/2}$ with the higher J value being the upper state. Because of the difference in statistical weights of the multiplet substates, the components differ by almost 0.3 in $\log gf$ and therefore have different optical depths at any column density.

The most important property of absorption lines arising from the diffuse interstellar medium is that they are seen in absorption *against* something external. This is not as circular a statement as it might seem. A stellar or extragalactic, continuum source (or even line emitter) has no interaction with the intervening medium. Yet, unlike the effect of an optically thickness on the emission lines formed within a cloud, the ability to see the illuminating object implies that one sees completely through the absorbing medium. If the distance to the background source is known, this provides an unambiguous measure of the mean volume density within the resolution of the spectrum. Low density clouds have such low optical depths, however, that the highest velocity portions of the profile can frequently be missed.

To quantify the total (absorption plus scattering) extinction fraction of the light removed by dark lines from any portion of the spectrum, a convenient measure is the *equivalent width*, W_λ, defined by

$$W_\lambda \equiv \int_{-\infty}^{\infty} \left(1 - \frac{I(\Delta\lambda)}{I(\lambda_0)}\right) d\Delta\lambda \qquad (2.26)$$

where λ_0 is the wavelength at line center. The integral is taken over the entire profile (or any band of the spectrum). Since the units are usually Ångstroms, the integral is written in wavelength units and the unit of W is Å although this choice of dimensions is purely conventional. Note that since the intensities are normalized to the continuum, W_λ is independent of distance. What is seen from the definition is that W_λ has several important limits depending on the line profile. Writing $\tau_\nu = \tau_0 \Phi_\nu$ where $\Phi(\Delta\lambda)$ is the (normalized) line profile function and τ_0 is the optical depth at line center,

$$W_\lambda = \int_{-\infty}^{\infty} (1 - e^{-\tau_0 \Phi(\Delta\lambda)}) d\Delta\lambda. \tag{2.27}$$

Even without precisely connecting the opacity with a specific transition, in the limit of a nearly transparent medium—when τ_0 is very much smaller than unity—the equivalent width varies as

$$W_\lambda \sim \tau_0 \sim N_i \tag{2.28}$$

since τ_0 is proportional to the column density of the lower state N_i. For a Gaussian line profile with full width at half maximum (FWHM), $\Delta\nu$,

$$\tau_i = \left(\frac{ln\, 2}{\pi}\right)^{1/2} \left(\frac{2}{\Delta\nu}\right) \frac{\lambda^2}{8\pi} A_{ji} \frac{g_j}{g_i} N_i \left[1 - exp\left(-\frac{h\nu}{kT_{ex}}\right)\right], \tag{2.29}$$

and the column density for a weak line scales as

$$N_i = 1.1 \times 10^{20} \lambda^2 f_{ij}^{-1} W_\lambda(\text{Å})\, \text{cm}^{-2}. \tag{2.30}$$

Here λ is in Å. With increasing optical depth the residual intensity at line center approaches zero but, because the profile is not a δ-function, there is always additional absorption from beyond line core, depending on the form of Φ. This is the *saturated* rate of growth of the profile, when the equivalent width increases more slowly than linearly with increasing column density. For a Lorentzian dominated line, also called a *damped absorption line* (as in the cosmological case of the Ly α forest), the curve of growth yields $W_\lambda \sim N_i^{1/2}$, as seen from Fig. 2.2. The effect of *gf* on the line strength is shown in Fig. 2.3.

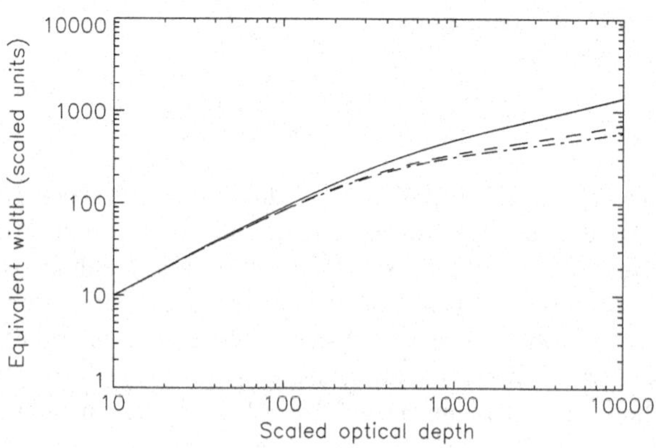

Fig. 2.2 Sample curve of growths in scaled units for a= 0.5 (*solid line*), 0.1 (*dashed line*), and 0.05 (*dash-dot line*)

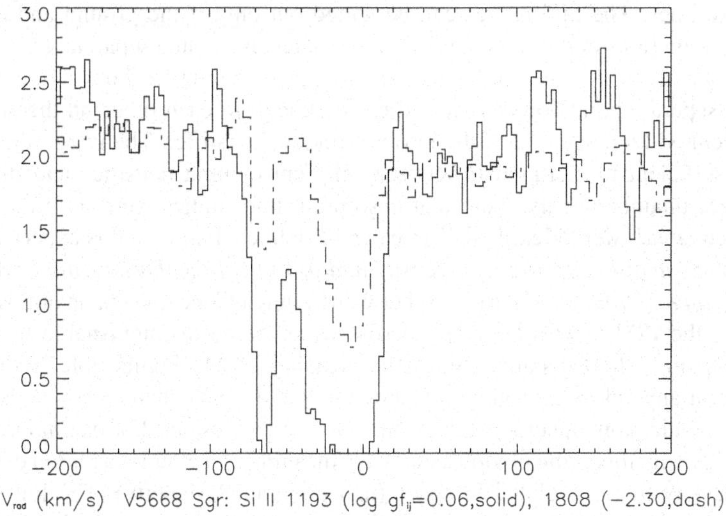

Fig. 2.3 Absorption lines along the line of sight to the nova V5668 Sgr 2015. The *solid line* is Si II at 1193 Å and the *dot-dashed line* is the 1808 Å line. This figure illustrates the effects of log (gf)

The lower level of an observed transition need not necessarily be the ground state, even under interstellar conditions. Fine structure states, because of their very small separations, can be collisionally excited even in the warm medium. Notwithstanding that such events are extremely rare, the induced populations may suffice to produce detectable absorption. Thus, the kinetic temperature can be obtained from the comparison of column densities derived for different transitions with different lower state energies. For example, the ratio of the ground state population to that of a fine structure state gives the excitation temperature, T_{ex}. This is well illustrated by the comparison of the two CII UV resonance lines (1334, 1335 Å). In a sequence, lines arising from different ground state multiplets of the same ion provide an ensemble of estimates of the electron temperature (or radiative excitation by the fine structure transitions). Since the latter is quite unlikely, given the low transition probabilities for those lines, the electron excitation dominates.

For the same ionic species, the set of available lines all have different gf values. If they are all resonance lines, each should have its own rate of saturation so taken together they provide a curve of growth. This is a simple means for visualizing the dependence of the equivalent width on the total τ. By including lines from excited states from the same ion, the line strengths are modulated by the term $\Delta E_i/kT_{ex}$. Multiplets have the same T_{ex}. The resulting integrated line strength is, however, very sensitive to spectral resolution. Along the line of sight, different clouds have different temperatures. This has no effect on the ground state lines but for fine structure lines, The differences can be sorted out only if the ground and excited state contributions in the same velocity interval can be treated separately.[9]

Thus, comparing two lines of sight through the same medium (should there be two separate point sources viewed close enough to each other on the sky) the equivalent width may differ if the intrinsic line widths differ. The integral measure is not sufficient to distinguish between different column densities and different dynamics unambiguously. Note that absorption line studies were the first means by which *clouds* were identified. These can be defined, based on this experience, as *coherent structures of enhanced density and/or optical depth discernible within a limited range in radial velocity*. In other words, much like clouds in an atmosphere, those in the ISM have a boundary across which some distinguishable feature(s) change(s) in a short distance. Since the meaning of "distinguishable" requires a change (at any velocity) of the optical depth, the physical change can be complex (density, ionization, opacity mechanism, velocity dispersion). For example, for the volume density this is the obvious, everyday meaning of a cloud—a region of higher local density that is isolated in space (hence in radial velocity). The line width, by construction, is due to the internal dynamics of the medium. If they follow a Gaussian distribution, whether kinetic or microscopic motions, the value is given by the sum of the individual contributions in quadrature. If, however, there are either multiple components along the line of sight or a more complicated velocity spectrum (such as turbulence, see Chap. 11), this line width will give a wrong estimate of the column density.

2.2.5 Obtaining Column Densities from Emission Lines

Emission lines from an optically thin transition do not follow a curve of growth (they don't saturate) but much of the analysis is similar to absorption lines. We can treat emission lines in the low frequency (Rayleigh-Jeans) limit by assuming that the excitation and kinetic temperatures are the same and ignoring collisional de-excitation. The emissivity is then $J_\lambda = \tau_\lambda B_\lambda(T_{ex})$ in approximate LTE. The conventional quotation of emissivity uses the brightness temperature integrated over the line width in velocity units so, over the FWHM ΔV, the column density of the *upper* level, N_j is

$$N_j = (T_{ex} - T_{bg})^{-1}\left[exp\left(\frac{h\nu}{kT_{ex}}\right) - 1\right]^{-1} \frac{4\pi^{3/2}}{(ln\ 2)^{1/2}} \frac{1}{\lambda^3} \frac{1}{A_{ji}} T_B dv. \quad (2.31)$$

[9]As an example, Jenkins (1996), Cardelli et al. (1989), Fitzpatrick and Massa (1990, 2007), and Savage and Sembach (1996) use the τ in individually chosen velocity intervals based on the ground state line to determine individual line of sight cloud properties.

Replacing $\Delta T_B \Delta V$ with the integral $0.935 \int T_B dv$ assuming again a composite Gaussian profile, we obtain the scaling law for the column density in terms of the velocity integrated brightness temperature

$$N_j = 24.1 (T_{ex} - T_{bg})^{-1} \left[exp\left(\frac{h\nu}{kT_{ex}} \right) - 1 \right]^{-1} \frac{1}{\lambda^3} \frac{1}{A_{ji}} \int \Delta T_B dv \quad . \tag{2.32}$$

The only novelty here is a change in vocabulary. Usually at shorter wavelengths, from the IR through UV, fluxes are expressed in physical units rather than temperatures but otherwise optical ground state emission is treated in much the same way, at least for fine structure lines.[10] Except for the terminology, expressing the intensities as temperatures, the principals are identical with a few exceptions. The exponential for stimulated emission can be neglected since, except at cm wavelengths, $h\nu \gg kT_{ex}$ for typical interstellar conditions. In general, since the spontaneous transition rates are so high relative to the downward collision rates and the background flux also high, the lines are seen in absorption and the resulting optical depth comes from the *lower* state column densities. Whether the line is formed by collisions or scattering, the emission is irrelevant for the lower states except when formed by a recombination cascade. However, at radio wavelengths, the rotational levels have energy separations of order kT_{ex} so collisional excitation is possible, resulting in emission lines. Note that the difference in line formation is essential. The background sources at millimeter and centimeter wavelengths, other than the Galactic nonthermal emission (and the CMB in the case of H_2CO) are not sufficiently intense that their source function overwhelms the emission.

2.2.6 An Example: HI—The 21 Centimeter Line

The HI "spin-flip" transition at 21 cm or 1420.40575177 MHz was the first spectral line detected in the radio part of the spectrum in the 1950s (Ewen and Purcell 1951; Muller and Oort 1951). With the exception of maser emission lines, it is the strongest radio spectral line, often visible in an oscilloscope monitoring the observations. An excellent review of the observations of HI in the Galaxy that have been made with this transition is by Dickey and Lockman (1990). We will limit the discussion to general considerations that are directly relevant to the diffuse ISM.

In the ground state of neutral hydrogen ($1s^2S_{1/2}$) the proton and the electron each have an intrinsic magnetic dipole moment so the state splits between the

[10]Those transitions that are radiatively pumped or result from recombination cascades require a more complicated approach and we refer the reader to, e.g. Kaplan and Pikel'ner (1970), Osterbrock and Ferland (2006), or Draine (2011) for further discussions.

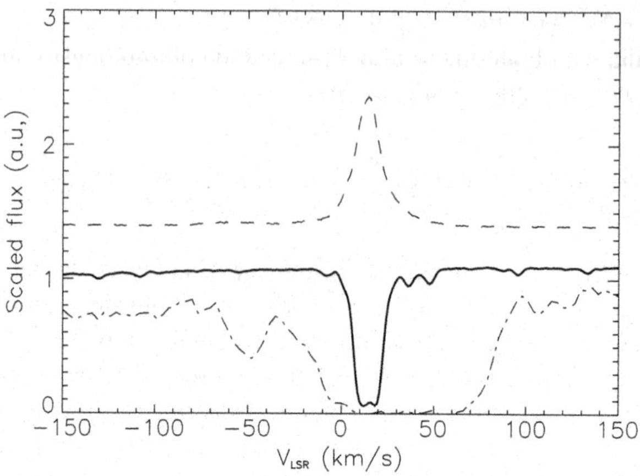

Fig. 2.4 An example of the neutral gas (NaI 5889 Å (solid black line) and OI 1302 Å (dot-dash-line)) compared with the 21 cm line emission profile (from the LAB survey—denoted by *dashed line*) along the line of sight toward the bright Galactic nova V339 Del. The clouds distinguished by the resonance metal absorption lines delineate a limited velocity range within the 21 cm emission, which is dispersed along the sight line beyond the star. The comparison of absorption and emission components in velocity constrain the E(B-V) through calibrations of the NaI D line absorption vs. reddening

parallel and antiparallel configuration with the latter having lower energy. These hyperfine sublevels are separated by 0.047 cm^{-1}, corresponding to 0.7 K, and have statistical weights of 3 and 1 for the upper and lower levels, respectively. Any radiative transition between the levels changes the spin, so it is forbidden and the transition probability, A_{10} is very small: 2.85 $\times 10^{-15}$ s^{-1}, equivalent to a lifetime of about 10 million years. The line is strong only because of the overwhelming abundance of atomic hydrogen in the ISM (see Fig. 2.4). Finally, the $n = 2$ or higher levels (the Balmer series) all lie more than 82000 cm^{-1} above the ground electronic state so their population in the neutral diffuse ISM is negligible (they are seen in emission only from recombination in HII regions or when the Lyman series is so optically thick that the levels are radiatively populated, which virtually never happens). Observations of the 21 cm line thus lead to very robust estimates of the atomic hydrogen column density. The excitation of the 21 cm transition is mainly by electron collisions. Excitation by the CMB can be ignored at typical interstellar densities and kinetic temperatures, Thus, in an example we cited earlier for collisional excitations, the excitation temperature for the 21 cm transition, also known as the *spin temperature*, T_s, is approximated by the gas kinetic temperature of the gas.

2.2.6.1 HI Emission

The line optical depth is $\tau \approx -ln(1 - \frac{\Delta T_B}{T_s})$ so when $\tau \ll 1$, the column density for the upper level is

$$N_u = 1.82 \times 10^{18} \int \Delta T_B dv \quad cm^{-2}. \tag{2.33}$$

If, instead, the line is optically thick, the column density is

$$N_u = 1.82 \times 10^{18} \int \Delta T_B \frac{\tau}{(1 - e^{-\tau})} dv \quad cm^{-2}. \tag{2.34}$$

As we discussed in Chap. 1, the diffuse ISM is partly atomic and partly molecular. Consequently, a complete view of the medium perforce requires molecular spectroscopic observations along with those at 21 cm.

The atomic gas not only envelops the molecular clouds (the outer parts of the PDRs), it is also mixed within the molecular gas. This neutral atomic gas is detected via "HI narrow self-absorption" lines or HINSA (Li and Goldsmith 2003), although the H^o density in the HINSA regions is about 1.5×10^{-3} that of the molecular gas (which is most likely around 10^3 cm^{-3}). The atomic component in dark clouds is produced by cosmic rays destroying H_2.

In emission, the problem of studying the HI distribution is primarily one of angular resolution. For example, at 21 cm, the Arecibo radio telescope has a relatively coarse resolution of $\sim 4'$. All-sky surveys of HI, such as the Leiden/Argentine/Bonn (LAB) survey (Kalberla et al. 2005) have even worse angular resolution, about 0.6 degrees. This is larger than the size of many diffuse molecular clouds and renders atomic-molecular comparisons approximate at best. Despite these difficulties, studies of the HI distribution around the lower-density molecular clouds can yield interesting results (e.g., Gir et al. 1994; Moriarty-Schieven and Wannier 1997). For example, in some instances the atomic gas envelops the molecular gas as in the case of MBM 40 (Shore et al. 2003), while in other cases the HI peaks to one side of the principal molecular emission (e.g., see Fig. 1.6, bottom right panel; Fig. 9.2, bottom right panel; and de Vries et al. 1987). However, improvements in resolution hold the key to understanding in detail the relationship between the gas components of the diffuse ISM.

Savage et al. (1977) established that in regions where E(B-V) $<$ 0.1 mag the hydrogen nucleons were almost entirely atomic and the relationship between N(H_{total}) and E(B-V) could be expressed as

$$N(H) = 5.8 \times 10^{21} E(B - V) \qquad cm^{-2}. \tag{2.35}$$

This relationship was derived from Lyα observations toward early-type stars and is used extensively by astronomers (see Sect. 6.3).

All-sky maps of the neutral hydrogen sky provide the column density of HI in any direction albeit usually at somewhat poor resolution. The LAB HI survey (Leiden/Argentine/Bonn Galactic HI survey (Kalberla et al. 2005).[11] Covers the entire sky at a resolution of $\sim 0.6°$. The survey merges the Leiden/Dwingeloo survey by Hartmann and Burton (1997) with that of the Instituto Argentino de Radioastronomia. Covering a velocity range of $\pm 400\,\text{km s}^{-1}$ at a velocity resolution of $1.03\,\text{km s}^{-1}$, the survey has an excellent rms noise in brightness temperature of 0.07–0.09 K and is corrected for stray radiation (see Sect. 4.2.5). With the most extensive coverage both spatially and kinematically and excellent sensitivity, this survey's only flaw is the relatively coarse spatial resolution. Another notable HI survey is HIPASS (the HI Parkes All-Sky Survey) with most of the data taken with the CSIRO 64-m Parkes Telescope. The survey covers the entire southern sky as well as declinations in the Northern Celestial Hemispher up to $+25°$. With a spatial resolution of $15.5'$, a velocity coverage of $-1280 < cz < 12700\,\text{km s}^{-1}$, and a velocity resolution of $18\,\text{km s}^{-1}$, the survey is best suited for extragalactic HI studies because the bandpass correction renders the Galactic signal virtually unusable for dynamics (each spectrum has the median signal from a declination strip $4°$ in length centered on the declination of the position subtracted from it). Finally, the Arecibo GALFA (Galactic Arecibo L-Band Feed Array) HI survey (Peek et al. 2011) provides the best resolution in a wide-field HI survey ($3.5'$), but the coverage is limited to what can be observed from Arecibo ($-2° \leq \delta \leq 38°$). Nevertheless, 13000 square degrees of sky will be covered once the survey is complete, at an excellent velocity resolution ($0.18\,\text{km s}^{-1}$) and with excellent sensitivity (80 mK rms in integrated $1\,\text{km s}^{-1}$ channels).

2.2.6.2 HI Absorption

Because HI can be found in every direction, there will be plenty of continuum sources that will show HI absorption. If we go back to the radiative transfer equation in the Raleigh-Jeans limit (equation 2.18) we can rewrite it as

$$T_B(\nu) = T_s[1 - e^{-\tau(\nu)}] + T_c e^{-\tau(\nu)} \tag{2.36}$$

where T_s is the 21 cm excitation temperature and the continuum temperature, T_c, replaces the background temperature, T_{bg}. Here $T_B(\nu)$ is just the on-source measurement and not $\Delta T_B(\nu)$. If we assume that the optical depth of the HI is small, then the above equation becomes:

$$T_B(\nu) = T_s\tau(\nu) + T_c[1 - \tau(\nu)] \tag{2.37}$$

[11] http:://lambda.gsfc.nasa.gov/product/foreground/LAB_HI_Survey_info.cfm.

If we denote the depth of the absorption line as $T_L(\nu) = T_c - T_B(\nu)$, we can rewrite equation 2.37 as

$$\tau(\nu) = \frac{T_L(\nu)}{T_c - T_s}. \tag{2.38}$$

Compare this formulation to that of equation 4.3 that allows for the determination of the optical depth of atomic hydrogen from on-off observations.

2.2.6.3 HI Absorption Continuum Measurements

There is yet another way to obtain information about the column density of neutral hydrogen, and also heavy elements such as the CNO group, using X-ray observations. Although still in development (e.g., Gatuzz et al. 2016) using *XMM* and *Chandra* X-ray band spectroscopy and *Swift* X-ray and UV photometry work rather well in discriminating continuum absorption intrinsic to the source and that from the gas along the line of sight. The idea is quite straightforward. Absorption above the Lyman limit decreases rather slowly with energy so even in the soft X-ray spectral range the neutral gas has an optical depth $\tau(E) \approx \tau_{LyC}(E/13.6eV)^{-3}$. Studies of explosive events, such as classical novae, show that the minimum N_H obtained from the X-rays equals that inferred from the Lyα and other UV and optical resonance transitions. From spectrophotometry, the absorption edges of heavy elements can be studied independent of the line excitation conditions. Finally, using X-ray sources, mainly supernova remnants and low mass X-ray binaries observed in the continuum, Güver and Özel (2009) derived an alternate relation between atomic hydrogen column density and extinction, $N_H = (2.21 \pm 0.09) \times 10^{21} A_V$ (see Chap. 6). The method is very close to that used for the relative abundances based on the Lyman series.

2.2.7 *Line Pumping by the Cosmic Background Radiation: The Optical CN Transitions*

There is one important case, however, where the occupation of the levels responsible for diffuse phase optical transitions are radiatively determined by the far infrared background. This was discovered when McKellar (1940) and Adams (1941) identified the two ground state rotational lines of CN in coude spectra. McKellar's study was made possible by Herzberg's contemporary laboratory studies, and his paper included a discussion where it was noted that the rotational excitation of the lowest CN states in absorption correspond to $T_{ex} = 2.3^{+1.1}_{-0.7}$ K, depending on the uncertainties in the R(1)/R(0) line ratios. In a lengthy section in his 1941 paper, McKellar included questions about the applicability of LTE to such an environment, but argued that the ubiquity of this anomaly indicates the presence of a universal exciting continuum and refers to Eddington's estimate of the mean radiation energy density resulting from a diffuse interstellar radiation field McKellar (1941).[12] When Alpher, Herman and Gamow were exploring the consequences of Big Bang cosmology for the production of the light elements, their predictions of possible residual emission from a primordial high temperature state were in the same range but they made no reference to the CN result.[13] The result re-appeared only after the detection of the cosmic microwave background by Penzias and Wilson in 1965. In back-to-back papers, Field and Hitchcock (1966a) and Thaddeus and Clauser (1966) explicitly cited the CN temperature as an indicator of the intensity of the CMB at 2.63 mm. Field and Hitchcock (1966b) went farther, showing that the CN supports the idea that the background radiation is not a manifestation of the DIRF (hence, finally, removing the conundrum raised by McKellar).

2.2.8 Collisional Excitation in a Nearly Collisionless Medium

Observed interstellar atomic transitions are either from the ground or fine structure states. The densities in the diffuse gas are so low that collisions are too infrequent to establish significant populations in the more highly excited states. The lines formed by absorption are therefore directly proportional to the column density of the ion. For molecular absorbers, however, this is more complicated. Rotational transitions of diatomic molecules have typical separations of a few hundredths of an eV, near the thermal velocity of the ambient particles (especially electrons). They are, therefore, potentially excited and the lower state populations reflect the thermal distribution of the gas, forming a rotational excitation ladder. In the absence of a strong radiation source, and in an optically thin medium, the populations of successively higher levels will relax to a Boltzmann distribution for which the excitation temperature approaches the kinetic temperature of the ambient perturbers so the logarithm ratio of column densities for the successive lines should be proportional to $\Delta E_{ij}/kT_{ex}$ where $\Delta E_{ij} \rightarrow \Delta E_{J,J+1} = (J+1)/I$ for dipole transitions from rotational levels J to $J+1$ and I is the moment of inertia. Vibrational states are more widely separated and may not be excited in the lower densities or by the cold gas. When the gas is optically thin, emission lines arise from the cascade to the ground state and the intensity ratios again follows the exponential distribution.

[12]In the last chapter of Eddington's 1926 tome *The Internal Constitution of the Stars*, he had speculated on the mean energy density of the space between the stars based on their diffuse radiation and computed an equivalent temperature for matter in thermal equilibrium that was similar to the excitation temperature required from the CN measurements.

[13]Ned Wright (http://www.astro.ucla.edu/~wright/CMB.html) called attention to the only pre-1965 reference to the possible cosmological implication of the CN detection. It is a review of one of Gamow's books by Hoyle (1950) that notes that the inferred relic radiation from the epoch of primordial nucleosynthesis was too hot compared to the CN excitation temperature. Nobody else noticed.

Collisions also establish a thermal distribution between fine structure levels if sufficiently frequent, as happens in the denser diffuse clouds, so the column density ratios are approximately the ratio of the statistical weights of the levels. This is not, however, always the case; for example, the ratio of the two fine structure states of hydrogen. For H_2, the ground state potential, $^1\Sigma_g$ comes in two forms, symmetric and antisymmetric in the nuclear spins, which are degenerate except for the rotational transitions. These are, alternately, positive for the odd J states and negative for even J. The ratio of these should be the ratio of the nuclear spin statistical weighs (the ortho- to para- ratio), which for $I = 1/2$ is either 1 or 0, hence the ratio of the statistical weights of 3:1 results when collisions produce a thermal statistical distribution. There are no radiative transitions between these. It is likely however, that the formation of H_2 leaves a mark on this ratio since it departs from the canonical value in dense media (see the discussion of astrochemistry in Chap. 3). Fine structure atomic transitions, of which the far infrared lines such as [CII] 158 μm is typical, are like the 21 cm line. They have very small transition probabilities and are not seen in absorption. But emission occurs when collisions excite the upper state and are not sufficiently frequent to de-excite the gas without emission of a photon. The same holds for the 21 cm line.

Line formation in the diffuse ISM is rather different from an atmosphere in another respect since there is no feedback from the radiative to mechanical properties of the gas. Emission following collisional excitations cools the gas but, when self-gravity can be safely ignored, the dynamics are unaffected by this (although the internal energy decreases, since pressure gradients are not responsible for structuring the equilibrium there is no dynamical consequence of this cooling). The electrons (and other perturbers) lose kinetic energy but this has no affect on the structure of the medium.

The diffuse gas is not strictly collisionless, as emission in neutral hydrogen at 21 cm demonstrates. The formation of this line is especially simple so we will use it as a prototype. The excited state has a very long lifetime, $A_{21} \approx 2.8 \times 10^{-15}$ s^{-1}. Since the B-value is proportional to this, the absorption coefficient follows immediately, $\kappa_{HI} \sim B_{21cm}$, the optical depth is linear in the column density of the ground state, and column densities of order 10^{21} cm^{-2} or higher are required for this to become optically thick. The transition is usually seen in emission, which might seem puzzling since it is not a scattering line like Lyα and with so low an absorption coefficient one would expect no radiative excitation. But there is an alternate, collisional channel for excitation. If the collision rate is $n_e C_{ij}$, where the rate coefficient depends only on temperature, then it suffices that the electron density be sufficiently low that detailed balance is not achieved to insure that a collisional excitation is followed by a radiative de-excitation. This condition gives the population ratio

$$n_j/n_i = n_e C_{ij}/A_{ji} \tag{2.39}$$

that, because of the extremely low thermal threshold, 0.1 K, guarantees collisional population of the upper state anywhere in the ISM. Thus, without some collisional excitation from electrons, the line would be observable only in absorption against Galactic and extragalactic nonthermal sources and then only very weakly given its minuscule oscillator strength.

The same reasoning can be applied to the fine structure of the resonance multiplets. Take, as an example, two of the dominant ions in the ISM, CII 1334, 1335 Å and MgII 2796, 2803Å. The former is a multiplet with the resonance line being $^2P^o_{1/2} -^2 D_{3/2}$ while the longer wavelength component is a $^2P^o_{3/2} -^2 D_{5/2}$ transition arising from a level 63 cm^{-1} above the ground state. The difference corresponds to the wavelength of the CII fine structure line, $^2P^o_{1/2} -^2 P^o_{3/2}$ at $157.7\mu m$. If there is no absorption to the excited level, without collisional excitation there will be no emission in this infrared line. On the other hand, electron collisions can easily excite the upper state, increasing the optical depth of 1335 Å relative to the 1334 Å line. The fine structure line has a transition probability of $A_{ul} = 2.29 \times 10^{-6}$ s^{-1} so it is collisionally suppressed for densities above about 10^3 cm^{-3} (depending on the temperature, see below), but at lower densities it is radiatively de-excited and observable in the far infrared. This also illustrates the important role these fine structure lines play in the cooling of the clouds and the CNM. Their optical depth is negligible so any collision in the diffuse medium will be followed by emission of a freely escaping photon. The contrast is with MgII.

The two ultraviolet lines of MgII arise directly from the ground state and the separation from the first excited level is too large to be accessible to collisional excitations (it requires about 4 eV). Thus, the ratio of the optical depth of the two components of the line depends only on the column density of Mg^+. The same holds for the principal optical doublet of NaI 5889, 5893 Å. The two components are both from the ground state and their ratio is used as an extinction proxy. These, and CaI 4226.7 Å a singlet, and the doublet KI 7664.5, 7688.9 Å, are the only neutral atomic species of sufficient abundance to leave a signature on the spectrum. The other important optical ionic transition, CaII 3933, 3968 Å, is also a pair of ground state lines.

Electron excitation also affects the optical CN rotational transitions R(0) ($J = 0-1$)and R(1)($J = 1-2$) . These are principally excited by the cosmic background radiation at 1.3 mm and 2.6 mm and should give an excitation temperature of $T_{CMB} = 2.725484$ K $\pm$ 570 μK Fixsen (2009). Gamow originally cited this line ratio as evidence for a FIR excitation by a cosmic radiative background. Ritchey et al. (2011) show, however, that the excitation exceeds what would be possible from radiation alone. For the diffuse cloud in the line of sight to HD 154368, they found an excess of about 63 mK. Since the exciting window is a minimum for Galactic thermal and synchrotron emission, even a difference of a few percent becomes significant. Harrison et al. (2013) show that electron collisions can account for the added populations within a relatively narrow density range in the diffuse medium, $n_e \approx 0.01$ to $0.06\,cm^{-3}$. This is on the low side for the mean density but quite in line with the expected range of fluctuations. The excitation cross section peaks at around

20 K . Similarly to the 21 cm line, weak emission has been detected in the hyperfine structure transition $(N = 1, F = 3/2, J = 5/2) - (0, 1/2, 3/2)$ at 113.49 GHz (Palazzi et al. 1990) with a brightness temperature of about 20 mK.

2.2.9 *Dynamical and Nonlocal Complications to Radiative Transfer*

2.2.9.1 Escape Probability and Photon Trapping

The equivalent width is the cumulant of all absorption along the line of sight to a source within the velocity range of the line profile. If the observed line is due to only one component, W_λ depends on a unique column density and the analysis proceeds as we have outlined. If, however, there are many individual absorbers, such as a string of clouds toward a background source each moving with a different center of mass velocity and unrelated to each other, the optical depth at each wavelength is the sum of all components capable of absorbing at the observed velocity. This is problematic for UV, optical, and IR spectroscopy where the resolutions are limited to about 0.5 km s^{-1}. For mm wavelengths this is not a problem since current autocorrelators easily achieve resolutions of less than 0.05 km s^{-1}. Unless we can separate the individual contributions, we do not know from a single profile what might be the differences from one absorbing system to another (see Jenkins 2009). It is now well established that above helium, the atomic phase is differentially depleted by adsorption on dust. If one only observes the line of sight total absorption, and the profile appears monolithic, only the total depletion can be measured. But depending on the local conditions, depletion is more effective in dense than rarefied clouds. So without being able to separate individual velocity coherent structures, even a comparison between lines of the same species may be misleading. To see this, think of two lines from the same ground state, one stronger than the other (they have different *gf* values, e.g. the doublets NaI D1 5895 Å, D2 5889 Å or CII 1334, 1335 Å). At large column densities the stronger component saturates first for a single constituent absorber (e.g., Wilson and Merrill 1937). But both line components may be unsaturated because of differential absorption in individual line of sight structures, some of which may be intrinsically more transparent and others less so. The component ratio is conventionally taken as a direct indicator of optical depth, hence it would normally be used to get N, the column density. The equivalent width is, however, an integral measure over the whole velocity range covered by the profile so the ratio of the components may yield the wrong optical depth. Recall that the line of sight contributors to τ add linearly, an ensemble of profiles produce, at some ν, a total

$$\tau_\nu = \Sigma_i \tau_{0,i} \Phi(\nu - \nu_{0,i}; \sigma_i) \tag{2.40}$$

where σ_i is the velocity dispersion of the ith component whose optical depth at line center is $\tau_{0,i}$.

The problem arises when either the velocity dispersion or the column and volumetric densities are different in each component. Densities of a particular species can change because of differential elemental depletion, while possible differences in internal dynamics change the dispersion and skew. The line *shifts* may be systematic, because of Galactic rotation separating clouds along the line of sight that are possibly distant from each other, or there can be large scale random motions of similar magnitude as the differential rotation, or the changes may come from superposition of small individual absorbers which are in the same dynamical complex but otherwise unrelated to each other. Hence a word of caution: all these effects are conflated in unresolved profiles.

Escape probability methods have the intuitive appeal of following a photon on its journey through the obstacle course of a medium. Although it is less important for treating line formation in diffuse, optically thin environments, when the column densities become sufficiently high escape probability makes the concept of *photon trapping* easier to grasp. As we discussed in Sect. 2.2, the penetration probability for a photon along an arbitrary line of sight is an exponential in the optical depth. But we sometimes want to know what the likelihood is that it will *escape* from some internal point, for instance when treating the cooling of a cloud or calculating the volumetric emissivity of a structure. Such questions involve integrating over all possible paths and are particularly important for correctly accounting for radiative losses from denser environments. Remember, "optical depth unity" is just another way of specifying that the photosphere is for a bounded structure. We are interested in the radiation escaping from a medium so the relevant quantity is the flux originating at some depth τ. Imagine, for illustration, a finite thickness slab for which the total optical depth is τ_0. Photons can exit at any angle so they see along any direction a scaled optical depth τ/μ where μ is the directional cosine relative to the surface normal. In a (horizontally) infinite slab, those trajectories parallel to the surface see ahead of them an infinite line of sight τ so they can only escape if deviated. Otherwise, at any point in the medium they will surely be absorbed. The escape probability is obtained by integrating over all directions

$$p_{esc,\nu} = \frac{1}{\tau_{0,\nu}}(1 - e^{-\tau_{0,\nu}}) \tag{2.41}$$

where now we explicitly include the frequency. In the limit of small τ_0, $p_{esc} \rightarrow 1$ and varies as τ_0^{-1} for large τ_0. Slabs are rather poor models for clouds, although they are good approximations to PDRs. For a sphere, the maximum optical depth in any line of sight through the sphere is $\tau_\nu = \tau\mu$ with $\mu = \cos\theta$ being the direction cosine. Then, if the center of the sphere has an optical depth τ_0, we can write

$$F_\nu \sim \int_0^1 T_{ex}[\exp(-\tau_{\nu,0}\mu)]\mu d\mu. \tag{2.42}$$

In the limit of infinite optical depth this quantity vanishes, when $\tau \rightarrow 0$ it goes to unity. This provides the *escape probability* for a photon emitted from some optical depth reaching the surface of the cloud. A systematic velocity gradient makes an enormous difference. When the linewidth given by what we can consider to be microscopic processes (whether thermal or small scale homogeneous turbulence) is smaller than the gradient of the macroscale motions along any line of sight (for instance, for an expanding medium) the optical depth decreases and the escape rate goes up.

For a cloud with an internal source function $S_\nu = j_\nu/\kappa_\nu$ where volume emissivity is j_ν, replacing κ_ν by the optical depth gives $S_\nu = j_\nu/(\tau_\nu/R)$ for a radius R. The escape probability is the ratio of the emerging flux from some optical depth to the total emitted from the volume so from equation 2.41

$$p_{esc,\nu} = \frac{3}{8\tau_\nu^3}[2\tau_\nu^2 + (1 + 2\tau_\nu)\exp(-2\tau_\nu) - 1]. \tag{2.43}$$

The important feature here is that the optical depth is the maximum through the sphere that depends on the column density. The formalism is therefore an integrated measure and valid at each frequency. The line profile now enters since p_ν must be averaged over the profile. Since for a velocity gradient the optical depth at line center decreases, the escape is enhanced. Instead, turbulence, while it decreases the opacity at line center, enhances that at large displacement from ν_0 so the effect (for fully developed microscale motions) is the same as a broadened Gaussian. A different spectrum, for instance a power law in scale as we will discuss in Chap. 11, changes this.

2.2.9.2 Velocity Gradients and Line Transfer

The monochromatic optical depth depends on the total line of sight absorption at *a specific point in the line profile*. For a static medium, the absorption line profile is symmetric around line center. If, instead, the medium displays velocities that are coordinated on longer distances than a photon mean free path, the centroid also shifts and the opacity at any point in the profile now depends on how these shifts occur along the line of sight. Take the example of a line that is thermally broadened on the microscale but that is formed in a uniformly accelerating mass flow. Since we have no constraint that an interstellar cloud is hydrostatic, there can be a net divergence in some volume. Alternatively, the flow can have a complex (but not chaotic) pattern in three dimensions (shears, twists, and the like). A photon is emitted at some frequency $\Delta\nu$ relative to line center at point A, the rate depending on $\Phi(\Delta\nu)$ as we discussed earlier. It then, statistically, goes straight ahead until it hits something, say a molecule in the ground state, at another point B. The mean free path would normally be $[n\kappa_0\Phi(\Delta\nu)]^{-1}$ were the profiles the same. But in a larger scale flow, there is a Doppler shift, $\Delta\nu_D = \nu_0(\mathbf{v}/c)\cdot\hat{\mathbf{k}}$ that displaces the B profile relative to A by some $\Delta\nu + \Delta\nu_D$ that may increase or decrease the line of sight opacity relative to the static case. Thus, the mean free path changes. The effect is easiest to see for a Gaussian, where the line width is σ. If $\Delta v_D \gg \sigma$, then the velocity shift *statistically* decreases the opacity and increases the mean free path. This is called the Sobolev or "on-the-spot" approximation. The effect of changing the mean velocity is the same as altering the column density but in a subtle way. If the flow changes its orientation relative to the observer, the optical depth can change abruptly.[14] If a molecular cloud is sufficiently dense it may have a "photosphere" in the far infrared but, in general—and especially for diffuse and translucent molecular clouds (see Chap. 9)—they are transparent in the continuum. Instead, the absorption is seen only against the line emitting region itself. Any part of the flow that is opaque along the line of sight leaves an imprint on some part of the profile at its specific velocity and self-absorption can occur because a change in the velocity or density produces an increase in the line of sight optical depth.

We will return to the issue of random flow fields when discussing turbulence in Chap. 11. But it is useful to make a remark here concerning the difference between a smooth and chaotic velocity field. In the disorder of turbulence, the macroscale is covered by a broad range of fluctuations in both density and velocity (and temperature, but for now we will assume isothermality). The optical depth in the line is simply redistributed over a broader bandpass and the line core becomes more transparent. Since in emission the photons are most likely to emerge from the core, turbulent reduction of the optical depth at line center increases the radiative losses. A diagnostic of the opacity is usually isotopic ratios. Since the chemistry cannot produce unlimited enrichments of rare species, for instance the ^{13}CO/^{12}CO ratio, the more abundant species is also the less transparent. Thus, if the optical depth increases, the ^{12}CO lines saturate first. Note that were there a continuum against which to absorb, the total velocity-corrected column density could be directly measured. But since the only source function is line emission, the broader the profile the less easily the opaque regions can be identified.

The comparison shown in Figs. 2.5, 2.6, 2.7 of two different lines of sight through the Galaxy illustrates how the optical depth depends on the excitation conditions. The lines are resonance transitions observed against two bright Galactic novae, V339 Del 2013 ($\ell, b \sim 62.2°, -9.4°$) and T Pyx 2011 ($\ell, b \sim 257.2°, +9.7°$). Both were unassociated with any star forming region, and were merely passive

[14]This is most familiar from stellar winds where, near the terminal velocity, the radial velocity gradient asymptotically vanishes. Despite the decreasing number density, the effective optical depth can produce saturated absorption up to nearly the maximum velocity of the flow. This is the familiar P Cygni profile characteristic of mass outflows. This effect is also strongly nonlocal since very distant zones of the flow can now communicate through absorption. In contrast to an interstellar flow, that in a wind is mainly radial with an optically thick surface at the lower boundary. Consequently, the absorption is always shifted to the blue and the emission, on the redward side of the line, is always optically thin. There is always a surface from which a broadband continuum emerges against which the absorption occurs. The relative intensity of absorption and emission (for complete coverage by a spherical outflow) is set by the ratio of the solid angle subtended by the outer regions of the wind to the photosphere.

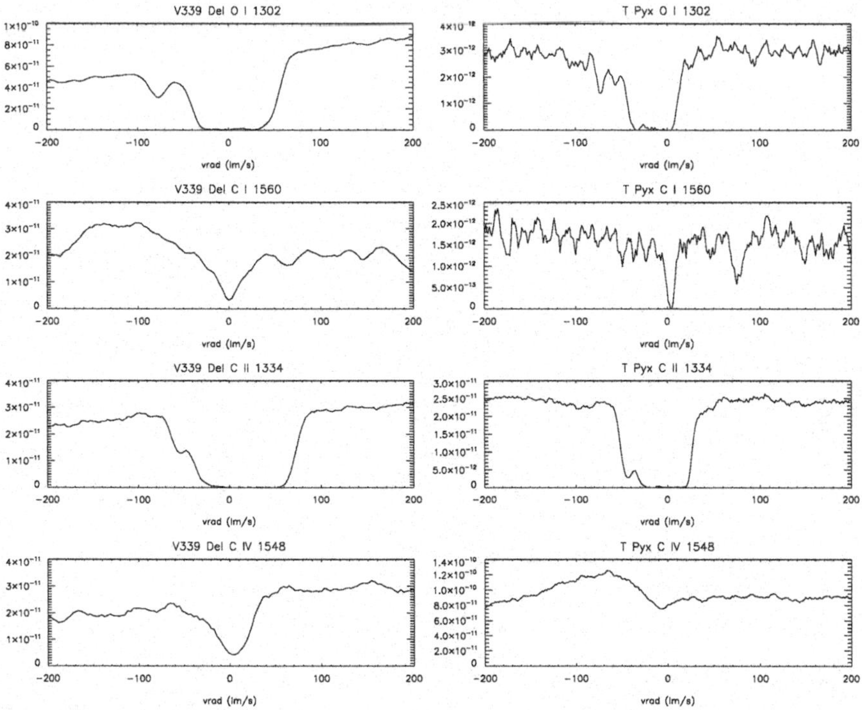

Fig. 2.5 Absorption lines along the line of sight to the nova V339 Del on the left and T Pyx on the right. The species and the wavelength (in Å) of the depicted transition are labeled above each spectrum. The spectra were taken with the Space Telescope Imaging Spectrograph (STIS) on the Hubble Space Telescope

background sources whose spectra were time variable. Hence, it was easy to distinguish the interstellar from stellar lines.

A number of important molecular species also absorb at optical and ultraviolet wavelengths, and were first identified long before millimeter wave astronomy became available for studying molecules in the diffuse phase. The observed lines, representing the species CN, CH, CH^+, HCO^+, and CO, are vibrational-electronic transitions. An example of such excitation are the H_2 bands formed by absorption from the ground state $X^1\Sigma_g^+$ to the two excited states $B^1\Sigma_u^+$ (Lyman band) and $C^1\Pi_u$ (Werner band) electronic states (Abgrall et al. 1993) which is, in fact, the *only* way to directly observe this essential species in the cold neutral medium in absorption (noting that the molecule is symmetric and has no dipole transitions, a point to which we will repeatedly return). The absorption is observable only in the vacuum ultraviolet (VUV), the Lyman band lies below 1108 Å and the Werner band is below 1008 Å, so until the 1990s, using spectra obtained with *FUSE*, only a few select lines of sight were available at high enough resolution to study the gas distribution (using *Copernicus* spectra, see Savage and Sembach

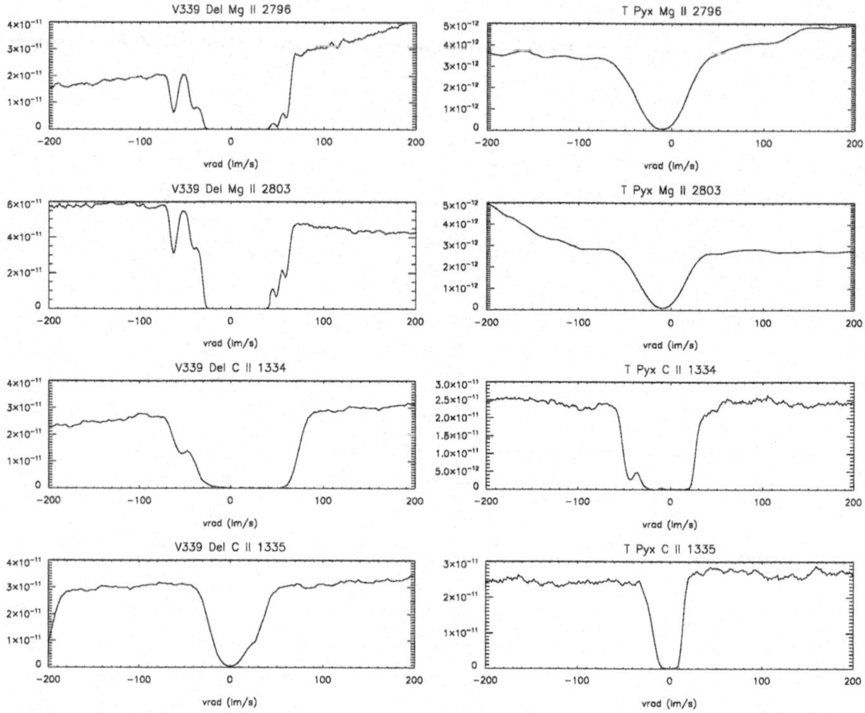

Fig. 2.6 Same as Fig. 2.5; additional absorption lines along the line of sight to the nova V339 Del on the left and T Pyx on the right

1996). The lines occur in emission by two mechanisms. If the colliding electrons have sufficient energy to excite the rotational lines of the $\mathrm{v} = 0 \rightarrow \mathrm{v} = 0$ system with $\Delta J = 0, \pm 2$ but this is not normally the case in the diffuse ISM. In shocked gas, where the velocities are above a few tens km s^{-1}, rotational lines from the ground state are excited. For $\Delta \mathrm{v} = 1$, requiring 0.5 eV, rotational lines can also be excited but the shocks must be much stronger (with velocities > 25 km s^{-1}). In all cases, the essential requirement is that the collisions not be sufficiently energetic to dissociate the H_2, which requires about 4.5 eV. Although this is far higher than the $\Delta \mathrm{v} = 1$ excitation, the destruction of the molecule is not balanced by rapid re-association. Alternatively, absorption in the neighborhood of strong UV sources pumps the excited states that decay radiatively. But, in general, the lines are observed in absorption.

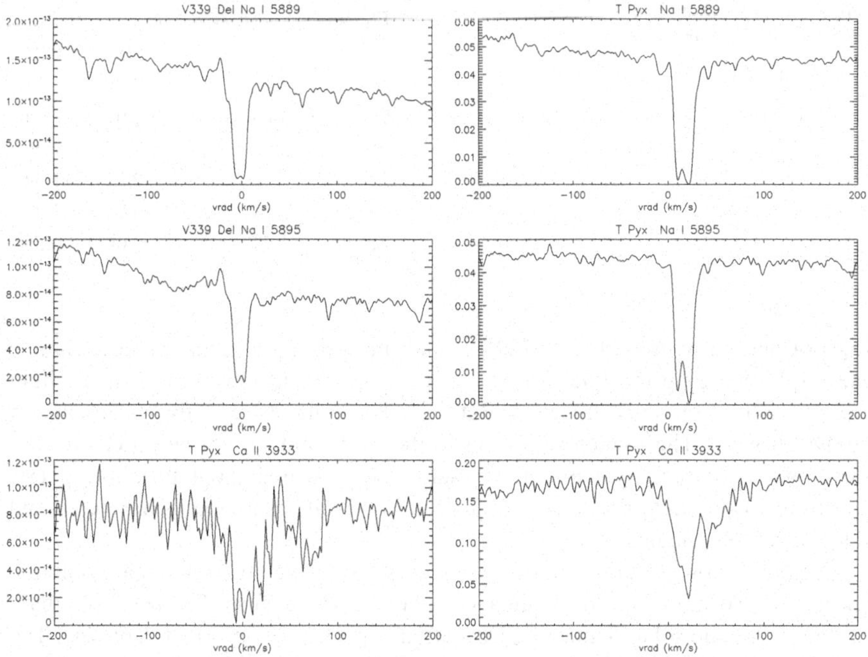

Fig. 2.7 Same as in Figs. 2.5 and 2.6; additional optical absorption lines along the line of sight to the nova V339 Del on the left and T Pyx on the right

2.2.10 Elemental Abundances and Depletion in the Diffuse Medium

Absolute elemental abundances are not the same as ion column densities. To achieve a comparison with stellar values, the ionization state must be known. In general, only one or two ions are observed for most species for the diffuse medium in currently available wavelength ranges. For instance, nitrogen has only N^0 and N^{+4}, oxygen shows only the neutral lines like Na, and the same applies to Mg and Ca. Carbon and silicon are better represented, and resonance lines are observable for the neutral through three times ionized states. This wide range is very important since their neutral state ionization potentials are lower than that of neutral hydrogen. As we will discuss, these are the major constituents of the solid interstellar phase, dust, so it is rather important to have some idea of how the abundances compare with those in stars. Stellar abundances are conventionally normalized to hydrogen—based mainly on the optical Balmer lines, but for the ISM, obtaining the hydrogen abundance is more difficult. For a start, the population of the parent state for the Balmer series is negligible in the ISM, unlike the far denser environment of a stellar atmosphere. The neutral hydrogen column density is available from the 21 cm line, but the emission line includes gas beyond the source. The resonance absorption

Table 2.1 Depletion factors in the diffuse medium for key elements (Jenkins 2009)

Element	$\log(X/H)_{\odot}$	f_{min}	f_{max}
C	8.46	−0.1	−0.21
N	7.40	−0.11	−0.11
O	8.76	−0.01	−0.24
Mg	7.62	−0.27	−1.36
Si	7.61	−0.22	−1.36
Fe	7.54	−0.95	−2.24
Zn	4.70	0.06	−0.55

lines of neutral hydrogen, Lyα 1215 Å and the rest of the Lyman series, sample the same gas along the line of sight as all other absorption lines toward a light source arising from the diffuse gas. But this measurement is not always possible or straightforward. The Lyman series lies in the VUV and *cannot* be observed from the ground. Additionally, hot luminous stars also show this line in the photospheric spectrum so sufficient resolution is required to permit separation of the stellar and interstellar components.

There are now several hundred lines of sight available from high resolution spectra. These show that the abundance pattern in the diffuse ISM is substantially different than the solar elemental distribution, a phenomenon called *depletion*. The dependence of the deficit on local conditions has been detailed by Jenkins (2009).

In Table 2.1, the depletion factors, measured as $\log(X/H) - \log(X/H)_{\odot}$ are indicated for different environments, defined by Jenkins (2009) as minimally and maximally depleted. The solar abundances are from a compilation by Lodders (2003) representing a best estimate of pre-solar system diffuse interstellar metallicities. We include Zn because it is well studied using strong resonance lines in the UV at 2054 Å and 2063 Å, analogous to the CII and MgII lines.

The ionization state of the gas is estimated by comparing the individual neutral and ion line profiles in velocity. Since the resolution is the ultimate limit to such studies, the ratio of column densities—hence, the ionization fraction for the heavy metal of interest—can be obtained within each component. The advantage is that the transitions are from the same element. While it is often impossible to determine the full ionization sequence, most of the doubly ionized species have resonance transitions in the FUV, the ionization state is mainly due to the neutrals and singly ionized species. A further feature of the absorption is that it can be used to check emission line studies for some diffuse regions. In HII regions, for instance, the column densities are sufficiently high that absorption from the diffuse gas toward the central stars is detectable in some systems. And it should be added that in addition to HII regions, supernova remnants (e.g. SN 1006) have been studied by this method in the ultraviolet.

To obtain the hydrogen column requires FUV observations of the Lyman and Werner bands for H_2 and those Lyman series lines for atomic hydrogen lying below 1100 Å that are available from archival FUSE or *Copernicus* spectra, and HI Lyα profiles from *HST*. The available Galactic directions are sufficiently broadly dis-

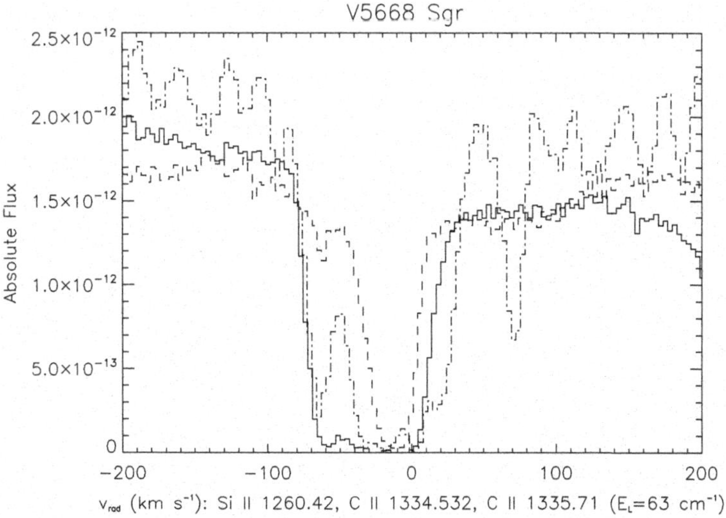

Fig. 2.8 Absorption lines along the line of sight to the nova V5668 Sgr 2015. The *solid line* is the CII 1334.532 Å line, the *dashed line* is the CII excited state, and the *dot-dash line* is the SiII. These spectra show the effects of effective excitation and were taken with the Space Telescope Imaging Spectrograph (STIS) on the Hubble Space Telescope

tributed that a representative abundances can be obtained. For the neutral hydrogen component, modeling the spectral energy distributions of X-ray sources provides estimates of N_H. Additional information has recently been derived from interstellar ionizing continuum absorption from C, N, and O but this is just beginning and is severely limited by the requirement that the hydrogen column densities cannot be too large. The advantage of this spectroscopy derives from the ability to separate individual components in radial velocity so the material conditions and column densities (and abundances) can be obtained even for individual clouds. Examples are shown in Figs. 2.8, 2.9, 2.10 for the line of sight to the classical nova V5668 Sgr 2015 ($\ell, b \sim 5.4°, -9.9°$). For instance, while SII is very strong in several transitions, there is no neutral sulfur along the line of sight, nor FeI. Neutral nitrogen and oxygen are quite strong but individual transitions are sensitive to different excitation conditions. The difficulty this poses for any abundance analysis is then evident: without the full set of available ionization states, only ionic column densities are available. Consider the case where the line of sight intersects several clouds. These may have nothing at all to do with each other yet contribute to the absorption in each velocity interval. Individual transitions of different optical depth help disentangling the components but this is critically dependent on the resolution and, therefore, can be confused with macroscopic broadening or turbulence.

Snow et al. (2002)—specifically for Fe along translucent lines of sight with FUSE—derive the mean number density by taking total H column densities and

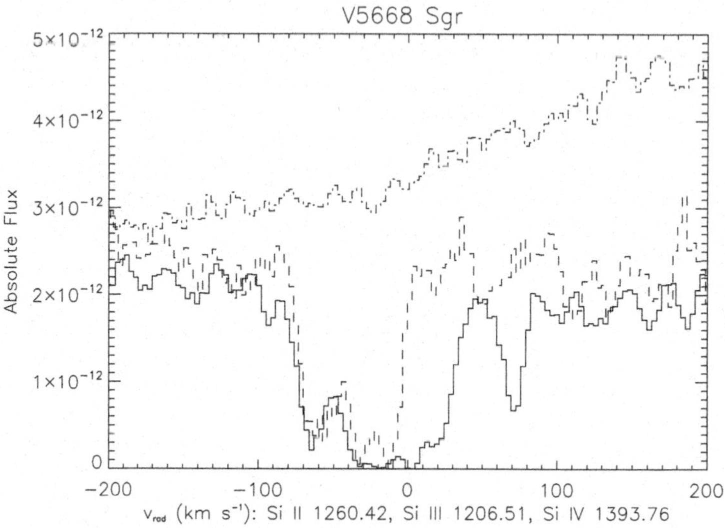

Fig. 2.9 Absorption lines along the line of sight to the nova V5668 Sgr 2015. The *solid line* is the SiII line, the *dashed line* is the SiIII line, and the *dot-dash line* is the nondetection of SiIV. These spectra show the effects of ionization differences in silicon and were taken with the Space Telescope Imaging Spectrograph (STIS) on the Hubble Space Telescope

dividing by distance, something that is possible for stars with known characteristics since these are absorption measurements.

2.2.11 *H_2 and Fine Structure Transitions*

One of the subtle conspiracies of nature is that the dominant species responsible for most of the mass of molecular clouds (helium atoms are not negligible in the mass budget of a molecular cloud) is intrinsically unobservable. Neutral molecular hydrogen has no permanent dipole moment so infrared emission from the lowest rotational and vibrational bands is strongly prohibited. There are, however, observable electronic transitions in the ultraviolet, in absorption, that can be viewed through translucent (but not dense) clouds: the Lyman and Werner bands.[15] This requires a background ultraviolet source (at this wavelength a relatively rare beast) that either has a long enough line of sight through the ISM or a high enough local density. In the former case, this requires distant massive stars or QSOs, although it is possible to observe. For nearby stars, hot subdwarfs, it requires high volume densities. This comes in the form of vibrational bands, all arising from the ground

[15]The Lyman bands are from the $B^1\Sigma_u^+$ to $X^1\Sigma_g^+$ and the Werner bands are from the $C^1\Pi_u$ to the $X^1\Sigma_g^+$ states, lying below 1100 Å (above 11 eV).

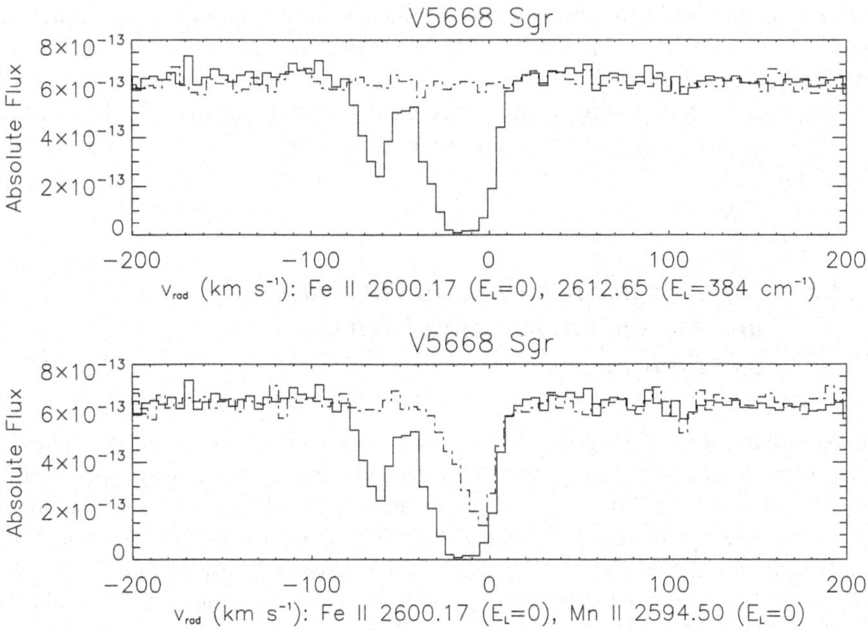

Fig. 2.10 Absorption lines along the line of sight to the nova V5668 Sgr 2015. In the top panel, the *solid line* is the FeII line at 2600.17 Å and the *dashed line* is the excited state. In the bottom panel, the *solid line* is again the FeII line at 2600.17 Å but the *dashed line* is the MnII 2594.50 Å line. The top spectrum illustrates excitation effects and the bottom differing abundances (both spectra arise from the same lower level). The spectra were taken with the Space Telescope Imaging Spectrograph (STIS) on the Hubble Space Telescope

state since the energy separation is about 0.1 eV ($T_{kin} \approx 10^3$ K for the vibrational states). The de-excitation, however, rather than leading to an emission band (again from the ground state) dissociates the molecule because of an intermediate energy, unbound $^3\Sigma_g^+$ state. This two step mechanism, first realized in the '60s (Stecher and Williams 1966) is the cornerstone for the formation of PDRs (see discussion in Sect. 1.7.1). But for our purposes, the most important point is that H_2 is nearly invisible—and certainly *is* so in the infrared and millimeter for the cold ISM. Thus, for molecular clouds, the presence of H_2 is an inference, not a direct observational detection, and its study depends critically on a chain of calibrations (see Chap. 8).

To further complicate matters, the interactions with CO are with a strongly dipolar molecule, that exists only in the neutral state. The interactions are, therefore, only weak and of a dipole-neutral type. The collision cross section is that of a pseudo-molecular state in which the perturber is weakly bound (the same as an autoionizing or pre-dissociating state).

This provides the excitation rate since there are few strong continuum sources at 1–3 mm in the Galaxy. Against extragalactic sources, there is a possibility of absorption but the flux densities are too low to balance the collision rates so that is seen as an absorption along a line of sight toward a point source (however observations of this type have led to some interesting conclusions—see Sect. 7.4.3). As for the 21 cm line, we have the same excitation condition: every collision with a rate $n_{H_2} C_{01} n_0$ is followed by a radiative de-excitation with rate $n_1 A_{10} h\nu$. hence, the ratio n_1/n_0 depends on the kinetic temperatures (in effect, on $T_{ex}/T_{kin} \approx 1$ for the transition).

2.2.12 *Diffuse Interstellar Features in Emission and Absorption: Radiative Excitations and Fluorescence*

Even in the diffuse ISM, radiative processes alone can determine the spectrum if the collisions are sufficiently rare. This depends, however, on the properties of the radiation source. Take the example of the infrared diffuse emission features. These are due to C-H and C-C bond stretching of complex species. Their precise identification remains elusive, the same for the optical diffuse interstellar bands (DIBs, broad absorption features whose width cannot be either thermal or dynamical in origin—see 3.5.3). Complex molecules, those almost massive enough to show solid state band structures, are very efficient absorbers in the UV. In the diffuse interstellar medium, the global higher-energy-than-infrared radiation field resembles a star with an effective temperature of about 10^4 K. This emission extends beyond 3 eV, and PAHs are especially stable up to about 10 eV. Absorption against this diffuse UV continuum can excite electronic transitions, without ionization, that de-excite through vibrational bands in the near IR. The emission will, therefore, drive the radiation temperatures far above those in thermal equilibrium, as expected for larger dust grains. The process conserves photon energy but not number and the band lifetimes are so short relative to the collision frequency that the levels are far from thermal equilibrium. The emission was first detected in the IRAS cirrus and in reflection nebulae near hot stars.

Another familiar example of photon coupling is from IR pumping of transitions that can produce population inversions, hence *maser* emission. The optical depth is $\tau(\Delta\nu) \sim n_i B_{ij} - n_j B_{ji}$ so if there is a strong pumping of a third level whose decay overpopulates the upper state of a resonance emission line, τ can be negative. The medium then becomes an amplifier. The emission comes out within the natural line width and only couples within a very narrow range in velocity around line center. Thus, the pumping radiation is channeled from a broadband absorption into a very narrow emission with a consequent increase in brightness temperature. In the laboratory, this is usually referred to as a negative temperature but, since we have been strict in defining temperature as kinetic, connected with the thermal motion of the gas, there is no reason for invoking any arcane description: population inversion and emission line amplification are the signature of NLTE. The emission is extremely fine-tuned, any turbulence induced fluctuation can spontaneously lead to desaturation of the maser along any line of sight. The same is true for large velocity gradients. So the region over which motion is coherent is the scale of the amplification length.

In 1920, Heger found a pair of weak, broad absorption lines in the spectrum of the B star ζ Per (and other B stars) at 5780 and 5797 Å that she suspected might be interstellar in origin (see Merrill 1938). The reasoning was the same as for the CaII and NaI lines: They appeared to have a different radial velocity than the photospheric absorption lines. Subsequent work by Merrill, Beals, and Greenstein among others added to the list and Beals added a number of such *diffuse interstellar bands* (DIBs) to the list, notably at 4430, 5778, 5780, 5796, 6177, 6196, 6284, and 6614 Å. Herbig (1975) attempted an analysis of possible carriers of the features based on their Galactic occurrence and wavelength distribution, going so far as to produce a tentative energy level diagram. But despite the best efforts over the past century, the identification of the responsible agent(s) remains an open problem (the most recent collection of studies, IAU Symp. 297 ends inconclusively).[16] The 5780, 5796 Å bands have also been observed in extragalactic environments, notably the LMC, M 31, and M 33 (e.g., Bailey et al. 2015). All surveys, whether of lines of sight within the Galactic or elsewhere, find a positive correlation between extinction (measured by E(B-V)) and the strength of the features but no single law for all of the bands. To further complicate matters, the individual bands do not all correlate with each other in strength. There are now over 400 cataloged features, some almost as narrow as atomic or molecular transitions, others having FWHM of many Ångstroms (see Tielens 2014) and varying in absorption depth from a few to almost thirty percent. An interesting feature is that they display a power law distribution for line strengths in single sources. The positive correlation with broad interstellar properties argues for a common atomic composition, mainly hydrogenated forms of C, N, and O bearing molecules but that is about all that is known. Only one species, the fullerene C_{60}^+ has recently been firmly associated with two DIBs, [9577/9633 Å] by Campbell et al. (2015).

However, there is a related component of the ISM, the polycyclic aromatic hydrocarbons (PAHs hereafter) that are associated with emission bands in the near infrared (Tielens 2008). The detection of excess FIR emission from reflection nebulae stimulated the search for a large molecular origin of the bands. The bands arise from generic transitions, C-H and C-C bond stretching and various bending modes, and the originating molecules are difficult to distinguish from these observations alone. There is even a question about whether one is even possible.

Radiative excitation is rare in the diffuse medium since the diffuse interstellar radiation field has far too low an energy density to produce significant emission by direct absorption. There is, however, one exception that relates directly to the physical meaning of the source function. Solids have such an enormous number of available internal degrees of freedom that one expects the absorption process

[16]See, *Proceedings*, IAU Symposium 279, *The Diffuse Interstellar Bands*, eds. J. Cami and N.L.J. Cox, Cambridge U. Press (2013).

to lead to thermalized emission. So for large ensembles, such as dust grains, the emissivity should be a Planck function at the radiative equilibrium temperature. Hence, if the absorption coefficient is $\kappa_{\nu,D} = Q_{abs}(\nu)\pi a^2$, where a is the grain radius and Q is the efficiency factor (see Draine 2003), the emission spectrum is $j_\nu = \kappa_{\nu,D} B_\nu(T)$ and T is determined by the net absorbed flux independent of the spectral distribution. If, however, the grains are in the gray area between very large molecules and small solids, the absorption occurring in bands is followed by cascades through the allowed transitions back to the ground state in a time that is short compared to the arrival rate for the individual photons. Instead of a modified blackbody with the same brightness temperature at all wavelengths, the emitted spectrum will be a complex of bands whose individual intensities are generally far higher than expected from simple thermal emission. The complex is actually a line spectrum, a down-conversion of the absorbed photon distribution through the cascade. If the lines are not individually resolved but, instead, photometrically folded through a broad bandpass, the emissivity will exceed that measured at longer wavelengths and appears as an anomalously hot component. This is how the small grain component of interstellar dust was first detected (e.g. Sellgren 1983; Tielens 2008) from near infrared emission (first in reflection nebulae in the vicinity of hot but not ionizing stars, then in cirrus diffuse 12 and 25 μm band emission from *IRAS*). Although the process can be described as non-equilibrium thermal fluctuations in the grains, hence that the specific heat is temperature dependent, the emission is a manifestation of NLTE. The level populations are independent of the collision rates and produced by radiative equilibrium.

The emission mechanism is an extreme sort of fluorescence. Recall that the simplest version of this is a radiative excitation followed by a down-conversion cascade through lower lying states. The cascade is only interrupted by collisions if the transition probabilities are sufficiently low (the same criterion as forbidden line emission). The pump can operate in any spectral interval above that of the target levels. The emission from small grains, particularly those like PAHs and perhaps fullerenes, is a more complicated version of the simple conversion process. Having so many vibrational modes, the redistribution is a pseudo-continuum with the strongest emission concentrated at the main bands due to C-C and C-H bond stretch. The bands are broadened because of the large number of independent modes in the molecule.

The extinction curve in the UV is distinguished by a broad absorption feature at 2175 Å. This, however, varies across the Galactic plane with the two extremes being ζ Oph type for a strong bump and σ Sco for the weak feature.

2.3 Radio Continuum Measurements of the Diffuse Medium

The diffuse medium also modifies radio frequency signals because it is partially ionized and magnetic. The continuum processes are refractive and depend on electron scattering. They are, therefore, only frequency dependent because the response of a medium (its characteristic frequency) is quite near the groundbased-accessible range. Centimeter and meter wavelength radio observations are also useful for probing the ionization and structure of the diffuse medium. The real part of the index of refraction for a radio signal (here we intend below 10 GHz or so) depends on the frequency and the plasma frequency, $\omega_p^2 = 4\pi e n_e/m_e$ as

$$n^2 = 1 - \left(\frac{\omega_p}{\omega}\right)^2 \tag{2.44}$$

Thus, across a bandpass $\Delta\omega$, the difference in the phase speed across the band differentially delays a periodic signal. This would normally not be an issue since the sources are not, in general very rapidly variable. But that is not the case for pulsars. For a pathlength ds the change in the arrival time is the same as a phase shift. Pulsars, magnetized rotating neutron stars with periods of from milliseconds to several seconds, are especially well suited to measure this. They have strong centimeter radio pulses and their intrinsic width is a fraction of the rotation period. The delay time can be determined using digital de-dispersing through the *dispersion measure*, DM:

$$DM = \frac{\Delta\omega}{c}\int nds \sim \int \frac{\omega_p^2}{\omega^3}ds \tag{2.45}$$

that is proportional to the electron column density along the line of sight. Fluctuations in n_e therefore introduce a variation in this measure and also are a driver of scintillation of Galactic and extragalactic radio sources (leading to finite angular diameters). Models based on the combined effects are discussed by Lazio and Cordes (2002) and Cordes and Lazio (2003). The propagation of polarized radio signals is also disturbed by the intervening magnetic field but, differently than the electron density, it is not isotropic and has several features that strongly differ. Across the same bandpass, right- or left-handed polarized continuum have different indices of refraction. The intervening electrons, tied to magnetic field lines, have a definite helicity depending on the orientation of the field. Calling the background field $\mathbf{B}_0$, oriented along the line of sight, the electron Larmor frequency (cyclotron frequency) is $\omega_L = eB_0/m_e c$. The *rotation measure*, measured in rad m^{-2} is the angle through which the plane of polarization is rotated,

$$RM = \frac{C}{\omega^2}\int \mathbf{B}\cdot\hat{k}n_e ds \tag{2.46}$$

which defines the *Faraday depth*

$$\phi(s_0) = \frac{e^3}{2\pi m_e^2 c^4}\int_{s_0}^{0} n_e(s)\mathbf{B}(s)\cdot\hat{k}ds \tag{2.47}$$

(Opermann et al. 2012), where the viewing direction is along $\hat{k}$. The result is the mean line of sight projected magnetic field and that is the core problem. This is *not* the field strength averaged over the electron column density. Instead, it is the *projected* field that can produce either sign of the rotation angle, depending on the relative orientation of the field lines toward the observer. Therefore, in a turbulent, or merely randomly oriented, field, the line of sight component may average to zero despite the energy density in the field possibly being large. It does not matter how closely spaced the frequencies are within the bandpass chosen for the observation. The only way out is to choose a scale for multiple lines of sight that is considerably larger than a single beam to capture some of the large scale structure. With the exception of a few high latitude clouds, the field in the diffuse medium can only be derived on a rather large scale, of order tens of parsecs or larger, because of the sparse sampling afforded by radio sources. This has significantly improved in recent years (see Opermann et al. 2012; Planck Collaboration, Plank 2015 results, paper XXI)), but it remains limited to line of sight only and requires substantial modeling efforts to disentangle orientation from field strength (e.g., Biermann and Kronberg 2004).

References

Abgrall et al. 1993, J,. Mol. Spec., 157, 512
Adams, W.S. 1941, ApJ, 93, 11
Athay, R.G. 1972, *Research in Radiative Transfer at NCAR*, Reidel Publishing Co.
Bailey, M., van Loon, J.Th., Sarre, P.J., adn Beckman, J.E. 2015, MNRAS, 454, 4013
Biermann, P. and Kronberg, P. P. 2004, JKAS, 37, 527
Campbell, E.E.B., Holz, M., Gerlich, D., and Maier, J.P. 2015, *Nature*, 523, 322
Cardelli, J.A, Clayton, G.C., and Mathis, J.S. 1989, ApJ, 345, 245
Chandrasekhar, S. 1950, *Radiative Transfer*, Oxford University Press
Cordes, J. and Lazio, J. 2003, http://arxiv.org/abs/astro-ph/0301598
Cowley, C.R. 1970, *The Theory of Stellar Spectra*, Gordon and Breach
Crutcher, R. 2012, ARAA, 50, 29
Crutcher, R. 2014, 69th International Symposium on Molecular Spectroscopy, June 16–20, at the University of Illinois at Urbana-Champaign, Talk TF15
de Vries, H.W., Heithausen, A., and Thaddeus, P. 1987, ApJ, 319, 723
Dickey, J.M. and Lockman, F.J. 1990, ARAA, 28, 215
Draine, B.T. 2003, ARAA, 41, 241
Draine, B.T. 2011, *Physics of the Interstellar and Intergalactic Medium*, Princeton University Press
Ewen, H.L. and Purcell, E.M. 1951, *Nature*, 168, 350
Field, G.B. and Hitchcock, J.L. 1966, PRL, 16, 817
Field, G.B. and Hitchcock, J.L. 1966, ApJ, 146, 1
Fitzpatrick, E.L. and Massa, D. 1990, ApJS, 72, 163
Fitzpatrick, E.L. and Massa, D. 2007, ApJ, 663, 320
Fixsen, D.J. 2009, ApJ, 707, 916
Gatuzz, E., García, J. A., Timothy R. Kallman, T. R. and Mendoza, C. 2016, A&A, 588, A111
Gir, B.-Y., Blitz, L., and Magnani, L. 1994, ApJ, 434, 162
Goldreich, P. and Kylafis, N.D. 1981, ApJ, 243, 75
Griem, H.R. 1964, *Plasma Spectroscopy*, McGraw-Hill, New York

Griem, H.R. 1997, *Principles of Plasma Spectroscopy*, Cambridge Unversity Press
Güver, T. and Özel, F. 2009, MNRAS, 400, 2050
Harrison, S., Faure, A., and Tennyson, J. 2013, MNRAS, 435, 3541
Hartmann, J. 1904, ApJ, 19, 268
Hartmann, D. and Burton, W.B. 1997, *Atlas of Galactic Neutral Hydrogen*, Cambridge University Press
Herbig, G. 1975, ApJ, 196, 129
Hoyle, F. 1950, Observatory, 70, 194
Hubeny, I. and Mihalas, D. 2014, *Theory of Stellar Atmospheres*, Princeton University Press
Jackson, J.D. 1999 *Classical Electrodynamics*, 3rd ed., John Wiley & Sons
Jenkins, E. 1996, ApJ, 471, 292
Jenkins, E.B. 2009, ApJ, 700, 1299
Kalberla, P.M.W. et al. 2005, A&A, 440, 775
Kaplan, S.A. and Pikel'ner, S.B. 1970, *The Interstellar Medium*, Harvard University Press
Lazio, J. and Cordes, J. 2002, http://arxiv.org/abs/astro-ph/0207156v3
Li, D. and Goldsmith, P.F. 2003, ApJ, 585, 823
Lodders, K. 2003, ApJ, 591, 1220
McKellar, A. 1940, PASP, 52, 187
McKellar, A. 1941, PDAO, 7, 251
Merrill, P.W. 1938, *The Nature of Variable Stars*, The Macmillan company
Muller, C.A. and Oort, J.H. 1951, *Nature*, 168, 357
Moriarty-Schieven, G.H. and Wannier, P.G. 1997, ApJ, 475, 642
Opermann, N., et al. 2012, A&A, 542, A93
Osterbrock, D.E. and Ferland, G.J. 2006, *Astrophysics of Gaseous Nebulae and Active Galactic Nuclei*, 2nd ed., University Science Books
Palazzi, E, et al. 1990, ApJ, 357, 14
Peek, J.E.G., et al. 2011, ApJS, 194, 20
Planck Collaboration XXI, Plank 2015 results, A&A, 5765, 106
Ritchey, A.M., Federman, S.R., and Lambert, D.L. 2011, ApJ, 728, 36
Rybicki, G.B. and Lightman, A.P. 1979, *Radiative Processes in Astrophysics*, John Wiley & Sons, Inc.
Savage, B.D., Bohlin, R.C., Drake, J.F., and Budich, W. 1977, ApJ, 216, 291
Savage, B.D. and Sembach, K.R. 1996, ARAA, 34, 297
Sellgren, K. 1983, AJ, 88, 985
Shore, S.N., Magnani, L., LaRosa, T.N., and McCarthy, M.N. 2003, ApJ, 593, 413
Snow, T.P., Rachford, B.L., and Figovski, L. 2002, ApJ, 573, 662
Spitzer, L. Jr. 1978, *Physical Processes in the Interstellar Medium*, John Wiley & Sons, Inc.
Stecher, T.P. and Williams, D.A. 1966, ApJ, 146, 88
Thaddeus, P. and Clauser, J.F. 1966, PRL, 16, 819
Tielens, A.G.G.M. 2008, ARAA, 46, 289
Tielens, A.G.G.M. 2014, *The Diffuse Interstellar Bands, Proceedings of the International Astronomical Union, IAU Symposium*, Volume 297, 399
Wilson, O.C. and Merrill, P.W. 1937, ApJ, 86, 44

Chapter 3
The Diffuse ISM from the Ground: Chemistry and Tracers

You look at the floor and see the floor. I look at the floor and see molecules

—Dan Aykroyd

Abstract In this chapter we discuss some of the radio and optical transitions which are used to observe the denser, colder portions of the diffuse ISM. We begin with a brief history of molecular observations, from the optical detections of CH, CN, and CH^+ in the late 1930s, to the rise of radio astronomy in the 1950s and 1960s. A brief discussion of astrochemistry and some of the basic reaction networks for oxygen, carbon, and nitrogen chemistry is included before individual molecular species of particular importance to studies of the diffuse ISM are described. We then discuss optical absorption line observations, the diffuse interstellar bands, and the red luminescence phenomenon. Finally, a brief mention of observations of the diffuse ISM in a distant galaxy is included.

3.1 Introduction

In this chapter, we discuss the chemistry and composition of the diffuse ISM concentrating on the most important probe species. The molecular component typically ranges in temperature from 5–100 K depending on the density, and may be gravitationally bound or unbound. The non-self-gravitating portion, with its associated atomic gas can be a part of the CNM, the WNM, or the WIM, which were discussed in Chap. 1. At the lower temperatures, rotational and hyperfine transitions of various molecular species are populated by collisions and give rise to emission lines at centimeter through sub-millimeter wavelengths. Optical absorption lines can be detected readily if an O or B star is located behind a diffuse molecular cloud (the cloud can't be too dense, otherwise the dust associated with the gas would obscure the background star). We will examine in some detail how to trace the atomic and molecular gas via ground-based radio and sub-millimeter wave observations of rotational and hyperfine transitions and optical absorption lines of electronic transitions.

L. Magnani, S.N. Shore, *A Dirty Window*, Astrophysics and Space Science Library 442, DOI 10.1007/978-3-662-54350-4_3

One of the great ironies of nature is that H_2, the most plentiful molecule by at least 4 orders of magnitude, is virtually unobservable with ground-based telescopes on account of its rotational transitions in the near-IR and its electronic transitions in the far-UV, as discussed in the last chapter. Other molecules must instead be used to trace the diffuse molecular gas. We will discuss them, in turn, and also briefly discuss some general aspects of the interstellar chemistry of the diffuse ISM. For recent general reviews see Tielens (2013) and van Dishoeck (2014). We begin by looking at the history of molecular observations of the ISM.

3.2 The First Detections of Interstellar Molecules

The discovery of the diffuse *atomic* interstellar medium was an accident but the means is the clue to its interpretation. Observations of the radial velocity variations of several unresolved binary star, especially δ Ori, led Hartmann in 1904 to identify stationary absorption lines with the presence of unbound gas interposed between the stellar system and the Earth. The centroid velocities of these features were neither that of the center of mass of the binary nor directly related to any of the neighboring stars. They were also specific, two absorption lines that were normally not seen in the spectrum of the hot stars being studied.

The two main optical atomic absorption features are the doublets of NaI 5889, 5895 Å and CaII 3933, 3968 Å. Importantly, both are resonance lines of their respective ions and both are doublets. The NaI D lines, $2p^6 3s^2 S_{1/2} - 2p^6 3p^2 P^o_{3/2,1/2}$ share a common lower state and CaII H and K, similarly, have almost identical behavior, $3p^6 4s^2 S_{1/2} - 3p^6 4p^2 P^o_{3/2,1/2}$ with the higher J value being the upper state. Because of the difference in statistical weights of the multiplet substates, the components differ by almost 0.3 in log gf and therefore have different optical depths at any column density. The most important property of absorption lines arising from the diffuse ISM is that they are seen in absorption *against* something external. This is not as circular a statement as it might seem. A stellar or extragalactic, continuum source (or even line emitter) has no interaction with the intervening medium. Yet, unlike the effect of an optically thick medium on the emission lines formed within a cloud, the ability to see the illuminating object implies that one sees completely through the absorbing medium. If the distance to the background source is known, this provides an unambiguous measure of the column density integrated within the resolution of the spectrum.

The first detection of molecular gas also occurred very early in the "modern era" in the late 1930's as Dunham and Adams (1937a,b) identified absorption lines of CH electronic transitions along the lines of sight to distant stars. The molecular absorption is produced by what we would call today diffuse molecular clouds, although this was not clear at the time. Detection of interstellar absorption from CH^+ and CN followed shortly thereafter but studies of the distribution of molecular gas were stymied by the lack of suitable background stars to provide a continuum against which to observe interstellar spectral lines. It was clear by the 1950s that molecular rotational lines might be observable with further improvements in radio technology. The atomic hydrogen "spin-flip" transition line was first detected in 1951 by Ewen and Purcell, but it was not until 1963 that Weinreb *et al.* detected the hyperfine ground state transitions of the OH radical at radio frequencies in GMCs. This breakthrough discovery was quickly followed by the detection of NH_3 (Cheung et al. 1968), H_2O (Cheung et al. 1969), H_2CO (Snyder et al. 1969), and, most importantly, CO (Wilson et al. 1970), and molecular astrophysics was born.

The most abundant molecule, H_2, was not directly detected in the cold ISM until 1970 because its relevant electronic transitions occur in the ultraviolet part of the spectrum, necessitating rocket-borne instruments to circumvent the opacity of the Earth's atmosphere (Carruthers 1970; Smith 1973). Unfortunately, direct observations of the rotational transitions of the most common ISM molecule are difficult (see Sect. 2.2.11), and are restricted either to absorption lines in the far-ultraviolet regime if the H_2 is cold (via ro-vibrational electronic transitions—see Fig. 3.1), or emission lines in the near-infrared if the H_2 is heated by shocks or

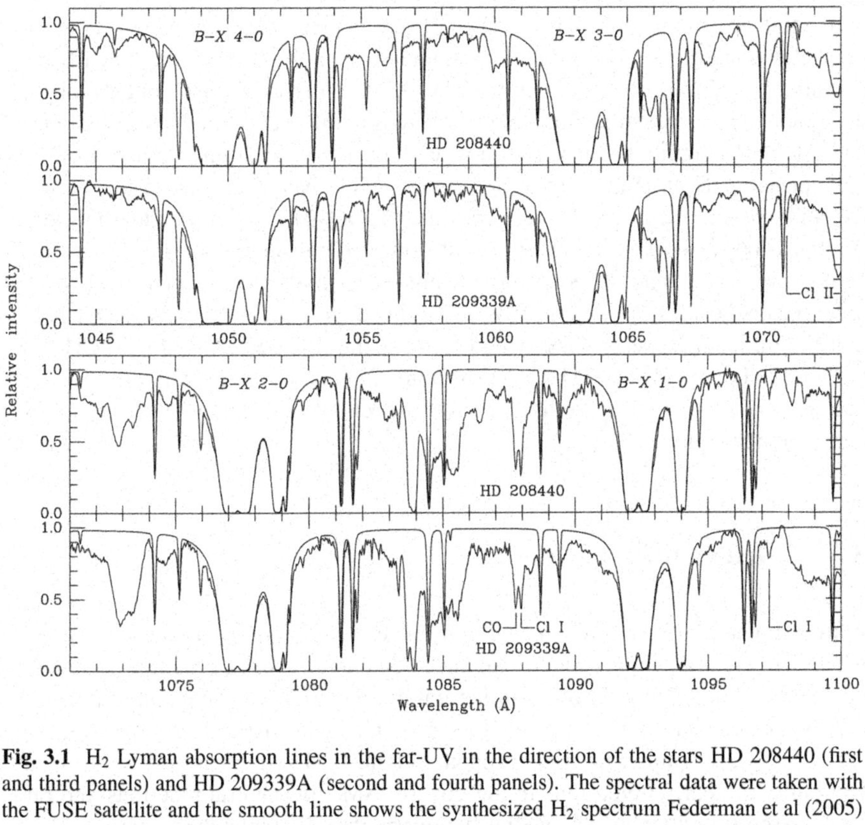

Fig. 3.1 H_2 Lyman absorption lines in the far-UV in the direction of the stars HD 208440 (first and third panels) and HD 209339A (second and fourth panels). The spectral data were taken with the FUSE satellite and the smooth line shows the synthesized H_2 spectrum Federman et al (2005) for details). Stellar features and other interstellar lines account for the remaining absorption. In addition to the four noted H_2 bands, one of the CO Rydberg bands is noticeable at 1088 Å as well as CI and CII lines. The figure is from Federman et al. (2005)

an intense ultraviolet radiation field (Shull and Beckwith 1982; Black and van Dishoeck 1987). The former method is very limited because suitable background stars that provide the continuum against which the absorption lines can be seen are scarce (*e.g.*, Spitzer and Jenkins 1975; Savage et al. 1977); the latter requires heating the gas. In light of these difficulties, surrogate tracers for molecular gas in the ISM had to be found.

In 1968, Carl Heiles detected OH emission lines at 18 cm from dense ($n \geq 10^3$ cm^{-3}), small ($\lesssim$ 10 pc) interstellar clouds (Heiles 1968). This discovery was quickly followed by the detection of H_2CO *absorption* at 6 cm Palmer et al. (1969) and CO emission at 2.6 mm Penzias et al. (1972) from these objects. Given its abundance and with advances in microwave technology, the CO(1-0) and (2-1) rotational transitions soon became the tracers of choice for molecular gas in the ISM. Surveys of the important astrophysical molecular species in dark clouds (as opposed to surveys in the larger GMCs) included OH by Cudaback and Heiles (1969) and Crutcher (1973); H_2CO by Dieter (1973); and CO by Dickman (1975). Concurrently, the great surveys to map the Galactic plane in CO and identify the distribution of GMCs began (these are discussed in Chap. 7).

It soon became apparent that nearly all regions with internal extinction greater than a few magnitudes showed molecular emission or absorption lines. In particular, Dickman (1975) reported observations of CO(1-0) emission from 63 of the 64 Lynds dark clouds he surveyed, including all 4 sources at $|b| \geq 25°$: If an interstellar cloud had $A_V \gtrsim 2$ magnitudes, it was almost certain that at least emission from CO at 2.6 mm and OH at 18 cm would be present. Moreover, the relationship between CO and extinction was quite linear with the ^{13}CO(1-0) transition being the species of choice because of its relatively small optical depth in all but the densest dark cloud cores (Dickman 1978). As molecular astrophysics developed in the 1970s other tracers besides CO were used with varying degrees of success. We argue below that the molecular species best suited to trace the diffuse ISM are CO, CH, OH, H_2CO, C_3H_2, and HCO^+. The relevant transitions and reasons why are discussed in Sect. 3.4. First, we briefly discuss some aspects of interstellar chemistry in low-density environments.

3.3 Astrochemistry

Astrochemistry differs fundamentally from terrestrial chemistry because typical “dense” interstellar environments like the CNM and molecular clouds are about 15–20 orders of magnitude lower than laboratory densities. Consequently, molecules and processes that have no effect in terrestrial chemistry become important in the ISM. For example, the radical methylidyne (CH) which is not stable in the laboratory is nearly ubiquitous in the diffuse ISM. Ion-neutral reactions and photoprocesses are critical in driving the chemistry of diffuse interstellar environments, and the abundances and importance of the various species in the ISM are very different from what is found in planetary atmospheres, for example. Our goal in this section is not to review astrochemistry [for that, see, e.g., the introductory text by Duley and Williams (1984) and the recent, comprehensive review by Tielens (2013)], but to highlight some of the processes that dictate the abundance of species which are tracers of molecular gas in the diffuse medium. In particular, we will look at how the oxygen, carbon, and nitrogen chemical reaction networks are thought to produce some of the molecules discussed later in this chapter.

3.3.1 General Considerations

The formation of molecules in the diffuse ISM proceeds through ion-neutral and neutral-neutral gas phase photon dominated reactions and surface reactions on dust grains (see Wakelam et al. 2010, for a recent review). Photo-processes are favored because of their low activation barriers and ionization, whether they are driven by photons or cosmic rays. The chemical reaction networks required to model the detailed time development of the species' relative abundances are a very large set of nonlinear coupled differential equations, many of which are poorly known, but neglect spatial terms explicitly involving transport and thermodynamical feedback on the environment. The destruction processes are a combination of ion-neutral reactions and photo-dissociations driven primarily by the UV radiation field produced by the stars. Table 3.1 shows the ionization potentials (IP) for some common atoms and diatomic molecules in the ISM. Atomic species with IP less than 13.6 eV are going to be ionized. Molecular species, which may be ions, dominate in regions where self-shielding and dust opacity are far greater and so the calculation is more complicated.

The diffuse molecular component is overwhelmingly diatomic (but with some important exceptions that we will discuss below), so the processes are particularly simple. For gas phase radiative association either ionic or van der Waals interactions determine the collision cross section. These rates also depend on the internal excitation of the individual atoms; they are usually in the ground state since there are only weak collisions at these low densities and the diffuse radiation field cannot maintain substantial population in the excited states. During this formation process,

Table 3.1 Ionization (IP) and dissociation energies (D_0^0) of key atoms and diatoms

Atom	IP (eV)	Diatom	D_0^0 (ev)	IP (eV)
H	13.598	H_2	4.478	14.426
He	24.587	HD	4.514	15.445
C	11.260	CH	3.465	10.64
N	14.534	OH	4.392	12.90
O	13.618	CN	7.76	14.17
Si	8.152	CO	11.092	14.014
S	10.360	CS	7.355	11.335
Ca	6.113	C_2	6.21	12.15

two atoms (A and B) come together to form a diatom AB in an excited state; thus excess energy must be removed from the system. The de-excitation may pass to a dissociative state and the molecule just breaks up again, or the radiative de-excitation arrives at a stable state after which rotational and vibrational transitions can remove the residual energy. Ionic reactions are generally faster, having greater cross sections, but then some external source is required to ionize the diffuse gas. Once formed, charge exchange can neutralize the molecule through $AB^{\pm} + C \rightarrow AB + C^{\mp}$ and stabilize AB, always depending on the state after the collision. However, there are many more possible outcomes for any molecular reaction than atomic. A radiationless dissociation, an analogy to atomic autoionization (e.g., Fano 1961), is produced when a bound state of one electronic configuration spatially overlaps with a continuum state of another electronic configuration.

Gas phase reactions, that dominate the cold dense ISM, are rare in the diffuse phase but when they are exothermic the reaction can be very efficient. An alternative pathway, the one that governs the formation of the dominant molecule, H_2, cannot occur without some additional preconditioning of the hydrogen atoms. In the neutral medium, the excited electronic state of H_2 de-excite dissociatively. The most intense DIRF ultraviolet emission is longward of 2000 Å and photons below 912 Å are effectively removed by ionization of H^0 from the ground state. Shortward, however, below 1200 Å, absorption in the Lyman and Werner bands excite the molecule to a stable state, from $X^1\Sigma_g^+ \rightarrow C^1\Pi_u$ after which decay is possible to vibrational states above the ${}^1\Sigma_g^+$ dissociation limit that has a vibrational quantum number, $v \geq 14$. Any direct absorption to the ${}^3\Sigma_u^+$ state requires a spin flip and is strongly forbidden. The timescales are extremely long for the radiative processes, of order 10^6 sec or longer, since they require quadrupole transitions. This process is driven by radiative excitation within a couple of relatively narrow electronic vibrotational bands so its rate is strongly dependent on the H_2 column density, as for any absorption line. But because the dissociation is driven only through these bands and is not a continuum process, if the radiation at the critical wavelengths (i.e. the Werner and Lyman bands) is suppressed by absorption, the dissociation rate also drops. This is called self-shielding. For a continuum process, any radiation above a threshold can ionize or dissociate the species. Because the excitation is from the ground state its is almost completely insensitive to temperature.

The inverse process—formation of this molecule by a two-body radiative association reaction—is extremely inefficient in removing the excess energy since the de-excitation is very slow and interactions on the surface of dust are the favored mechanism (see Chapter 6 of Duley and Williams 1984; van Dishoeck 2014). In this case, energy can be removed from the system and transferred to bulk heating of the grain, resulting in formation of a stable molecule. The current uncertainty is how the hydrogen atoms migrate between grain sites and then how the molecule detaches from the grain. Both of these require that the dust is not cold. Tunneling and hopping are the main stochastic processes that allow the accumulated H atoms to bond (Charnley 2001). An important clue is that the ortho- to para- ratio is not thermalized (see e.g., Le Bourlot 2000; Pagani et al. 2011). The two nuclei are identical but in combination their nuclear spins are either I = 1 (ortho) or I = 0 (para) so the ratio of the statistical weights should be 3. In homonuclear molecules these correspond to different rotational levels, those for I = 0 are linked to even J (2, 4, ...) and those with I = 1 to the odd states (1,3,...). This causes the line ratios to differ, if thermalized because of the ratio of the statistical weights.[1] The energy difference is 170 K between the two levels (J = 0 to J = 1), but radiative processes do not mix these states and collisions are insufficient to alter the spin state populations in the diffuse medium with $n_{H_2} < 10^3\ \mathrm{cm}^{-3}$ (see Habart et al. (2005) for a comprehensive review of molecular hydrogen in the ISM). The departures from the canonical equilibrium ratio can, however, be achieved by proton collisions on grains and is the signature of the formation site and thus also correlates with the total extinction, A_V (see Chap. 6).

This feature of band driven resonant absorption explains the sensitivity of the formation and excitation of other diatomic molecules, especially CO, to the H_2 column densities. Since radiation impinges on a cloud *from the outside*, the penetration depends on the integrated column density in the absorption lines. The transition probabilities are very low so large pathlengths and number densities are needed to achieve effective self-shielding or self-absorption. But as we discussed in Chap. 2, this is not anything special; once the column density of *any* species is large the resonance lines become opaque and they inhibit both cooling and heating.

The same holds for the photo-destruction of other homonuclear molecules of importance for interstellar chemistry, e.g. N_2 and O_2. All require a radiative excitation in series of vibrotational absorption bands between allowed electronic states. Because none of them has a dipole moment, all radiative de-excitations are slow and quadrupolar. The same sequence of excitation followed by de-excitation to a vibrationally unstable lower level accounts for the photon-driven dissociations that are prerequisites for more complex associative reactions. Molecular hydrogen is merely more abundant so these others, especially N_2, are not so easily optically thick in their dissociating transitions. It further depends on the details of the potentials. For example, for C_2 [see Herzberg (1989) for the potentials of the bound states] there are a number of accessible electronic states whose bound states allow for internal conversion.

In Table 3.2 we list typical reaction rates for common processes of interstellar chemistry. For details on the processes themselves see Duley and Williams (1984).

An important difference between atomic and molecular processes is also discernible from Table 3.2. Like the networks handling nuclear reactions, especially nuclear statistical equilibrium (see Rolfs and Rodney 1988), an input channel may have a range of outcomes. For triatomic and more complex molecules, such as H_2O or NH_3, the branching ratio between exit channels dominates the input to the network and drives the system out of strict equilibrium. The extremely low densities and temperatures in the diffuse phase virtually insure this since, often, the reverse

[1]For a beautiful discussion of how this was used to determine the spin of the proton see Chap. 4 of Tomonaga, S-I. 1997, *The Story of Spin* (Chicago: Univ. of Chicago Press).

Table 3.2 Characteristic molecular rates

Characteristic molecular rates (van Dishoeck 2014)		
Process	Schematic reaction	Rate ($cm^3\ s^{-1}$)
Radiative association	$A + B \rightarrow AB + \gamma$	$10^{-17} - 10^{-14}$
Grain surface reactions		10^{-17}
Associative attachment	$A + B \rightarrow AB + e$	10^{-9}
Dissociative recombination	$AB^+ + e \rightarrow A + B + e$	$10^{-7} - 10^{-6}$
Ion-molecule exchange	$A^+ + BC \rightarrow AB^+ + C$	$10^{-9} - 10^{-8}$
Charge transfer	$AB^+ + C \rightarrow AC + C^+$	10^{-9}
Neutral-neutral	$A + BZ \rightarrow AB + C$	10^{-11}–10^{-9}
Collisional dissociation	$AB + Q \rightarrow A + B + Q$	10^{-26}
Photodissociation	$AB + \gamma \rightarrow A + B$	$10^{-11} - 10^{-9}$

reaction is three-body. Take, for instance, the interaction of OH^+ with H_2. Among the possible outcomes, depending on energy, one might have $H_2O + H^+$ or $OH + H^+ + H_2$ or $OH + H_2^+$. Each final state has an associated binding energy and activation energy. The delicate business of measuring the low temperature reaction rates has been developing apace in the last decade (see Snow and McCall 2006) and empirical branching ratios and cross sections are now available for most of the main reaction sequences that follow.

In contrast to molecules, dust grains are enormous, and through the adsorption by sticking and diffusion their surfaces become covered over time with accreted atoms that can form ice coatings and deplete the diffuse medium. This depends, unfortunately, on one of the more poorly known parameters in molecular networks, the atomic sticking fraction for grains of different compositions. The abundances in the gas are *not* solar under most conditions because over the mean grain lifetime, about 10^7 yr, the refractory species (e.g. C, Mg, Si, Fe) accrete. Absorption line studies are required to determine these factors (Jenkins 2009) but the advantage is that with the low excitation of the atoms the equivalent width measurements suffice to give column densities with no T_{ex} uncertainties. As we described in Chap. 2, this makes the diffuse medium, in effect, an enormous PDR in the sense that photon dominated effects on the grains and in the gas phase drive the chemistry that produces the molecular species. These, in turn, act as tracers of the environments in which they form.

Cosmic rays permeate and ionize all interstellar clouds, but in translucent clouds they are not the only means for producing atomic ions; some ionization is still provided by penetrating UV radiation (in diffuse clouds UV radiation dominates the ionization).[2] The tracers are, therefore, simpler than those found in GMCs and dark cores where the very large optical depths and higher densities lead to much more

[2]Translucent molecular clouds have $1 \leq A_V \leq 5$ mag and diffuse clouds have $A_V < 1$ mag. See Chap. 7.

complex molecules. In addition to the diatomic species, some simple polyatomic species are detected in the diffuse gas, especially HCO, HCO^+, and H_2CO. Each requires a different, but related, chemical path that includes photodissociation of precursor species followed by gas phase and grain surface reactions.

Determined mainly by their abundances (relative to hydrogen), C, C^+, O, N, Mg, Mg^+, Si, Si^+, S, and S^+ are important reacting partners with hydrogen in the CNM, WNM, and WIM. None of the these species reacts exothermically with hydrogen, however, so the starting reactions of the oxygen, carbon, and nitrogen networks are not obvious. We briefly discuss each network below. As a rule, astrochemical modeling has concentrated on denser, colder environments as tracers of star formation. The diffuse medium is, however, also a factory in which these tracers of density and excitation conditions can be exploited to determine local properties of even translucent environments.

An uncertainty, however, remains regarding the chemical pathways for molecular production in the diffuse gas. Dust is the gorilla in the room, yet its role in the chemical networks is still uncertain; as are the laboratory rates for radiative processes for some of the critical reactions (Snow and Bierbaum 2008; Smith 2011). For tabulations of observed abundances see Turner (2000) and Turner et al. (2000) for translucent clouds and Lucas and Liszt (2000) for diffuse clouds.

3.3.2 The Oxygen Network

Ionized oxygen is required to start the oxygen network, since reactions with O and H^+ or H_2^+ are endothermic. To form O^+, cosmic rays are necessary to dissociate H_2 and ionize H to produce a sufficient abundance of H^+ and to ionize O. Because the ionization potential energy of H and O are similar, but O will be neutral where H is ionized, O^+ can form by charge exchange reaction with hydrogen,

$$H^+ + O \rightleftharpoons H + O^+ \tag{3.1}$$

This permits a series of gas phase hydrogenization reactions with H_2,

$$O^+ + H_2 \rightarrow OH^+ + H \tag{3.2}$$

that sets the stage for a variety of end states. For instance

$$OH^+ + H_2 \rightarrow H_2O^+ + H \quad \text{and} \quad H_2O^+ + H_2 \rightarrow H_3O^+ + H \tag{3.3}$$

The ions H_2O^+ and H_3O^+ can then undergo dissociative recombinations with ambient electrons to form OH and H_2O. Alternatively, H_2^+ formed from cosmic ray reactions produces H_3^+ and then

$$H_3^+ + O \rightarrow OH^+ + H_2 \tag{3.4}$$

to yield OH^+, thereafter proceeding according to the sequence in equation 3.3. The difference in these two initiating steps is the hydrogen ion with which charge exchange occurs. The ISRF is sufficiently strong that some fraction of the atomic hydrogen is ionized, as shown by observations of the diffuse Hα background (see Chap. 1). In the colder medium, ionization by cosmic rays insures that there is a reservoir of H_2^+ available to maintain the H_3^+ ion.

3.3.3 The Carbon Network

Of the six, easily detectable, nearly ubiquitous species that trace the diffuse molecular ISM mentioned in Sect. 3.2, five are part of the carbon reaction network (CH, HCO^+, CO, C_3H_2, and H_2CO). We can see below where they form in the astrochemical network, a clue as to which portions of the PDR structure of the medium they best trace.

The problem with this network is that C^+ cannot react exothermally with H_2 to give CH^+. So, to get H_2 involved, the radiative association reaction

$$C^+ + H_2 \rightarrow CH_2^+ + h\nu \tag{3.5}$$

must occur as the entryway into the network, followed by reactions such as

$$CH_2^+ + e \rightarrow CH + H \tag{3.6}$$

$$CH_2^+ + H_2 \rightarrow CH_3^+ + H \tag{3.7}$$

$$CH_3^+ + e \rightarrow CH + H_2 \quad \text{or} \quad CH_2 + H \tag{3.8}$$

and so on, to give the basic hydrides and their ions. We note here CH forms as first generation molecule and so it should be one of the best candidate for tracing the region where H_2 may be present as "dark molecular gas" (see Sect. 8.4). The formation of CO follows from that of HCO^+

$$C^+ + OH \rightarrow CO^+ + H \tag{3.9}$$

$$CO^+ + H_2 \rightarrow HCO^+ + H \tag{3.10}$$

$$HCO^+ + e \rightarrow CO + H \tag{3.11}$$

If we consider OH a first generation molecule (along with CH), then the distribution of both HCO^+ and CO should be more restricted to denser regions of the PDR compared to OH and CH. However, the astrochemical argument for this must be balanced with excitation considerations, which greatly complicate the issue. The main formation pathway to H_2CO also begins with CO by successive addition of hydrogen atoms, through the sequence

$$CO \rightarrow HCO \rightarrow H_2CO(\rightarrow H_2COH \rightarrow CH_3OH) \tag{3.12}$$

where the last steps depend on the hydrogenation of the grains (see below). Continuing reactions depend on the physical properties of the grains and on the radiation in which they are immersed.

The nearly ubiquitous C_3H_2 molecule, a pyramid structure, forms from successive hydrogenation of C_3H^+ and dissociative recombination

$$C_3H^+ + H_2 \rightarrow C_3H_3^+ + e \rightarrow C_3H_2 + H \tag{3.13}$$

Production of the precursor appears to require dust chemistry.

Diffuse phase turbulent dissipation, driven in concentrated vortices, has promise as a viable thermal source (Godard et al. 2009, 2014). The required energies are comparable to those available from shocks (e.g., Sonnentrucker et al. 2002). These dissipative structures can account for the widespread presence of specific ions, notably CH^+, SH^+, and HCO^+, and the higher excitation rotational lines detected for H_2

The CN molecule, one of the first three molecules detected in the ISM, forms primarily from the neutral-neutral gas phase exchange reaction:

$$CH + N \rightarrow CN + H \tag{3.14}$$

We emphasize the important organic molecule H_2CO. A precursor of amino acids and easily detected at 4.8 GHz in absorption against the CMB (the transition is the 1_{11}-1_{10}—see below), formaldehyde is most likely formed on grains rather than in the gas phase by a hydrogenation sequence (Federman and Allen 1991). This is accomplished by adding two hydrogen atoms to CO, which will have already formed on the grain or accreted (if the environment is dense and cold enough). However, this is also a relatively slow three body reaction that can be interrupted by, for example, HCO formation. Finally, these carbon-bearing species may also be connected with the formation of the carriers of the diffuse interstellar bands (DIBs, see below). Tielens (2013) notes that an important feature connecting the hydrogenated series is that dissociative recombination of CH_5 proceeds through $CH_5^+ + e \rightarrow CH_3 + H_2$ instead of forming CH_4. The reactions between O and the organic hydrides then lead to CO while those with C^+ end in acetylene and long carbon chain species.

3.3.4 The Nitrogen Network

For nitrogen chemistry the problem is again the endothermicity of the initiating reactions. The neutral nitrogen ionization potential is 14.5 eV so charge exchange does not work. Instead, neutral exchange reactions with CH produce CN, and with OH produce NO, both of which end with N_2. Subsequent reactions with the omnipresent H_3^+ leads to N_2H^+ and then dissociative recombination leads again to N_2 with a small ($\approx 7\%$) channel for NH formation. Nitrogen ions can be produced by cosmic rays, either directly, or indirectly by interactions with He^+ (which is produced by cosmic ray ionizations). Once a population of N^+ is produced, the reactions

$$N^+ + H_2 \rightarrow NH^+ + H \quad (3.15)$$

$$NH^+ + H_2 \rightarrow NH_2^+ + H \quad (3.16)$$

$$NH_2^+ + H_2 \rightarrow NH_3^+ + H \quad (3.17)$$

occur fairly rapidly. The next reaction in the sequence:

$$NH_3^+ + H_2 \rightarrow NH_4^+ + H \quad (3.18)$$

is slower, but occurs frequently enough to lead to the production of ammonia by dissociative recombination. In the same way, NH and NH_2^+ are produced from NH_3^+. For those reactions that involve grain surfaces, it is now thought that the nitrogen bearing species, since they are shielded by hydrogen, are lost from grains when attached to molecules with low binding energies.

For production of the cyanogen and CN-bearing species in the diffuse medium, Liszt and Lucas (2001) give the following example (recalling, however, our discussion at the start of this section). Beginning with the associative gas phase reaction

$$C^+ + NH \rightarrow CN^+ + H \quad (3.19)$$

followed by

$$CN^+ + H_2 \rightarrow HCN^+ + H \quad (3.20)$$

leads to CN, HCN, HNC once the compounds undergo recombinations reactions, e.g.,

$$H_2CN^+ + e \rightarrow HCN + H \quad . \quad (3.21)$$

Such sequences have been described by van Dishoeck and Black (1986a), Federman et al. (1984), and Crawford and Williams (1997). Thus, even some of the most abundant simple radicals require complex networks, with several formation pathways and must be treated through extensive chemical networks.

It is becoming clearer, as more species are observed, that the astrochemistry of diffuse clouds is more complex than originally thought (see, e.g., Liszt et al. 2013). The fact that Liszt and Pety (2012) routinely detect CO(1-0) emission at the 10 K antenna temperature level in regions where A_V is less than 1 magnitude is a clear indication that much conventional thinking about diffuse molecular clouds is incorrect. Although most astrochemistry studies and reaction networks focus on the denser molecular regions where the most complex molecules are found, there are sure to be new surprises even when the gas density is less than $10^3\ \mathrm{cm}^{-3}$.

3.3.5 Cosmic Rays as Agents of Astrochemical Processing

Cosmic rays below a few hundred GeV are responsible for ionization of the densest phase of the ISM because of their high penetration depths. The penetration depth is governed by their energy dependent ionization losses traversing a neutral medium (see Bethe and Ashkin 1953). Figure 3.2 shows an example of the stopping power (range) in terms of mass column density of hydrogen for protons with energy less than 10 GeV. The average cosmic ray appears to have passed though $\sim 1\ \mathrm{gm\ cm}^{-2}$ of intervening material (Strong et al. 2007). The diffuse ISM has a baryonic volume density of about $1\ \mathrm{cm}^{-3}$ which is a column mass density of $1\ \mathrm{g\ cm}^{-2}$ for a pathlength of $\sim$ 100 kpc, which is larger than the size of the Galaxy. For protons with energies in the MeV range, the gyroradius is substantially smaller. Thus, they must scatter within the tangled magnetic field of the Galaxy instead of freely streaming and remain trapped within the disk and halo. The trapping is more effective than the energy losses and the cosmic rays random walk within the turbulent ISM.

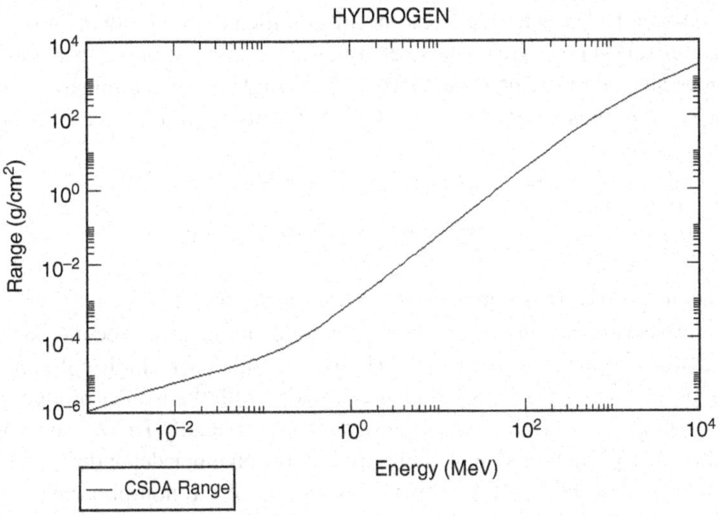

Fig. 3.2 Stopping power (range) for relatively low energy cosmic ray protons, up to 10 GeV, in hydrogen gas. Graph made with data from NIST database using the interface at their site: http://www.nist.gov/physics-portal.cfm

Two related issues arise when treating interstellar chemistry, both relating to the diffuse medium. The interstellar photon distribution, G_0, is derived from modeling the cumulant spectrum weighted by the population of the UV stellar continua that governs surface and low extinction environments (the outer regions of PDRs; see Sect. 1.7.1). Whatever the optical depth of the cloud in the visible, it will be higher in the ultraviolet. Porosity is consequently important for determining the penetration since any photon harder than about 15 eV is capable of ionizing light elements and shifting the driver of the chemistry toward ionic interactions. Cosmic ray ionization is, however, independent of these uncertainties.

The other issue is the cosmic ray ionization rate, ζ_{CR}, or when referring to the cosmic ray ionization rate of molecular hydrogen, ζ_2. This is obtained from modeling reaction networks in which, for dense clouds, G_0 does not suffice to explain the ionization fraction of molecular species. The problem with both is their derivation, based on statistical assumptions that may differ dramatically depending on environment. While the diffuse UV Galactic radiation measurement is complicated by extinction, the CR fraction can also be estimated from a combination of Faraday rotation and synchrotron emissivity, but this depends on the energy distribution of only the electrons and local magnetic field strengths.

Indriolo and McCall (2012) provide an excellent example of how the cosmic ray ionization rate is observationally constrained. The crux of the method is that the presence of one molecule, H_3^+, is uniquely dependent on the ionization produced by protons in the diffuse medium and that the absorption lines from this molecular ion, in the near infrared around 3–4 μm determine the column density. Primary losses for protons at low energy are by ionization of the background ions, with $\Delta E \sim$20–40 eV per ionization. The released electrons are significantly hotter than those released by photoionization processes, whose energies are usually an order of magnitude or more lower since they are predominantly at threshold for the neutral gas. They argue that since *every* cosmic ray proton that ionizes H_2 produces H_3^+ by the sequence:

$$H_2 + CR(p) \rightarrow H_2^+ + e + CR'(p') \tag{3.22}$$

$$H_2^+ + H_2 \rightarrow H_3^+ + H \quad , \tag{3.23}$$

the abundance of H_3^+ tells you the cosmic ray ionization rate.

The destruction mechanisms for H_3^+ include dissociative recombination and recombination in the gas phase with H. Among others, in high column density regions, $H_3^+ + CO \rightarrow H_2 + HCO^+$, which divides with the isomer product HOC^+, and $H_3^+ + O \rightarrow H_2 + OH^+$. All of these species, especially HCO^+, are important interstellar constituents and that their production channel depends on gas phase ion chemistry and that leads to their use as tracers of the non-photonic ionization rate. The cosmic ray ionization rate, ζ, is obtained from the H_2/H_3^+ ratio since the production of the ion is directly proportional to ζ. The recombination rates, whether associative or destructive, are functions of the kinetic temperature through the normal chemical rate coefficients and the H_3^+ is very easily destroyed. The ratio $N(H_3^+)/N(H_2)$ is observationally accessible (although to be precise, the H_2

is obtained by an ultraviolet measurement because the molecule, lacking a dipole moment, is infrared inactive.

Using the absorption column densities that distinguish individual filaments and/or clouds in velocity resolved profiles, the $N(H_3^+)/N(H_2)$ ratio can be measured along the same sight line. Then from the chemical equations the cosmic ray ionization rate follows immediately since the temperature is given by the population ratio from individual line observations. The atomic hydrogen column density, that provides an additional density constraint, comes from 21 cm line strengths and ultraviolet Lyα absorption. The result is that $\log \zeta \approx -16.5 \pm 0.5\ \mathrm{s}^{-1}$. Other tracers give slightly different values for the diffuse medium but all in the same range, between $1.7 \times 10^{-16}\ \mathrm{s}^{-1}$ and $1.1 \times 10^{-15}\ \mathrm{s}^{-1}$. The penetration depth depends on the number density, composition, and spectrum of the protons but these numbers have been obtained from a variety of environments. Since these measurements are based on absorption lines, the details of line formation are enormously simplified since these molecular lines are from the ground state or very low lying rotational states. The ionization is effected by the lower energy protons, those between 10 keV and a few MeV, hence the part of the energy distribution not discernible from Earth without (still debatable) corrections for heliospheric modulation. Recall that the main contributors to the ionization are protons below a few tens of MeV.

Some of the gas, however, seems almost immune to interaction. The absence of H_3^+ along some lines of sight and other chemical modeling indicates regions with $\zeta_2 < 10^{-17}\ \mathrm{s}^{-1}$ in regions for which the molecular hydrogen column density is $N(H_2) > 10^{21}\ \mathrm{cm}^{-2}$ that would be completely optically thick in H_2 and self-shielding (hence capable of producing a CO emitting translucent cloud).

3.4 Molecular Tracers of the Diffuse ISM in the Radio Regime

Although the CO(1-0) line is the traditional way to trace molecular gas on large scales, there are other transitions that can perform this and complementary tasks. Mapping the distribution of molecular gas on large scales basically involves trying to trace molecular hydrogen with densities in the 10^2–$10^4\ \mathrm{cm}^{-3}$ range. Higher density molecular gas is confined to the small, dense cores within the much-larger molecular clouds and is best traced using species such as CS or NH_3. We are primarily concerned with the lower density gas, so, we now examine in more detail the species and transitions that can be used to trace that matter.

3.4.1 The CO Rotational Transitions

The CO molecule has a simple rotational ladder structure in its ground vibrational state that is ideal for detecting molecular gas in cold clouds. The lowest rotational transition, the famous CO(1-0) line has a rest frequency of 115.2712 GHz (equivalent to a wavelength of 2.6 mm) and was first detected in the ISM in 1970 (Wilson et al.). Although optically thick, an empirical conversion factor known as the X-factor or X_{CO} is used to convert the line profile to the physically relevant quantity, $N(H_2)$. The many ways to calibrate this factor are discussed in detail in Chap. 8. In the first two decades after its discovery, this transition was used to trace the overall molecular distribution of the Galaxy. The GMCs along the Galactic plane were mapped out and the conventional wisdom at the time was that the CO basically traced virtually all the molecular mass. The principal reason for this assertion was the very rapid drop-off in CO emission at the edges of GMCs (Blitz 1979). It has become clear in the last decade that more molecular gas is present at the edges of GMCs and other, smaller clouds, than was traced by the early surveys. The CO(1-0) molecule may still be a relevant tracer in these regions, but it may not be the best tracer. This important issue is discussed in detail in Sect. 8.4. For the moment, we will consider the CO(1-0) line to be the most effective tracer of molecular gas in the denser regions of GMCs and in the so-called "dark molecular clouds" (see Chap. 7). In density regimes with $n \geq 10^2\,\mathrm{cm}^{-3}$ the CO(1-0) line almost always is detectable with relatively short integration times (a few minutes using current millimeter-wave receivers).

The energy difference between the first rotational state and the ground state is equivalent to a temperature of ~ 5 K, so collisions in molecular clouds can still produce a sizeable population in the first rotational state. This is also true of the other low rotational states so that observations of the CO(2-1) line at 230.5 GHz and the CO(3-2) line at 345.8 GHz were commonplace in the early days of molecular radio astronomy. With equivalent temperatures of 5.6, 17, and 33 K,[3] the lowest three rotational levels of CO are ideally spaced to probe the bulk of the molecular gas in molecular clouds. The CO(2-1) line is actually strong enough to allow significant survey work (e.g., Sakamoto et al. 1997) to be done. Interestingly, the CO(4-3) line at 461.04080 GHz (650 μm) is sometimes observed from the ground, but it is solidly in the submillimeter where the opacity of the atmosphere is beginning to be a serious problem. The higher rotational transitions are sometimes observed in denser regions where the relevant rotational levels can be sufficiently populated, but large-scale surveys are almost never made with transitions higher then the CO(3-2) line.

Of more importance than the higher CO rotational transition are the isotopes of $^{12}C^{16}O$, and in particular the $^{13}C^{16}O$(1-0) line at 110.20137 GHz and the $^{12}C^{18}O$(1-0) line at 109.782182 GHz. In the lower density molecular regions (i.e., those with

[3]Note that the energy difference between $\Delta E_{J,J+1} = (J+1)\hbar/I$. Here, I is the moment of inertia of the electronic state.

$n < 10^3\,\text{cm}^{-3}$), the $^{13}C^{16}O$(1-0) line is optically thin so that the column density of $^{13}C^{16}O$ is readily obtainable following the procedure outlined in Sect. 2.2.5. However, by the time the density reaches $10^3\,\text{cm}^{-3}$, the line becomes optically thick and $^{12}C^{18}O$ must be used. Many studies of these isotopic species were made in molecular clouds in the 1970s and 1980s in order to obtain the abundance of CO with respect to H_2 (e.g., Dickman 1978; Frerking et al. 1982).

In addition to the rotational transitions, CO can be detected via ultraviolet electro-vibrational transitions (see Chap. 5). Once the characteristics of the CO(1-0) transition had been established, the large scale surveys of the molecular gas along the Galactic plane began. We will discuss their results in Chap. 7.

3.4.2 *The OH 18 cm Lines*

The hydroxyl radical, OH, is the third most abundant molecule in the ISM, though it is significantly (at least two orders of magnitude) less abundant than CO. Hydroxyl has a $^2\Pi$ ground state with an unpaired electron whose spin couples with the electronic angular momentum, **J**, to split the rotational levels into two ladders, $^2\Pi_{1/2}$ and $^2\Pi_{3/2}$. Because the OH molecule is inverted (the constant describing the interaction between the orbital angular momentum and the electron spin of the system is negative), the $^2\Pi_{3/2}$ state has lower energy than the $^2\Pi_{1/2}$ state and the former is thus the ground state. Each rotational level in each ladder is again split by Λ-doubling which arises from the degeneracy of states with the same absolute value of Ω, the projection of the total angular momentum of the electrons onto the internuclear axis ($\Omega = \Lambda + \Sigma$, where Λ is the projection of the orbital angular momentum on the internuclear axis and Σ is the total spin angular momentum). Each Λ-doubled state is again degenerate because the unpaired spin of the proton, **I**, changes the electronic angular momentum to $\mathbf{F} = \mathbf{J} + \mathbf{I}$ (see Herzberg 1989, for details). This hyperfine interaction splits the Λ-doubled levels again so that each rotational state has four levels. The energy-level diagram for the lowest three rotational states in both $^2\Pi$ ladders is shown in Fig. 3.3.

The four transitions of the $^2\Pi_{3/2}$ ground state have the following frequencies: 1612.231, 1665.401, 1667.385 and 1720.533 MHz; these correspond to wavelengths of 18.61, 18.02, 18.00, and 17.44 cm, respectively. The main lines correspond to the $\text{F} = 2 - 2$ (1667 MHz) and the $\text{F} = 1 - 1$ (1665 MHz) transitions. In thermal equilibrium the ratio of the 1667 to the 1665 MHz lines is 9:5 and the 1667 MHz line to the satellite lines is 9:1. However, maser emission from the 18 cm lines has been detected in various astronomical environment (star-forming regions, circumstellar shells, extragalactic nuclei). Masers require high column densities with nearly collisionless environments and strong radiative pumping (see Elitzur (1992) for details). Neither condition is typical of the diffuse ISM.

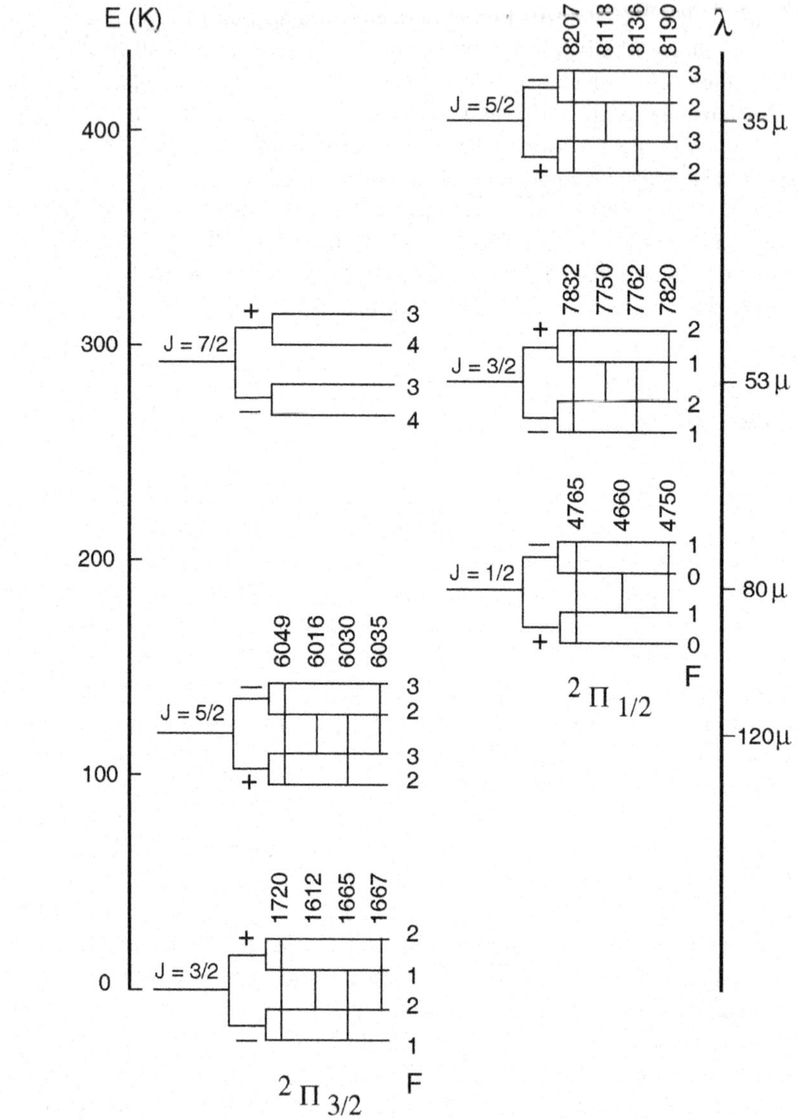

Fig. 3.3 Energy-level diagram of the 3 lowest OH rotational transitions in each $^2\Pi$ state. The frequencies of the 18 cm lines are noted in the $^2\Pi_{3/2}$ $J = 3/2$ term. The scale on the left is in Kaysers (cm^{-1}) and the spacings of the lambda-doubled and hyperfine transitions are not to scale. The figure is from Pihlström et al. (2008)

Once one or more of the 18 cm lines have been detected, column densities of OH can be determined following the method described in Chap. 2. Basically, for each line (denoted by *i*) the antenna temperature and line widths (FWHM) are related to

T_{ex}, τ, and N(OH) by

$$T_{B,i} = \eta_B f\ (T_{ex,i} - T_{BG})(1 - e^{-\tau_i}) \tag{3.24}$$

$$N(OH) = C_i\ T_{ex,i}\ \Delta V_i\ \tau_i \tag{3.25}$$

where η_B is the efficiency of the telescope (see Sect. 4.2.6), f is the filling factor of the source in the main beam, and C is a constant that differs for each line (C equals 22.21, 4.30, 2.30, and 20.82 for the 1612, 1665, 1667, and 1720 MHz transitions, respectively). Assuming all four lines are detected, the above eight equations have 9 unknowns, so that some assumptions about the excitations temperatures are usually made. However, for those lines of sight where all four lines are successfully detected, the sum rule

$$\nu_{1612}/T_{ex,1612} + \nu_{1720}/T_{ex,1720} = \nu_{1665}/T_{ex,1665} + \nu_{1667}/T_{ex,1667} \tag{3.26}$$

can be used to provide an extra equation to solve for N(OH), $T_{ex,i}$ and τ_i. The sum rule is derived readily from the Boltzmann distribution.

Usually, the satellite lines are too weak to be detected or radio-frequency interference (RFI) wipes out observations of the 1612 and/or 1720 MHz line so that only data on the main lines is available. In those instances, the usual procedure it to assume an excitation temperature (usually, the assumptions is $T_{ex} \sim T_{kin}$) and then solve for N(OH). Once N(OH) is determined, $N(H_2)$ can be readily derived because the $N(OH)/N(H_2)$ ratio is fairly constant at 1×10^{-7} (e.g., Weselak et al. 2010).

The 18 cm transitions were the first molecular spectroscopic lines detected in emission in the radio portion of the spectrum (Weinreb et al. 1963). Early studies by Cudaback and Heiles (1969) and Crutcher (1973) established that these lines were easily detectable in dark molecular clouds and Wouterloot (1981) mapped the Taurus/Auriga dark clouds in these transitions. Although supplanted by CO as the principal molecular tracer, the 18 cm lines were always acknowledged to be superb tracers of low-density molecular gas. Kazés et al. (1977), Crutcher (1979), Dickey et al. (1981), Magnani et al. (1988), and Magnani and Siskind (1990) all reported OH 18 cm detections in diffuse and translucent molecular clouds. Astrochemistry arguments (e.g., Viala 1986) make it clear that in PDR regions, the OH molecule forms before CO and has its greatest abundance at the edge of a cloud. Wannier et al. (1993) even claimed that the 1665 and 1667 MHz lines extend beyond the CO-emitting region in five molecular clouds encompassing both dark clouds and GMCs.

The results of these early studies were that the OH 18 cm main lines are readily detectable in diffuse and translucent clouds and the strengths of the various transitions were consistent with thermal excitation. Because these studies focused on the cloud cores as identified by CO mapping, with the notable exception of the work by Peter Wannier and B-G Andersson (see, e.g., Wannier et al. 1993), the question of OH emission at the cloud edges was put aside for many years.

Wennerstrom (2007) studied OH main lines as tracers of low-density molecular gas at the edges of a translucent cloud, MBM 16, and two diffuse molecular clouds: MBM 3 and 40 (some of her results were published by Cotten et al. 2012). Even moderate integrations (less than an hour) revealed a signal in the envelopes of diffuse molecular clouds. Figure 3.4 shows the locations of the OH observations in the diffuse cloud MBM 3. The spectra were obtained with the Arecibo 305 m radio telescope along two directions; one oriented N-S, and the other E-W. Hydroxyl main line emission is detectable up to the edge of the CO boundary of the cloud (denoted by a dashed line in the figure) where E(B-V) is ≈ 0.12 mag according to the Schlegel et al. (1998) dust maps. More significantly, beyond the edge of the CO(1-0) map of the cloud, the OH 1667 MHz main line is likely present. Figure 3.5 shows the composite spectrum of the 2 lines of sight on each side of the western dashed line shown in Fig. 3.4. This is clearly diffuse molecular gas showing E(B-V) level ≈ 0.09 mag.

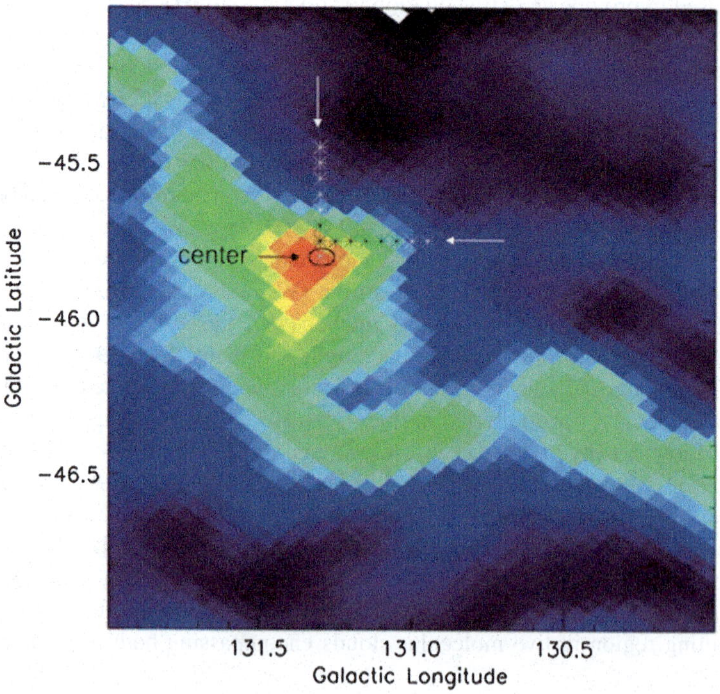

Fig. 3.4 Locations of the OH 18 cm main lines observed by Wennerstrom (2007) in MBM 3. The observations were made with the Arecibo 305 m radiotelescope at a resolution of 3′. The *asterisks* represent the observed lines of sight. *Black asterisks* represent the extent of the molecular cloud as mapped by the CO(1-0) line. The positions are numbered from the nominal center of the cloud (*circled*) in each direction, so that the position W5 would be the fifth "x" to the right of center (the westernmost black asterisk)

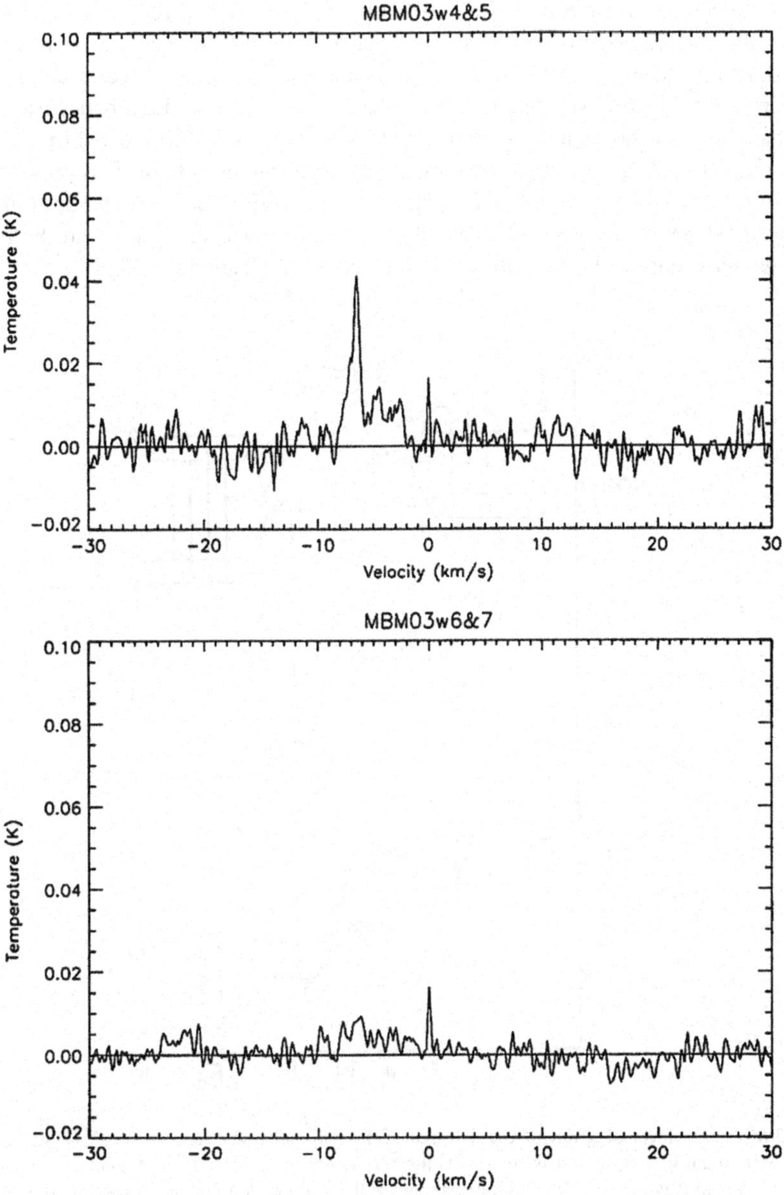

Fig. 3.5 Top spectrum: Composite OH 1667 MHz spectrum for positions W4 and W5 in Fig. 3.4 (Wennerstrom 2007). The 1667 MHz emission from the translucent high-latitude cloud MBM 3 extends from −2 to −9 km s^{-1}. The small chirp at 0 km s^{-1} is due to the autocorrelator and is seen here only because of the great sensitivity of the observations (rms ∼ 5 mK). Bottom spectrum: Composite OH 1667 MHz spectrum from positions W6 and W7 in Fig. 3.4. There is a hint of an emission line at the correct velocity for MBM 3. The lower spectrum arise from a region that has E(B-V) ≈ 0.09 mag. If emission is indeed present from this outermost region of the cloud, it would indicate that the OH 18 main lines are very effective tracers of dark molecular gas (see Sect. 8.4)

More recently, Barriault et al. (2010) noted that, in some directions towards the molecular clouds around the North Celestial Pole, the OH lines are significantly broader than those of CO(1-0) along the same lines of sight. Like Wannier and Andersson, they proposed that the OH 18 cm main lines may be a better tracer of H_2 than the CO rotational transitions. Can CO(1-0) emission be detected from these regions or is the OH 1667 MHz line tracing the PDR regime where H_2 is present but CO column densities are too low for detection at 115 GHz? Before that question can be resolved, more sensitive CO(1-0) observations are needed. We revisit the issue of high-sensitivity CO observations at cloud edges in Chap. 8.

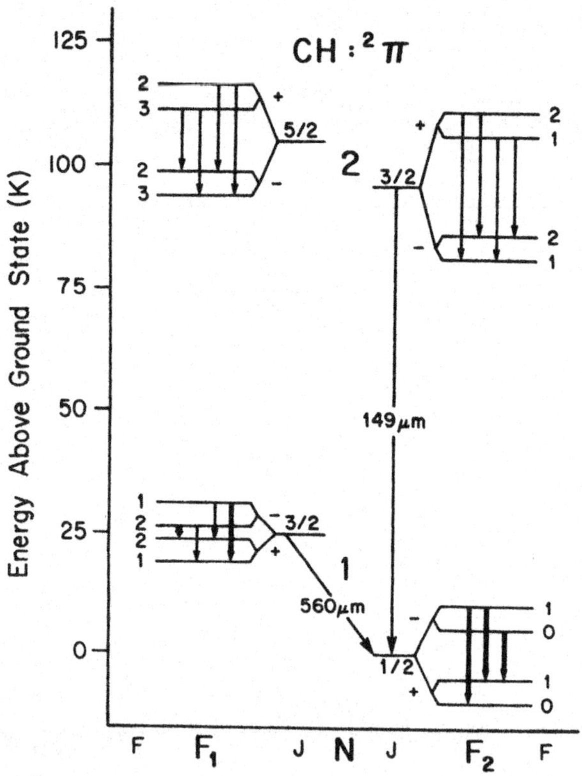

Fig. 3.6 Energy-level diagram of the 3 lowest CH rotational transitions in each $^2\Pi$ state. The 3335 MHz main line discussed in the text is the $^2\Pi_{1/2}, J = 1/2, F = 1 - 1$ transition. The scale on the left is in Kaysers (cm^{-1}) and the spacings of the lambda-doubled and hyperfine transitions are not to scale. The figure is from Ziurys and Turner (1985)

3.4.3 The CH 3.3 GHz Lines

The CH molecule (methylidyne) has a $^2\Pi$ ground state like OH. However, the rotational levels that are split into the two ladders, $^2\Pi_{1/2}$ and $^2\Pi_{3/2}$, are inverted with respect to OH, so that the ground state is a $^2\Pi_{1/2}$ state (see Fig. 3.6). Like the OH ground state, Λ-doubling and hyperfine splitting produces four levels, but because the ground state is a $^2\Pi_{1/2}$, the $F = 0 - 0$ transition is forbidden. Thus, there is only one main line, at 3335 MHz, and two satellite lines at 3264 and 3349 MHz. In thermal equilibrium, the ratio of the main to satellite lines is 2:1.

The CH main line is an excellent tracer for H_2 at low densities ($n \leq 10^3$ cm^{-3}). This was established early on for dark clouds (Sandell et al. 1981, 1980) and was confirmed for translucent clouds by Magnani et al. (1989) and Magnani and Onello (1993). There was speculation that the peculiar excitation characteristics of the CH ground state (Rydbeck et al. 1976; Bertojo et al. 1976 and Turner 1988) produced weak maser behavior which might allow the detection of the CH main line even from gas at number densities too low to excite the lower rotational transitions of CO. There are reports in the literature of CH main line detections from regions without associated CO(1-0) emission (Federman and Willson 1982; Sandell et al. 1987; Magnani and Onello 1993). But the brutally long integration times required to register the few mK rms levels necessary to find this gas made these results difficult to reproduce. In light of the recent work by Cotten and Magnani (2013), it seems longer than usual integration times for the CO(1-0) line (i.e., $\sim$ tens of minutes rather than the usual 1–2 minutes) will detect molecular gas where CH emission has also been found. In summary, like deep CO(1-0) and OH 1667 MHz integrations, the CH 3335 MHz line is sensitive to low-density molecular gas at the edges of clouds. However, both CO(1-0) and the OH main lines are more useful tracers because they require much lower integration times. So why should anyone ever observe the CH 3335 MHz line? The answer lies in the ease with which CH 3335 MHz emission leads to N(CH) and the linearity between N(CH) and N(H_2).

The problem with CO(1-0) observations is that the line is optically thick in virtually all astrophysical environments, so the conversion of the observed line parameters to N(H_2) is problematic. The CH main line is optically thin and so directly provides N(CH) (see §2.2.5 and Rydbeck et al. 1976). Once N(CH) is obtained, the conversion to N(H_2) for diffuse and translucent molecular gas is straightforward given the linear relationship between the two column densities for $A_V < 5$ mag. The evidence for this constant ratio of CH/H_2 in low-density gas has been confirmed over the last three decades by many observational studies: Sandell et al. (1981); Federman (1982); Danks et al. (1984); Mattila (1986); van Dishoeck and Black (1989); Magnani and Onello (1995); Liszt and Lucas (2002); Sheffer et al. (2008); Weselak et al. (2010). The CH/H_2 ratio in these studies varies from 3.5–4.5 $\times$ 10^{-7}, which sets a limit on the accuracy of N(H_2) determinations. We will discuss the technique and results of using the CH 3335 MHz line to calibrate the CO-H_2 conversion factor in Chap. 8, here we are concerned primarily with how well CH can trace molecular gas.

As stated above, CH goes away in high-density, high-extinction regions, but is fairly reliable for regions below 5 mag of visual extinction. How far down in density and extinction can this line be employed to trace molecular gas? An intriguing result was presented by L.E.B. Johansson in an Onsala preprint (1979). He described a poorly sampled CH survey of the Galactic plane whose results implied that the radial distribution of N(H_2) was between the distribution determined from CO(1-0) observations and HI 21 cm observations. The exact quote was

More significant is the observed excess CH abundance (compared with CO) at galactocentric distances larger than about 7 kpc, which may indicate that the radial distribution of CH to some extent is an intermediate case between those of CO and HI. Such an interpretation can be considered in the light of the different region contributing to the molecular and atomic emission. The CO emission originates from cool and compressed regions where atomic hydrogen has been converted to molecular hydrogen (see e.g. Burton 1976). The CH emission, on the other hand, is observed from the whole spectrum of "clouds", HI to molecular clouds, and thus is likely to be due to a mixture of the regions revealed by CO and HI observations.

The difference in distribution between CH and CO also manifests itself on the small scale. High resolution CH observations from Arecibo (angular resolution at 3.3 GHz $\sim 3'$) showed a positional offset between the CO(1-0) emission and the CH 3335 MHz line (see Fig. 3.7). This phenomenon was confirmed in velocity since the CH line profile is distinctly different from that of ^{12}CO and ^{13}CO (see Fig. 3.8). An extended blue-shifted wing is noted in the composite profile and does not appear in either CO line. CH and CO seem to arise in different parts of the cloud with the CH more strongly associated with the peripheral regions of the cloud. In this sense, the CH might be tracing molecular gas "invisible" using CO. Since composite profiles are used in comparing the different species, this is not a sensitivity question as the rms of the CO(1-0) line is only a few millikelvin. This ability of CH to probe the outermost molecular regions of a cloud was predicted theoretically by Federman (1982), but in some sense the debate is moot in that the necessary integration times to bring out this level of emission using CH at 3335 are prohibitively long. However, if a linear tracer of H_2 is required for regions with $A_V \lesssim 5$ mag, then it is worthwhile to remember the 3335 MHz CH line.[4]

3.4.4 The H_2CO 4.8 GHz Line

At first glance, based only on elemental abundance, the most abundant molecules after H_2 would be diatomic combinations of H, C, O, and N. Carbon monoxide, methylidyne and hydroxyl certainly fall into this category, and while CN, NO, and NH, C_2, O_2 and N_2 have all been detected in the ISM, for a variety of

[4]Something the designers of the NRAO 100 m Green Bank Radio Telescope forgot—no upper S-band receiver was ever built for it.

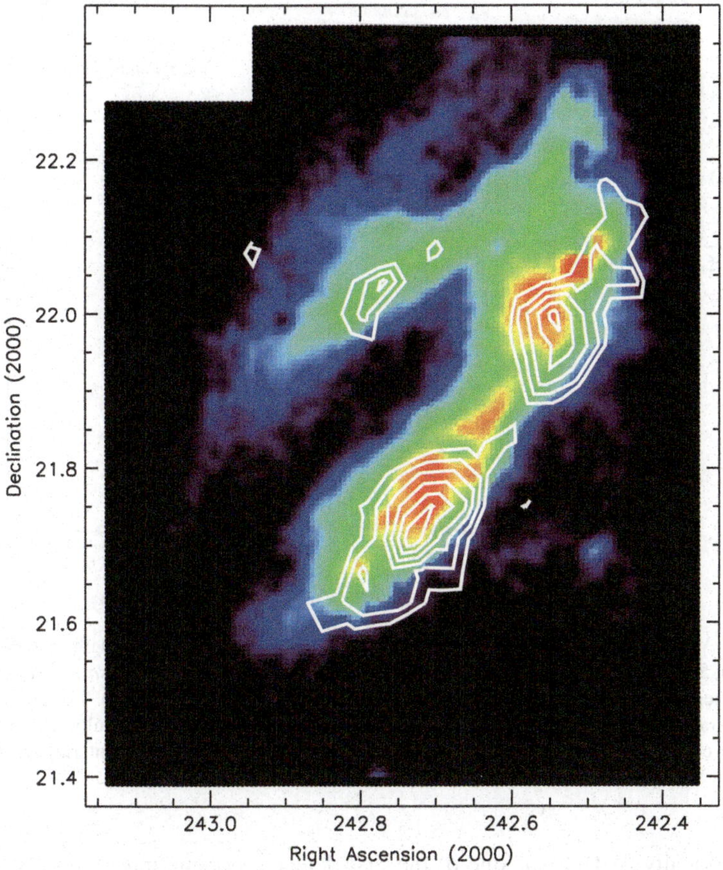

Fig. 3.7 Positional offset between CH and CO emission in MBM 40. The color image represents CO(1-0) emission from Chastain (2005) . The contours are the velocity integrated CH 3335 MHz line profiles with white contour levels set at 167, 195, 223, 251, 279 and 307 mK km s^{-1}. The lowest color level (violet) is at about 1 K km s^{-1}, so there is lower intensity CO emission throughout the region. This figure demonstrates the offset between the *peak* CO and CH emission, indicating that the two species are tracing different molecular regimes

reasons (excitation, chemistry, lack of favorable radio transitions) none of these molecules are good markers of low-density gas in the radio regime. However, several polyatomic molecular transitions do trace that material. Foremost of these is formaldehyde, H_2CO, with its hyperfine transition at 4830 MHz. The molecule is a near-symmetric rotator with one moment of inertia much larger than the other two producing a splitting of the levels into K doublets (Herzberg 1966). Additionally, because the two H atoms can have parallel or antiparallel spins, the energy levels divide into two separate states, ortho and para. This gives a rich spectrum of transitions at cm and mm wavelengths.

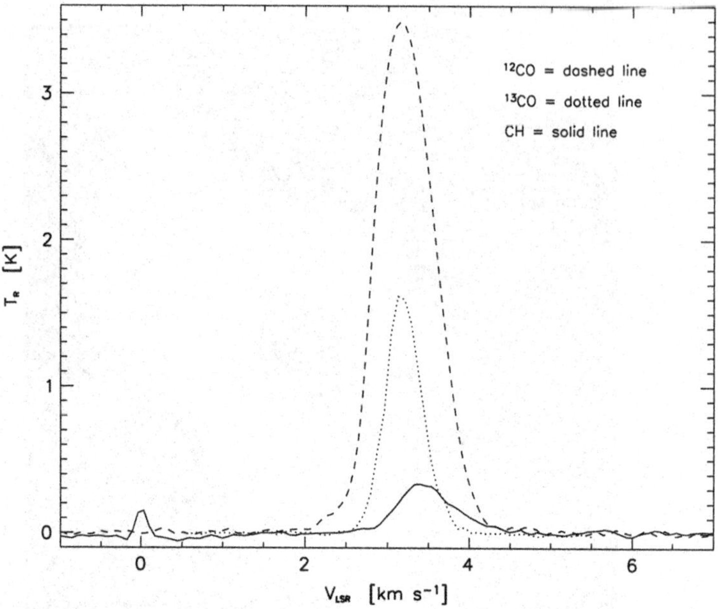

Fig. 3.8 Velocity offset between CH and CO emission in MBM 40 (compare with spatial offset in Fig. 3.7). Spectral profiles for CO (*dashed line*), ^{13}CO (*dotted line*), and CH (*solid line*) averaged over the region of ^{13}CO observations in MBM 40 described by Chastain (2005). The two averaged CO spectra are plotted at their actual values, while the CH spectrum has been multiplied by a factor of 10 for clarity. The spike in the CH spectrum at 0.0 km s^{-1} is produced by internal interference and should be ignored

Historically, H_2CO was one of the first detected organic interstellar molecules (Snyder et al. 1969) and its 1_{11}-1_{10} transition at 4830 MHz (6 cm) and, to a lesser extent the 2_{12}-2_{11} line at 14.5 GHz (2 cm), were used early on to probe cloud envelopes. The excitation of both transitions is anomalous. For regions with densities less than 10^6 cm^{-3} the lines appear in absorption, even when no obvious source of continuum radiation lies behind them. To produce detectable absorption lines requires overpopulating the lower level of the 2 cm doublet (2_{12}) relative to LTE, so $T_{ex} < T_{BG}$—see Chap. 2. Townes and Cheung (1969) proposed a "refrigeration pump" initiated by collisions with H_2 that preferentially populate the lower level of the doublet, rapidly followed by radiative decay to the lower level of the 6 cm doublet (1_{11}). Both levels therefore have greater than equilibrium populations in their lower states. This pump functions at relatively low ISM densities, so the collisions with H_2 are more frequent and the levels thermalize. For densities $> 10^6$ cm^{-3}, both lines go over into emission. The behavior of the H_2CO 2 and 6 cm lines as a function of densities was explored in a classic paper by Evans et al. (1975). Note, however, that there is an ambiguity when the lines are seen in absorption. A weak H_2CO absorption line may indicate the presence of very low-density gas ($\sim 10^2$ cm^{-3}), or, come from significantly higher density gas ($\sim 10^5$ cm^{-3}) as the line passes from absorption to emission.

For the 4830 MHz line, mapping of diffuse and translucent clouds shows that this transition picks up the low-density core regions but not the outer envelopes (at least at integration times of less than an hour per point—Magnani et al. 1993, Magnani et al. 1996 for MBM16 and 40, respectively). Not much mapping of cloud envelopes has been done so far with the 6 cm transition, but recently renewed interest in observing the 2 and 6 cm lines has come about as robust density probes of the dense cores in GMCs (e.g., Mangum et al. 2013).

Gas-phase chemistry models have always had a difficult time reproducing the observed column densities of H_2CO under diffuse and translucent conditions (e.g., Prasad and Huntress 1980). Federman and Allen (1991) suggest that, like H_2, H_2CO is formed on grain mantles and ejected from them into the low-density gas phase of molecular clouds on timescales of 10^5 yr.

3.4.5 The HCO^+(1-0) Transition

Another polyatomic molecular transition, HCO^+(1-0) absorption at 89188.526 MHz is widely seen in absorption in the diffuse ISM. This species is a precursor of CO. Lucas and Liszt (1996) detected HCO^+ absorption in 30% of the continuum sources in their sample which had HI absorption. In contrast, HCO^+(1-0) emission is rarer and the lines are rather weak compared to the CO(1-0) transition. Its large molecular dipole moment, 3.89 Debye, is mitigated by the significantly lower abundances compared to CO. The typical abundance of HCO^+ with respect to H_2. along diffuse and translucent lines of sight is $2\text{–}3 \times 10^{-9}$, and the abundance ratio with respect to CO [i.e., N(HCO^+)/N(CO)] is about an order of magnitude higher in diffuse regions compared to regions with $A_V > 1.5$ mag (Turner 1995). Antenna temperatures for emission lines of the $J = 1 - 0$ transition are typically tens of mK so that mapping of this species over large areas is unfeasible despite the relative ubiquity of the species in diffuse molecular regions. The transition may trace molecular gas in regions with $A_V < 1$ mag that is not observable via the CO(1-0) transition (Lucas and Liszt 1994; Hogerheijde et al. 1995). Liszt and Lucas (1998) point out that in a sample of interstellar absorption lines detected towards compact extragalactic objects, the CO(1-0) line widths are narrower than those of the HCO^+ lines (0.75 km s^{-1} vs. 0.85 km s^{-1}). This is important in that absorption line observations are more sensitive to low column densities than emission line studies. Note that the neutral molecule HCO has not been detected in either diffuse or translucent clouds Muller et al. (2011), although it is present in dense molecular regions.

3.4.6 C_3H_2, C_2H, and C_3H^+

The propenylidine molecule, C_3H_2, exists in the ISM in both linear form (ℓ-C_3H_2) and the more stable cyclic conformer state (c-C_3H_2, called cyclopropenylidine). The symmetry of the hydrogen atoms bonding to two of the carbon atoms produces ortho and para states in the energy level structure. Additionally, the two unpaired electrons on the other carbon atom give C_3H_2 a very large dipole moment ($\mu = 3.43$ Debye); consequently, the 1_{10}-1_{01} ortho-line is one of the strongest molecular lines in the ISM at cm wavelengths. The molecule is surprisingly ubiquitous in interstellar clouds (Matthews and Irvine 1985) with the 1_{10}-1_{01} and 2_{20}-2_{11} lines at 18343.145 and 21587.4 MHz having been detected in diffuse clouds in absorption (Cox et al. 1988). Many of the regions observed by Cox et al. have $A_V \sim 1$ mag and so this species can be used to probe low-density diffuse gas. The linear form of the molecule is also detected in the ISM but at much lower column densities (factor of 100 - Liszt et al. 2012), so its value as a low-density molecular tracer is minimal.

The c-C_3H_2 18.3 GHz ortho-line in emission and the 21.6 GHz $2_{20} - 2_{11}$ para-transition in absorption have also been detected in a sample of dark clouds spanning a broad range of physical conditions (Madden et al. 1989; Cox et al. 1989). Additionally, the $2_{12} - 1_{10}$ rotational ortho-transition at 85339 MHz has been detected in emission in dark clouds at fairly high antenna temperatures (a few K). In these clouds the abundance of C_3H_2 is related to that of HC_5N supporting a correlation between cyclopropenylidenes and cyanopollynes. Lucas and Liszt (2000) observed the molecule in a sample of clouds in front of compact, extragalactic mm sources and derived an abundance of 2.3×10^{-9} consistent with astrochemistry models. In the same paper, the ethynil radical (C_2H) was also reported in nearly all the observed directions making it also an excellent tracer of diffuse molecular gas. The transition used was the $N = 0 - 1$ line in absorption at 87 GHz. Lucas and Liszt (2000) and Gerin et al. (2011) found a relative abundance of $N(C_2H)/N(c\text{-}C_3H_2)$ of about 20 in diffuse molecular gas which decreases by an order of magnitude in dark clouds. C_2H is by far the most abundant of the pollyynes (C_nH molecules) and of the small polyatomic hydrocarbons. Liszt et al. (2012) determine an abundance of C_2H of (4–7) $\times 10^{-8}$, similar to that of CH. Both the cyclic and linear forms of C_3H have also been detected in translucent clouds.

As a side note, the detection of c-C_3H_2 in the ISM raised the possibility that other ring molecules might be found in molecular clouds. Planar ring molecules such as furan (C_4H_4O), pyrrole (C_4H_5N), and imidazole ($C_3H_4N_2$) are all biologically important, have been looked for in the ISM, but have not been detected. As of this writing, the only other cyclic molecules besides C_3H_2 that have been detected are silicon dicarbide, SiC_2, cyclopropynylidyne, C_3H, cyclopropenone, H_2C_3O, and ethylene oxide, C_2H_4O.

3.4.7 *HF, a Tracer of Molecular Hydrogen*

A most unlikely proxy has recently been found for H_2, the diatom HF (Gerin et al. 2016). Although the solar system abundance of flourine is $10^{-7.6}$ relative to hydrogen (Maiorca et al. 2014), far lower than the CNO group (e.g., the oxygen abundance is $10^{-3.3}$), *the diatom is unique in forming from an exothermic neutral gas phase reaction* $H_2 + F \rightarrow HF + H$ (Tizniti et al. 2014). The only destruction channels are photodissociation and reactions with C^+, both of which are relatively slow (see Table 3.2). The molecular was first observed by Neufeld et al. (1997) at 121.7 μm with ISO and more recently Indriolo et al. (2013) have detected the 2.499 μm ro-vibrational band in absorption against several embedded stellar sources.

3.5 Optical Manifestations of Diffuse Molecular Gas and Dust

Although radio observations in the last three decades have revolutionized our view of the diffuse ISM, the first observations were made in the visible portion of the spectrum. After the first detections in the late 1930s, C_2 was seen in absorption nearly four decades later. Other species which have detectable optical transitions are not discussed as they have limited roles in studies of interstellar clouds. The diffuse interstellar bands are discussed because their origin is still mysterious and they appear to be molecular in origin. Finally, we discuss the possibility that some of the emission visible in the red part of the optical spectrum may not be reflection from dust grains in the clouds but, rather, photoluminescence.

3.5.1 *CH, CH^+, CN*

As discussed in Sect. 3.2, the first identification of molecules in the 1930s involved absorption lines from CH, CH^+, and CN. For a quarter century, they were the only way to probe the molecular component of the ISM, and because background stars had to be visible through the cloud, sampled mainly diffuse molecular gas. Each of the species is interesting in its own right and we will briefly discuss them in turn.

The dominant CH absorption line is the $A^2\Delta$-$X^2\Pi$ transition at λ4300.3 Å . There are also three transitions from the $B^2\Sigma - X^2\Pi$ system at λ3878, λ3886, and λ3890 Å , and, in the near-UV, from the $C^2\Sigma^+ - X^2\Pi$ system at λ3137, λ3143, and λ3146 Å . Lien (1984) describes the term structure of these lines in detail and resolved a long-standing problem with the oscillator strength. Because of the strong linear correlation between N(H_2) and N(CH)—reviewed by Mattila (1986), observations of the λ4300 Å line have been made for hundreds of lines of sight traversing diffuse clouds. Before UV observations in the 1970s, this CH line was one of, if not the most, important transition for characterizing the molecular properties of the low-extinction clouds. The formation and observed abundances of this species are readily reproduced by steady-state chemistry models. At first, there was little communication between the optical astronomers who studied CH at $\lambda 4300$ Å and the radio astronomers who studied the 3.3 GHz transition described above. A notable exception was the work by Federman and Willson (1982) that compared column densities obtained by the two techniques. With the dearth of upper S-band radio receivers at radio observatories throughout the world, and the availability of time on relatively large optical telescopes, more complete surveys of the CH $\lambda 4300$ Å transition would be fruitful.

The CH^+ ion is detected in absorption from the $A^1\Pi - X^1\Sigma^+$ (0-0) transition at $\lambda 4232.5$Å. Lambert and Danks (1986) observed the $\lambda 4232.5$ Å transition in 65 lines of sight towards bright O and B stars for the $\lambda 4232$ Å line. The abundance of this ion in the ISM remains a problem for gas-phase chemistry models (e.g., van Dishoeck and Black 1986b). To date, all steady state models have failed by several orders of magnitude to reproduce the observed abundances in diffuse molecular clouds. Exothermic reactions are usually invoked to explain the enhanced abundance (e.g., between C^+ and vibrationally excited H_2). Bates and Spitzer (1951) suggested CH^+ formation by sublimation of CH_4 from dust grains followed by photodissociation and photoionization. Alternatively, shocks are invoked to produce the kinetic energy necessary to drive the reactions favoring its creation. The observed velocity shift between CH^+ and CH may indicate the formation in different parts of a shock flow. However, even these models have not yet been able to reproduce the observed CH^+ abundances (see Dalgarno 1976 and Hartquist et al. 1990).

The CN radical was discovered via its three interstellar lines from the 0-0 band of the violet ($B^2\Sigma^+ - X^2\Sigma^+$) system (the R(0), R(1), and P(1) lines at $\lambda 3874.6$, $\lambda 3874.0$, and $\lambda 3875.8$ Å, respectively. Interestingly, the excitation temperature of the J $= 0 \rightarrow 1$ rotational transition at 2.64 mm can be determined from the strengths of the R(0) and R(1) lines. In the absence of local excitation, this excitation temperature is due to cosmic background radiation and constitutes an excellent way of measuring the CMB brightness temperature at this wavelength (Field and Hitchcock 1966; Shklovsky 1966; Thaddeus and Clauser 1966). Federman, Danks, and Lambert (1984) established that the CN column density increases rapidly with H_2 in diffuse clouds, thereby making this transition most sensitive to the denser portions of these clouds. Lambert et al. (1990) observed all three species using the $\lambda 4300$, $\lambda 4232$, and $\lambda 3874$ Å lines in the direction of ζ Ophiuchi. Their results confirm that the CN profile is narrowest ($< 1\,\mathrm{km\,s^{-1}}$), the CH profile has both a narrow and broad component (0.7 and $1.3\,\mathrm{km\,s^{-1}}$, respectively); and the CH^+ is broad ($3.5\,\mathrm{km\,s^{-1}}$). In this way, the core and envelope regions of a cloud can be studied simultaneously. Despite the lack of area coverage in such studies of diffuse molecular clouds, an all-sky absorption line survey towards suitable stars in these three species would yield tremendous information about diffuse molecular clouds.

3.5.2 C_2

After the discovery of CH, CN, and CH^+ in the late 1930s, nearly four decades passed before the next molecule, C_2 was found in the visible in the ISM (Souza and Lutz 1977). Like H_2, C_2 has no permanent dipole moment and so its excited rotational levels have long radiative lifetimes. This means that, unlike CH, CN, and CH^+, C_2 can have excited rotational levels with relatively high populations. For example, van Dishoeck and de Zeeuw (1984) observed seventeen absorption lines of C_2 from the (2-0) Phillips system (near 8750 Å in the direction of χ Oph. The lines arise from several rotational levels in the v $= 0 - 2$ band of the first excited electronic state $A^1\Pi_u$. Its rich level structure allows the molecule to be used a density and temperature probe in diffuse clouds. Absorption lines can also be detected in the (3,0) Philips system near 7720 Å . The C_2 in absorption yields abundances in diffuse clouds of C_2/H_2 of about 0.5–1.5 $\times$ 10^{-8} (Gredel 1999). This can increase by an order of magnitude in more opaque clouds.

3.5.3 *The Diffuse Interstellar Bands*

One absorber in the ISM is in a completely different category, the *diffuse absorption bands* or DIBs (see Snow and McCall 2006 for a general survey of the properties, and Gredel et al. 2011 for a review of DIB attributions). A persistent mystery, not too strong a description, the DIBs were first discovered by Mary Heger (1922) and, to date, have increased in number and variety while remaining unidentified as far as their carriers. They are weak absorption features, never seen in emission, with low optical depths (less than about 0.2) and much broader (FWHM $\geq$ 140 Å) than can be accounted for by any thermal, or even turbulent, motion, of order a few Å. Their breadth has been interpreted as indicating that their carriers must be complexes with five or more atoms. They have little discernible fine structure, no obvious band appearance of single line cores, and no single form fits all of the absorbers. The most studied, and omnipresent, are at 4430 (the strongest), 5780, 5797, 6284, and 6613 Å, but extensive catalogs have been produced along many sight lines through the Galaxy (e.g., Jennisken and Desért 1994).

A clue to their origin is a strong correlation between the equivalent widths and extinction, in particular E(B-V). The O stars, having only a small range in intrinsic B-V, serve as the best background sources since the reddening is most easily specified for them and they have the advantage of being visible over the longest lines of sight through the Galaxy because of their intrinsically high luminosities (classical novae have the same advantage). The disadvantage in using O and B stars is that they possess strong UV continuum emission, hence they can modify their immediate environments through the formation of local photo-dissociation and ionization regions. Since the rest wavelengths of the individual DIBs are not well-known, their association with any particular structure along the line of sight is far more difficult than for the resonance atomic or molecular lines. Because DIBs do not have obvious associations with H_2 they are thought to reside in regions with low molecular fraction.

Platt (1956) suggested the presence of small dust particles in the ISM that would behave like a transitional structure between molecules and grains (Kroto 1988). Their link to the DIBs was, however, only made after the discovery of excess short wavelength infrared emission from reflection nebulae (Sellgren 1984) and the discovery of anomalous 12 μm and 25 μm emission from the diffuse medium. Leger and Puget (1984) and Omont (1986) proposed polycyclic aromatic hydrocarbons (PAH) as the carrier of both the bands and the source of the anomalous infrared emission (see (Leger and D'Hendecourt 1985; van der Zwet and Allamandola 1985). This identification was made by associating a set of broad *emission* features, seen in reflection nebulae and HII regions, with bond stretching modes of C-H bearing organic compounds such as coronene and pyroxene. The PAHs are especially stable under soft ultraviolet irradiation and are of sufficient mass that they satisfy the Pratt requirement. The excess IR is then thought to be produced by vibrational band redistribution of the energy of absorbed UV photons through a fluorescence process. Recently, the Buckminsterfullerene ion, C_{60}^{+} has been identified as the carrier of two DIBs (Campbell et al. 2015), confirming a conjecture by Harold Kroto in 1988. Given the presence of C_{60} in soot, it is likely that other carriers of the DIBs may be found there.

Related to the DIBs is the Anomalous Microwave Emission (AME); an excess of microwave flux over that from the CMB, the thermal Galactic bremsstrahlung and non-thermal synchrotron backgrounds, and the thermal emission backgrounds. It is most easily observed in the 10–60 GHz range and was first recognized in the COBE data as a diffuse, all-sky background correlated with dust emission (Kogut et al. 1996; De Oliveira-Costa et al. 1997; Leitch et al. 1997). The connection with DIBs is from the putative mechanism for producing AME: spinning dust. The correlation between AME and 100 μm emission from infrared cirrus implies that the phenomenon is linked to some property of the dust. The grains are thought to be charged because of photoelectric heating; they are also thought to be spinning rapidly ($\omega \sim 10^5$–10^7 Hz) so that they can emit electric dipole radiation. In addition, if they are also magnetized they could also emits magnetic dipole radiation (Draine and Lazarian 1998).

3.5.4 Red Luminescence

Red Luminescence is seen from regions known by reflection nebulosity in the blue to be dusty. Sometimes known as *extended red emission* or ERE, the signature is a broad, featureless emission band most prominent from λ6100- λ8200 Å. In some nebulae it can contribute as much as 30–50% of the flux in the I band. PAHs may account for the red luminescence (d'Hendecourt et al. 1986; Chlewicki and Laureijs 1987), but no conclusive identification has been made

Fig. 3.9 SDSS green image of a 2°× 2° region centered on MBM 40. The structure of the molecular cloud is visible in the band centered on 4770 Å . Compare this figure to Fig. 5.5. The image was made with the Skyview Virtual Observatory

The red luminescence from nearby, high-latitude molecular clouds is especially prominent and played a role in the search for these objects first reported by Blitz et al. (1984). Candidate regions were originally identified by looking for regions of low-extinction on the POSS prints, but it soon became clear that these regions are accompanied by faint, low-level emission on both the blue and the red prints. The blue emission is simply reflection by the dust component of the integrated starlight of the Galactic plane but, surprisingly, a faint emission of similar extent and morphology was evident on the red plates. This is a manifestation of red luminescence. An example of this can be seen in Figs. 3.9 and 3.10 which show a 2°× 2° region centered on the high-latitude cloud MBM 40 in both the red and the blue SDSS data. Note that the red low-level emission is just as evident as the blue emission. Most high-latitude molecular clouds that show blue reflection nebulosity also show red nebulosity which is likely to be red luminescence.

Fig. 3.10 SDSS red image of a 2°× 2° region centered on MBM 40. The structure of the molecular cloud is actually more pronounced in the red band centered on 6231Å than on the green band shown in Fig. 3.9. Compare this figure to Fig. 5.5. The image was made with the Skyview Virtual Observatory

3.6 The Molecular Medium in a Cosmological Line of Sight: PKS 1830-211

One of the most remarkable advances in the last decade is the observation of the diffuse medium in external galaxies using absorption line measurements against cosmologically distant background sources. The most complete survey to date is against the multiple images of the quasar PKS 1830-211 (Muller et al. 2011, 2014) that is lensed by a foreground galaxy at a redshift of $z = 0.89$. At such large distance, the lines are displaced by a factor of nearly two. The survey was performed in the 32–47 GHz band that is relatively free from atmospheric contamination. Table 3.3, based on that paper, gives the relative column densities compared to molecular hydrogen (as inferred from H I and CO observations, see Wiklind and Combes (1998); and references therein).

Table 3.3 Relative abundance of molecular species in different environments [$\epsilon(X)/\epsilon(H_2) \times 10^8$]

Molecule	Diffuse	Translucent	TMC 1	FG0.89SW
C_2H	2.6	6.6	7.5	6.24
HCN	0.3	3.6	2.0	1.52
H_2CO	0.4	0.6	2.0	1.20
HCO^+	0.2	0.2	0.8	0.87
CH_3OH	–	1.8	0.2	0.85
HCN	0.06	0.25	2.0	0.51
SiO	0.01	0.01	–	0.04
SO	0.15	3.20	0.50	0.13

References

Barriault, L, Joncas, G., Lockman, F.J., and Martin 2010, P.G. MNRAS, 407, 2645

Bates, D.R. and Spitzer, L. Jr. 1951, ApJ, 113, 441

Bethe, H. and Ashkin, J. 1953, in *Experimental Nuclear Physics*, ed. E. Segré, (New York: Wiley), 253

Bertojo, M., Cheung, A.C., and Townes, C.H. 1976, ApJ, 208, 914

Black, J.H. and van Dishoeck, E.F. 1987, ApJ, 322, 412

Blitz, L. 1979, Ph.D. Thesis, Columbia University

Blitz, L., Magnani, L., and Mundy, L. 1984, ApJ, 282, L9

Burton, W.B. 1976, ARAA, 14, 275

Campbell, E.K., Holz, M., Gehrlich, D., and Maier, J.P. 2015, *Nature*, 523, 322

Carruthers, G.R 1970, ApJ, 161, L81

Charnley, S.B. 2001, ApJ, 562, 99

Chastain, R.J. 2005, Ph.D. thesis, University of Georgia

Cheung, A.C., Rank, D.M., Townes, C.H., Thornton, D.C., and Welch, W.J. 1968, Phys. Rev. Let., 21, 1701

Cheung, A.C., Rank, D.M., Townes, C.H., and Welch, W.J. 1969, *Nature*, 221, 626

Chlewicki, G. and Laureijs, R.J. 1987, in *Polycyclic aromatic hydrocarbons and astrophysics*; Proceedings of the NATO Advanced Research and CNRS Workshop, Les Houches, France, Feb. 17–22, 1986, D. Reidel Publishing Co., 335

Cotten, D.L., et al. 2012, AJ, 144, 163

Cotten, D.L. and Magnani, L. 2013, MNRAS, 436, 1152

Cox, P., Guesten, R., and Henkel, C. 1988, AAp, 206, 108

Cox, P., Walmsley, C.M., and Güsten, R. 1989, AAp, 209, 382

Crawford, I.A. and Williams, D.A. 1997, MNRAS, 291, 53

Crutcher, R.M. 1973, ApJ, 185, 857

Crutcher, R.M. 1979, ApJ, 231, L151

Cudaback, D.D. and Heiles, C. 1969, ApJ, 155, L21

Dalgarno, A. 1976, in *Atomic Processes and Applications*, eds. P.G. Burke and B.L. Moiseiwitsch (Amsterdam: North-Holland), 109.

Danks, A.C., Federman, S.R., and Lambert, D.L. 1984, A&A, 130, 62

De Oliveira-Costa, A., et al. 1997, ApJ, 482, L17

d'Hendecourt, L.B., Leger, A., Olofsson, G., and Schmidt, W. 1986, A&A, 170, 91

Dickey, J.M., Crovisier, J., and Kazès, I. 1981, A&A, 98, 271

Dickman, R.L. 1975, ApJ, 202, 50

Dickman, R.L. 1978, ApJS, 37, 407

Dieter, N.H. 1973, ApJ, 183, 449

Draine, B.T. and Lazarian, A. 1998, ApJ, 494, L19

Duley, W.W. and Williams, D.A. 1984, *Interstellar Chemistry*, London; Academic Press
Dunham, T., Jr. and Adams, W.S. 1937*a*, Publ.Am.Astr.Soc., 9, 5
Dunham, T., Jr. and Adams, W.S. 1937*b*, PASP, 49, 26
Elitzur, M. 1992, *Astronomical Masers*, Springer-Science+Business Media
Evans, N.J.,II, Morris, G., Sato, T., and Zuckerman, B. 1975, ApJ, 196, 433
Fano, U. 1961, Phys. Rev., 124, 1866
Federman, S.R. 1982, ApJ, 257, 125
Federman, S.R. and Willson, R.F. 1982, ApJ. 260, 124
Federman, S.R., Danks, A.C., and Lambert, D.L. 1984, ApJ, 287, 219
Federman, S.R. and Allen, M. 1991, ApJ, 375, 157
Federman, S.R., Sheffer, Y., Lambert, D.L., and Smith, V.V. 2005, ApJ, 619, 884
Field, G.B. and Hitchcock, J.L. 1966, Phys. Rev. Letters, 16, 817
Frerking, M.A., Langer, W.D., and Wilson, R.W. 1982, ApJ, 262, 590
Gerin, M. et al. 2011, A&A, 525, A116
Gerin, M., Neufeld, D.A., and Goicoechea, J.R. 2016, arXiv160102985G
Godard, B., Falgarone, E., and Pineau des Forêts, G. 2009, A&A, 495, 847
Godard, B., Falgarone, E., and Pineau des Forêts, G. 2014, A&A, 570, 27
Gredel, R. 1999, A&A 351, 667
Gredel, R., et al. 2011, A&A, 530, A26
Habart, E., et al. 2005, SSRv, 119, 71
Hartmann, J. 1904, ApJ, 19, 268
Hartquist, T.W., Flower, D.R., and Pineau des Forêts, G. 1990, in *Molecular Astrophysics*, ed. T.W. Hartquist, Cambridge U. Press, 99
Heger, M.L. 1922, Lick Observatory Bull., 10 (337), 146
Heiles, C. 1968, ApJ, 157, 123
Herzberg, G. 1989, *Molecular Spectra and Molecular Structure, Vol. 1 - Spectra of Diatomic Molecules*, 2nd ed., Krieger Pub. Co.
Herzberg, G. 1966, *Molecular Spectra and Molecular Structure, III (Polyatomic Molecules)*, van Nostrand
Hogerheijde, M.R., de Geus, E.J., Spaans, M., van Langevelde, H.J. and van Dishoeck, E.F. 1995, ApJ, 441, L93
Indriolo, N. and McCall, B.J. 2012, ApJ, 745, 91
Indriolo, N., Neufeld, D.A., Seifahrt, A., and Richter, M.J. 2013, ApJ, 764, 188
Jenkins, E.B. 2009, ApJ, 700, 1299
Jennisken, P. and Desért, F.X. 1994, A&AS, 106, 39
Johansson, L.E.B. 1979, Research Lab. Electronics and Onsala Space Obs. Research Rep. 136 (Onsala: Onsala Space Obs.)
Kazés, I., Crovisier, and Aubry, D. 1977, A&A, 58, 403
Kogut, A., et al. 1996, ApJ, 460, 1
Kroto, H. 1988, *Science*, 242, 1139
Lambert, D.L. and Danks. A.C. 1986, ApJ, 303, 401
Lambert, D.L., Sheffer, Y., and Crane, P. 1990, ApJ, 359, 19
Le Bourlot, J. 2000, A&A, 360, 656
Leger, A. and Puget, J.L. 1984, A&A, 137, L5
Leger, A. and D'Hendecourt, L. 1985, A&A, 146, 81
Leitch, E.M., Readhead, A.C.S., Pearson, T.J., and Myers, S.T. 1997, ApJ, 486, L23
Lien, D.J. 1984, ApJ, 284, 578
Liszt, H.S. and Lucas, R. 1998, A&A, 339, 561
Liszt, H. and Lucas, R. 2001, A&A, 370, 576
Liszt, H. and Lucas, R. 2002, A&A, 391, 693
Liszt, H.S. and Pety, J. 2012, A&A, 541, 58
Liszt, H.S., Sonnentrucker, P., Cordiner, M., and Gerin, M. 2012, ApJ, 753, L28
Liszt, H., Lucas, R., Pety, J., and Gerin, M. 2013, Proceedings, IAU Symposium No. 297, J. Cami and N.L.T. Cox, ed., Cambridge U. Press, 163

Lucas, R. and Liszt, H.S. 1994, A&A, 282, L5
Lucas, R. and Liszt, H.S. 1996, A&A, 307, 237
Lucas, R. and Liszt, H.S. 2000, A&A, 358, 1069
Madden, S.C., Irvine, W.M., Swade, D.A., Matthews, H.E., and Friberg, P. 1989, AJ, 97, 1403
Maiorca, E. et al. 2014, ApJ, 788, 149
Magnani, L., Blitz, L., and Wouterloot, J.G.A. 1988, 326. 909
Magnani, L., Lada, E.A., Sandell, G., and Blitz, L. 1989, ApJ, 339, 244
Magnani, L. and Siskind, L. 1990, ApJ, 359, 355
Magnani, L., LaRosa, T.N., and Shore, S.N. 1993, ApJ, 402, 226
Magnani, L. and Onello, J.S. 1993, ApJ, 408, 559
Magnani, L. and Onello, J.S. 1995, ApJ, 443, 169
Magnani, L. et al. 1996, ApJ, 465, 825
Mangum, J.G., Darling, J., Henkel, C., and Menten, K.M. 2013, ApJ, 766, 108
Matthews, H.E. and Irvine, W.M. 1985, ApJ, 298, L61
Mattila, K. 1986, A&A, 160, 157
Muller, S., et al. 2011, A&A 535, 103
Muller, S., et al. 2014, A&A 566, 112
Neufeld, D.A., Zmuidzinas, J., Schilke, P., and Phillips, T.G. 1997, ApJ, 488, 141
Omont, A. 1986, A&A, 164, 159
Pagani, L., Roueff, E., and LeSaffre, P. et al. 2011, ApJ, 739
Palmer, P., Zuckerman, B., Buhl, D., and Snyder, L.E. 1969, ApJ, 156, L147
Penzias, A.A., Solomon, P.M, Jefferts, K.B., and Wilson, R.W. 1972, ApJ, 174, L43
Pihlström, Y., et al. 2008, ApJ, 676, 371
Platt, J.R. 1956, ApJ, 123, 486
Prasad, S.S. and Huntress, W.T. 1980, ApJ, 239, 151
Rolfs, C.E. and Rodney, W.S. 1988, *Cauldrons in the Cosmos: Nuclear Astrophysics*, U. Chicago Press
Rydbeck, O.E.H., et al. 1976, ApJS, 31, 333
Sakamoto, S., Hasegawa, T., Handa, T., Hayashi, M., and Oka, T. 1997, ApJ, 486, 276
Sandell, G., Hoglund, B., and Friberg, P. 1980, A&A, 83, 226
Sandell, G., Johansson, L.E.B., Rieu, N.Q., and Mattila, K. 1981, A&A, 97, 317
Sandell, G., Stevens, M.A., and Heiles, C. 1987, A&A, 179, 255
Savage, B.D., Bohlin, R.C., Drake, J.F., and Budich, W. 1977, ApJ, 216, 291
Schlegel, D.J., Finkbeiner, D.P., and Davis, M. 1998, ApJ, 500, 525
Sellgren, K. 1984, ApJ, 277, 623
Sheffer, Y. et al. 2008, ApJ, 687, 1075
Shklovsky, I.S. 1966, Astron. Circ., No. 364 (Acad. Sci. USSR)
Shull, J.M. and Beckwith, S. 1982, ARAA, 20, 163
Smith, A.M. 1973, ApJ, 179, L11
Smith, I. W. M. 2011, ARAA, 49, 29
Snow, T.P. and McCall, B.J. 2006, ARAA, 44, 367
Snow, T.P. and Bierbaum, V. M. 2008, Ann. Rev. Phys Chem, 1, 22
Snyder, L.E., Buhl, D., Zuckerman, B., and Palmer, P. 1969, Phys. Rev. Let., 22, 679
Sonnentrucker, P., Friedman, S. D., Welty, D. E., York, D. G., and Snow, T. P. 2002, ApJ, 576, 241
Souza, S.P. and Lutz, B.L. 1977, ApJ, 216, L49
Spitzer, L. Jr. and Jenkins, E.B. 1975, ARAA, 13, 133
Strong, A.W., Moskalenko, I.V., and Ptuskin, V.S. 2007, Ann.Rev.Nucl.Part.Sci., 57, 285
Thaddeus, P. and Clauser, J.F. 1966, Phys. Rev. Letters, 16, 819
Tielens, A.G.G.M. 2013, Rev. Mod. Phys., 85, 1021
Tizniti, M. et al. 2014, Nat. Chem., 6, 141
Townes, C.H. and Cheung, A.C. 1969, APJ, 157, 103
Turner, B .E. 1988, ApJ, 329, 425
Turner, B.E. 1995, ApJ, 449, 635
Turner, B.E. 2000, ApJ, 542, 837

Turner, B.E., Herbst, E., and Terzieva, R. 2000, ApJS, 126, 427
van der Zwet, G.P. and Allamandola, L.J. 1985, A&A, 146, 76
van Dishoeck, E.F. and de Zeeuw, T. 1984, MNRAS, 206, 383
van Dishoeck, E.F. and Black, J.H. 1986, ApJS, 62, 109
van Dishoeck, E.F. and Black, J.H. 1986, ApJ, 340, 273
van Dishoeck, E.F. and Black, J.H. 1989, ApJ, 340, 273
van Dishoeck, E.F. 2014, Faraday Discuss., 168, 9
Viala, Y.P. 1986, A&A, 64, 391
Wakelam, V., et al. 2010, SSRv, 156, 13
Wannier, P.G., et al., 1993, ApJ, 407, 163
Weinreb, S., Barret, A.M., Meeks, M.L., and Henry, J.C. 1963, *Nature*, 200, 829
Wennerstrom, E. 2007, MS Thesis, Univerisity of Georgia
Weselak,T., Galazutdinov, G A. Beletsky, Y., and Kreowski, J. 2010, MNRAS 402, 1991
Wiklind, T. and Combes, F. 1998, ApJ, 500, 129
Wilson, R.W., Jefferts, K.B., and Penzias, A.A. 1970, ApJ Let., 161, L43
Wouterloot, J.G.A. 1981, PhD Thesis, Leiden University
Ziurys, L.M. and Turner, B.E. 1985, ApJ, 292, 25

Chapter 4
Observing the Diffuse ISM: Making Sense of the Radio Observations

My dear Kepler, what would you say of the learned here, who, replete with the pertinacity of the asp, have steadfastly refused to cast a glance through the telescope? What shall we make of this? Shall we laugh, or shall we cry?

—Galileo Galilei

Abstract The diffuse interstellar medium is studied primarily via radio spectroscopic techniques as far as the colder atomic and low-density molecular structures are concerned. In this chapter, we discuss some of the practical considerations involved in making and analyzing these observations. Some of the topics discussed here should be useful in preparing observing proposals or reading observational papers on this medium. Topics covered include basic radio spectroscopic and mapping techniques, Stokes parameters, antenna temperature conventions, radio frequency interference, virial analysis, principal component analysis, and statistical analysis of the velocity fluctuations.

4.1 Introduction

The diffuse ISM is unusually beautiful! Its vast and diverse structures produce stunning visual images (see book cover and Fig. 4.1). Beauty aside, from radio waves to gamma rays, observations of the diffuse ISM are critical to characterizing and understanding it. Although observations at all wavelengths contribute to our knowledge of the many different manifestations of the ISM, our focus is on the denser substructures. Since these are mostly molecular, they are cold, so it is not surprising that observations in the radio and infrared regimes are the most important. Because of the transparency of the atmosphere at low frequencies, radio observations can easily be made from ground-based observatories. In particular, spectroscopy of the molecules discussed in Chap. 3 is mainly carried out at single-dish radio telescopes. In this chapter we focus on the some of the more relevant aspects of radio astronomical observations necessary for understanding the literature of the CNM and molecular clouds.

L. Magnani, S.N. Shore, *A Dirty Window*, Astrophysics and Space Science Library 442, DOI 10.1007/978-3-662-54350-4_4

Fig. 4.1 The diffuse ISM in and around ρ Ophiuchi. NASA Picture of the Day for 2012, August 28. A combination of dust, gas and young stars produced a reflection nebula (*blue, top middle right*), an emission nebula (*reddish, lower middle right and dark clouds* (center). The bright star Antares illuminates a yellowish dust cloud (*bottom center*). To the right of Antares is the globular cluster M4. Image courtesy of Tom O'Donoghue; website: astrophotography.ie

4.2 Radio Observations: Practical Considerations

There are several references available that describe the ins and outs of radio astronomy. Three that immediately come to mind are *Essential Radio Astronomy* by Condon and Ransom (2016), *Tools of Radio Astronomy* by Wilson et al. (2010) and *Radio Astronomy* by John Krauss (1966). We will not repeat the material covered in those texts but will confine our discussion to some practical considerations which are often skipped in the more general introductions to the subject. Although interferometers are becoming more useful in studying the low-density ISM (e.g., the series of papers by Liszt and Lucas on molecular species along diffuse lines of sight discussed in Chap. 7), the workhorse instruments for detecting and studying the diffuse molecular gas are still the large radio single dishes and spectroscopy is the principal way to determine the structure, composition, and physical parameters of the clouds in question. We saw in the preceding chapter which species and transitions are the most useful in tracing low-density molecular gas and one or more combinations of them will be used to tackle almost any question pertaining to the diffuse ISM. Once the tracer has been chosen, what considerations must then be addressed to successfully obtain radio spectroscopic data for a given project?

4.2.1 The Radiometer Equation

In radio astronomy the noise power per unit frequency bandwidth (P_ν) from any source (astronomical, electronic, interference, etc.) is compared to the temperature of a resistor which would produce the same power per unit bandwidth, i.e., $P_\nu = kT_N$, where T_N is the noise temperature and k is the Boltzmann constant. The antenna temperature, T_A, is the power output per unit frequency of a receiving antenna (we will define this quantity more carefully in Sect. 4.2.6). We can thus write $T_A = P_\nu/k$. A single-dish radio telescope typically uses a parabolic collecting surface to focus radio or microwave radiation into a feed horn that converts the electric fields to voltages which are then amplified and converted to power levels in the receiver.[1] The received signals are digitized almost immediately after the feed horn to avoid noise contamination and maintain maximum fidelity. A source with a flux density, S, can be expressed in terms of the antenna temperature as $T_A = AS/2k$ where A is the effective collecting area of the antenna.[2] Details can be found in Sect. 4.2.6 and Krauss (1966).

Probes at the base of the feed split the radiation into either two orthogonal linear polarizations or right- and left-handed circular polarizations (RCP and LCP, respectively). The signals follow separate paths and are amplified at various stages in the "front end" of the system. The "back end" uses one of several techniques to convert the voltages as a function of time into a spectrum (power as a function of frequency). The most common way is to use an autocorrelator which multiplies the voltage at one time with a time-lagged voltage, sums over all the lags, and finally takes the fast-fourier transform to produce a spectrum. Setting the maximum lag fixes the limits of the band, and choosing filters produces the bandpass, a signal as a function of frequency which drops to zero power at frequencies higher and lower than the bandpass edges. At this point, unless the spectral line is the 21 cm line of hydrogen or some strong maser, the line is buried in the noise of the bandpass that is generated by the electronics. An example of a 1.25 MHz bandpass centered on 1420 MHz is shown in Fig. 4.2. The autocorrelator power levels are converted to antenna temperature for extended sources and flux units of power per Hz per area called janskys[3] for point sources, by either periodically injecting a known power level into the system or by calibrating against a point source with a known

[1]A few radio telescopes (e.g., the Five hundred meter Aperture Spherical Telescope (FAST) and the Arecibo 305-m radio telescope) use spherical collecting surfaces.

[2]The flux density is the integral of the intensity over a source: $S = \int\int I_\nu(\theta, \phi)\, d\Omega$, where θ and ϕ are sky coordinates and Ω is the solid angle subtended by the telescope beam (equivalent to its *pdf*).

[3]A jansky is a flux unit of the spectral flux density, or spectral irradiance. It has a value of 1×10^{-23} erg [s cm^2 Hz]$^{-1}$. The power is assumed to be completely contained within the *psf* and therefore the flux and not the surface brightness is being measured.

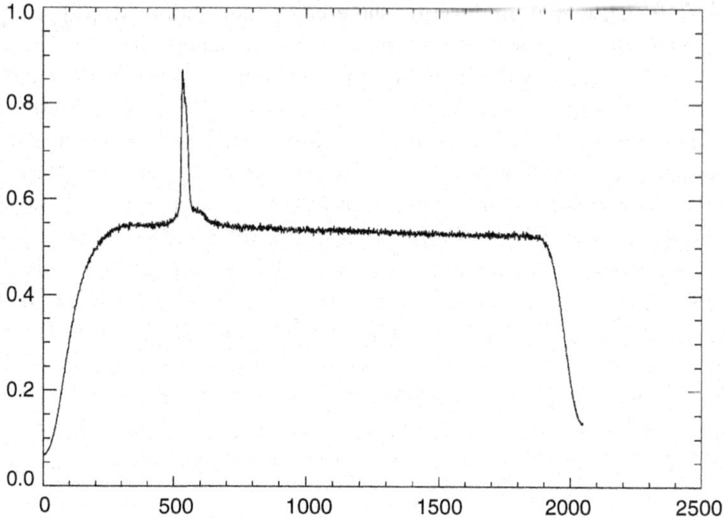

Fig. 4.2 Spectral bandpass of 1.25 MHz over 2048 channels, centered at 1420 MHz. The data represent a 4 minute integration using the Arecibo 305 m radiotelescope. The ordinate is the power in autocorrelator units and the broad pedestal at a level of about 0.55 units represents the system temperature of the electronics plus a minor contribution from the sky. The HI line at 1420.406 MHz is clearly evident

(and constant) flux level. Correcting for the losses at the reflecting surface and the spillover and other effects produced by the feed can be complicated but at all radio observatories the parameters are well-known and can be used to convert the observed T_A into the physically relevant brightness temperature, T_B, at lower frequencies (centimeter or larger), and the main-beam brightness temperature, T_{mb} at higher frequencies (millimeter and submillimeter). At some telescopes (e.g., the Arizona Radio Observatory 12-meter telescope), the output antenna temperature is given as T_R^*, where this quantity is related to the radiation temperature T_R by T_R^*/η_c, where η_c is the main beam efficiency corrected for forward spillover losses. Definitions of the factors necessary to describe the losses due to the various antenna/feed properties can be found in the classic paper by Kutner and Ulich (1981).

Regardless of how autocorrelator units are converted to antenna or radiation temperature, Fig. 4.2 is not the final spectrum. We still need to eliminate the instrumental response or spectral gain shape (the broad pedestal upon which the HI line is sitting) which dominates the spectral distribution. The height of this pedestal is a measure of the system temperature, T_{sys}, and represents the noise from the entire electronic system including the antenna and the sky. It is the sum of many contributions including the receiver noise, the atmosphere, the source, background sources, the CMB radiation, and any spillover of the power pattern of the antenna onto the ground.

The noise fluctuations, ΔT_{rms}, are determined by the system temperature, the frequency resolution of the back end (individual resolution elements of frequency width $\Delta\nu$ are called “channels”), and the integration time. The relationship is dictated by the radiometer equation which in its most direct form (i.e., not worrying about receiver or autocorrelator details) can be written as

$$\Delta T_{rms} = \frac{K_S T_{sys}}{\sqrt{t\Delta\nu}} \tag{4.1}$$

where ΔT_{rms} is the 1-σ noise in a channel, K_S is a weighting constant (of order unity), and t is the integration time. With modern autocorrelators it is possible to have of order 10^5 channels over the bandwidth. For Galactic work, this means the individual channels can be very narrow in frequency (tens of Hz—equivalent to an R of 10^8). In order to minimize ΔT_{rms}, the integration time and/or the channel width can be increased. Choosing wider channels in frequency is the most expedient strategy. As long as the spectral line is resolved over at least 4–5 channels, the channels can be made as wide in frequency as needed.

4.2.2 *Frequency Switching vs. Position Switching*

To produce a reduced, final spectrum, we must remove the spectral gain shape discussed in the previous section. This leaves behind a possible signal from the source embedded in the noise per channel, ΔT_{rms}. There are two main ways to extract the line, position switching and frequency switching.

In position switching, the telescope points and integrates on the source position for a chosen period of time (this is known as the “on” scan or the signal scan) and then is moved off the source for an equally long integration (this is known as the “off” scan or the reference scan). The difference of the two spectra divided by the off scan and then calibrated gives the final distribution of antenna temperature as a function of radial velocity. In this observing mode half the integration time is spent off source, but if the off position scan is taken with the antenna at the same elevation (ideally, the same altitude-azimuth track as for the on source scan) as the on scan, the baseline of the spectrum is usually linear and very stable and any variation can be removed by fitting a low-order polynomial. If the system is extremely stable, one off-position scan can serve as the off scan for a series of on-source scans. At frequencies less than 5 GHz at Arecibo the system was so stable in the early 2000s that off scans were not really necessary. One could remove the back-end systematics just by fitting a polynomial to the bandpass, if the line was narrow and there was no low-level wing emission. The complexity of the wings of the 21 cm line precluded using this technique and its most useful employment was in CH or H_2CO observations at 3.3 or 4.8 GHz.

Along the Galactic plane, finding a suitable off position free of molecular emission is challenging and can lead to significant slew times to shift between the on to the off position. For the 21 cm line this is actually impossible as there is no sight line free of emission. This problem also manifests itself in blind surveys for molecular gas. Given the correlation between gas and dust (see Chap. 6), an excellent way to find off positions for spectroscopic observations is to use the Schlegel et al. (1998) dust maps to locate regions of low dust column density near the desired target. This can be supplemented by channel 1 data from the Planck satellite mission (Sect. 5.2), which is effectively a snapshot of the CO(1-0) emission.

An interesting solution to this problem was devised by Dap Hartmann for the high-latitude blind CO surveys of the North and south Galactic hemispheres (NGH and SGH, respectively; see Chap. 9). In a blind survey of the high-latitude sky with one-degree sampling spacing, the off position for any on scan was located one degree north (in the NGH) or south (in the SGH) in latitude. If there is CO emission in the off scan, there is a danger of oversubtracting the background emission from the on scan. However, given the size and velocity structure of the high-latitude molecular clouds, the chances of complete cancellation were small and, most likely the velocities of emission in the on and off scans were different enough to lead to a final spectrum with both an upward and downward going emission line (see Fig. 4.3). This technique was efficient enough to indicate where the detections were; actual line parameters were obtained by re-observing the lines of sight in question in frequency-switched mode (described below).

The other way to remove the systematics is frequency switching. Here the telescope is not moved. Instead, the center frequency is switched electronically by some amount (the frequency throw) every second or fraction of a second. The data for the respective center frequencies are collected in separate registers and then combined in the same manner as for position switching. One tremendous advantage of frequency switching is that the central frequencies can be chosen so that the signal and the reference scans are both within the band (this is known as "in-band switching"; if the reference is switched out of the band, this is "out-of-band switching"). For example, for small bandwidths, the frequency for the signal scan can be shifted from the actual line frequency by 1/4 of the bandwidth towards lower frequencies, and the center frequency for the reference can be chosen so that the line frequency is shifted 1/4 of the bandwidth to the higher frequency side. This technique yields a signal in both the signal and reference scans, albeit separated by half of the bandwidth. In producing a final spectrum, the low-frequency half of the bandpass from the reference scan is subtracted from the signal scan, and the high-frequency half of the bandpass from the signal scan is subtracted from the reference scan. In our example, the two scans are shifted by 1/4 of the bandwidth in opposite directions to make the line frequency line up in middle of the band, then added together, divided by 2 (to account for the fact that the spectrum consists of two separate scans added together), and then calibrated to produce the final spectrum. When the two intermediate scans (after the subtraction) are shifted and

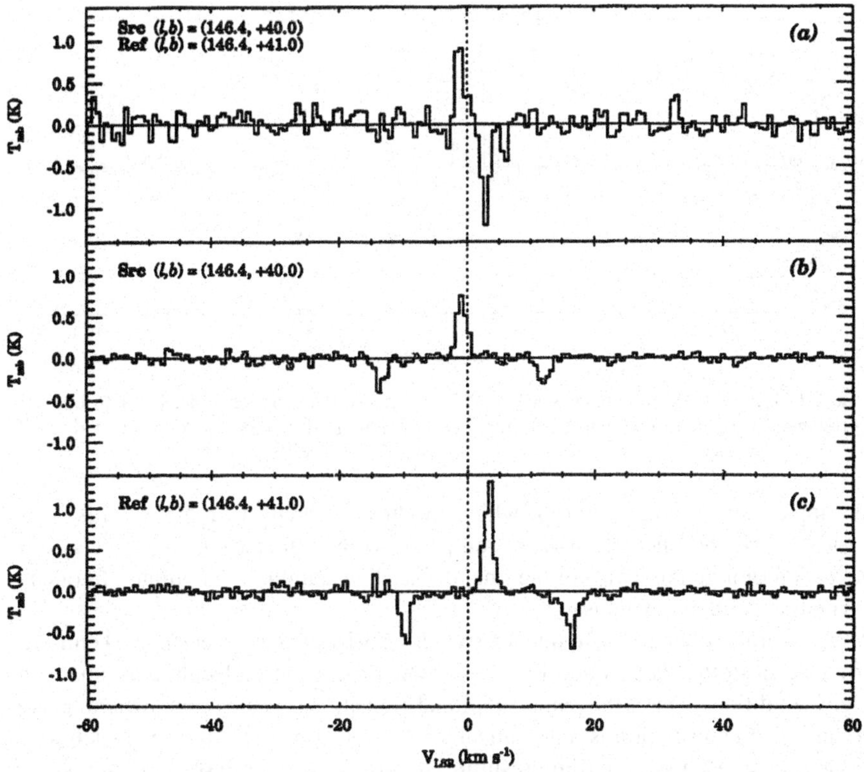

Fig. 4.3 Panel (**a**) shows the position-switched CO(1-0) spectrum from the NGH high-latitude CO survey (Hartmann et al. 1998). The on source position is designated Src in the figure and the off position (Ref) is one degree higher in latitude. A high-latitude molecular cloud (MBM 31) is detected in both the on and off positions but the velocity structure of high-latitude clouds is great enough that over 1 degree it is highly unlikely that the lines would be at the same velocity. Thus, the on position shows a normal CO emission line at -1 km s^{-1} and the off position data show a downward going CO line at $+5$ km s^{-1} as a consequence of the (on–off)/off data processing for position-switched spectra (see text). Panels (**b**) and (**c**) show frequency-switched spectra of both the on and off positions. The CO(1-0) line now appears in emission in both spectra. Note the downward-going "shadow" profiles at $\pm$ 13 km s^{-1} for (b) and -10, $+17$ km s^{-1} for (**c**). These are artefacts of the frequency-switching data reduction (see text for details). Figure from Hartmann et al. (1998)

added together, there will be a downward going signal on either side of the central emission line. These artifacts of the data reduction are called "shadows" of the signal and can be ignored. An example of the results of this procedure is shown in the bottom 2 panels of Fig. 4.3.

The great advantage of frequency switching is that no time is spent observing off source. The drawback is that you are not subtracting the same frequency bandpass from the equivalent on and off scans (each is shifted by a fraction of the bandwidth from the other), thus any frequency structure in the bandpass will cause fluctuations

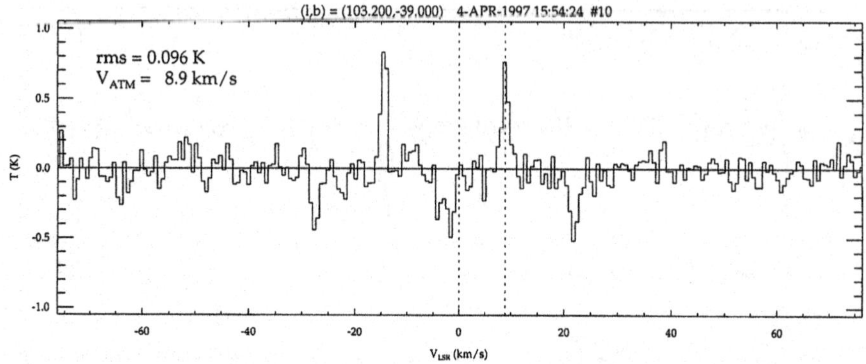

Fig. 4.4 Telluric CO(1-0) line at ∼ +9 km s^{-1} contrasted with an interstellar CO(1-0) line (at∼ −15 km s^{-1}) arising from the diffuse infrared excess cloud DIR 105-38 (see Magnani et al. 2000)

or ripples in the final spectrum which might be difficult to remove. This is not a problem if the lines are narrow compared to the bandpass but if there is low-level emission in the wings of the line, or baseline stability is important, frequency switching is often not advisable.

A separate problem for frequency switching arises when observing CO emission lines because the Earth's atmosphere contains the same molecule that, under the right conditions, emits the same rotational transitions. In position switching, this telluric CO contribution is automatically removed. But in frequency switching, it is present in each scan and the resulting telluric signal can look surprisingly like a weak interstellar signal (see Fig. 4.4). The telluric CO(1-0) line is produced in the thermosphere and mesosphere; lower lying telluric CO is pressure broadened to such an extent that it is not visible in the bandpasses typical of CO observations. There are seasonal variations in the telluric signal with stronger emission in the winter months (e.g., Solomon et al. 1985). If one is paying attention, the telluric CO signal is easy to identify as it occurs at the negative of the velocity corresponding to the Earth's motion with respect to the local standard of rest, but more than one novice observer has been fooled by the telluric CO line.[4]

4.2.3 *On-Off Radio Measurements*

The on-off procedure described above provides a variant of ΔT_B in equation 2.18 (see Sect. 4.2.6) . However, if the line is optically thin, there are two unknowns, T_{ex} and τ_ν. If one observes the on position for an interstellar cloud towards a background

[4]This includes one of the authors of this book who, as a graduate student, mistook the telluric line for a high-latitude interstellar cloud which found its way into the literature as MBM 10 (Magnani et al. 1985).

continuum source of sufficient flux, the equation of transfer discussed in Chap. 2 readily leads to the relationship between T_{ex}, the antenna temperature ($T_{A,on}$), the flux of the background source expressed in K, and the optical depth τ_ν:

$$\Delta T_{A,on}(V) = (T_{ex} - T_{src})(1 - exp(-\tau_\nu(V)) \tag{4.2}$$

where V is the velocity of any channel and we have ignored the contribution from the CMB. If the off position is measured a beam width away from the line of sight with the continuum source, it is easy to show that

$$exp(-\tau_\nu(V)) = 1 + [\Delta T_{A,on}(V) - \Delta T_{A,off}(V)]/T_{src} \tag{4.3}$$

So, the optical depth as a function of V can be determined and T_{ex} then follows from equation 4.2. The assumption is that T_{ex} does not vary significantly over a beamwidth requires that observations be made at the highest possible spatial sampling. A classic example of this technique is by Radhakrishnan et al. (1972).

4.2.4 Polarization Considerations

Just about all radio feed systems are equipped to detect two orthogonal polarizations of electromagnetic waves, either in linear or circular mode. In general, the polarization characteristics of the incoming signals are ignored and the two polarizations are run through the various mixers, amplifiers, and spectrometers separately only to be combined into a single signal. The most important exception to this is the study of interstellar magnetic fields via the Zeeman effect or Faraday rotation for spectropolarimetry—the two orthogonal polarizations are combined to form the Stokes parameters. We will concentrate on the Zeeman effect because it is the most important method for determining magnetic fields in molecular clouds.

The four Stokes parameters (I, Q, U, V) completely quantify polarized radiation (Chandrasekhar 1950; van de Hulst 1957; Born and Wolf 1959). In a Cartesian coordinate system in the plane of polarization, xy, they are defined as

$$I = \langle E_x E_x^* \rangle + \langle E_y E_y^* \rangle, \tag{4.4}$$

$$Q = \langle E_x E_x^* \rangle - \langle E_y E_y^* \rangle, \tag{4.5}$$

$$U = \langle E_x E_y^* \rangle + \langle E_x^* E_y \rangle, \tag{4.6}$$

$$V = -i[\langle E_x E_y^* \rangle - \langle E_x^* E_y \rangle], \tag{4.7}$$

where the angle brackets denote a time average of the electric field components and asterisks denote the complex conjugate of the electric field components in a given direction. In this formulation, $E_x(z,t) = E_{ox} exp[+i(kz - \omega t + \delta_x)]$ and

$E_y(z,t) = E_{oy}exp[+i(kz - \omega t + \delta_y)]$, and the relative phase δ is defined as $\delta_y - \delta_x$. Equation 4.7 defines the Stokes V parameter in the IEEE convention of IEEE right-handed circular polarization (RCP) minus IEEE left-handed circular polarization (LCP). The IEEE RCP convention follows the right-hand rule: The thumb points along the direction of propagation and the fingers curl in the E field direction. This is the standard sense of polarization that radio astronomers and engineers use, but it is important to note that physicists and optical astronomers tend to use the opposite convention.[5] With the IEEE convention and the definition of the electric field above, then E_y lags E_x by 90 degrees for IEEE RCP.

For ease of notation, we will refer to the Stokes parameters above as $\mathrm{I} = \mathrm{XX}^* + \mathrm{YY}^*$, $\mathrm{Q} = \mathrm{XX}^* - \mathrm{YY}^*$, $\mathrm{U} = \mathrm{XY}^* + \mathrm{X}^*\mathrm{Y}$, and $\mathrm{V} = i(\mathrm{XY}^* - \mathrm{X}^*\mathrm{Y})$, with the time average understood. The I Stokes parameter is the total intensity of the radiation, Q measures the tendency for the linear polarization to be aligned horizontally rather than vertically, U measures the tendency for the linear polarization to be aligned at $+45°$ to the horizontal, and V is the difference between RCP and LCP. For positive Stokes V, RCP is greater than LCP when using the IEEE conventions.

For magnetic field strengths in the diffuse ISM, the splitting of the RCP and LCP lines is not greater than the linewidth produced by the gas motions along the line of sight. Thus, the shift in the polarizations produces overlapping lines for which the Stokes V parameter shows as effectively the scaled derivative of the Stokes I parameter (see Fig. 4.5). Thus, in Zeeman measurements one must obtain well-calibrated cross-correlation products and then try to fit the typical down-up line profile pattern in the Stokes V spectrum at the central velocity of the spectral line as determined from the I spectrum.

While this curve-fitting is not a trivial problem, the real problem with Zeeman observations is that there is cross-talk between the circular polarizations. Effectively, each polarization "leaks" into the other so the cross-correlation products necessary for detection of the Zeeman signal Stokes V are contaminated. The various ways that the antenna feed and the motion of the source across the sky with respect to the feed contribute to this contamination are described in great detail by Heiles (1998). To remove this contamination, a calibration must be performed that involves observing a polarized continuum source (such as the quasar 3C286) across the sky. By following a source across the sky as the parallactic angle changes, the Mueller matrix for the antenna-feed system can be derived. The radio astronomical Mueller matrix specifies the relationship between the Stokes parameters obtained at a telescope and the conventionally defined Stokes parameters. The actual Mueller matrix can be determined from plots of the four calibrated correlator products as a function of the parallactic angle, ρ. If the antenna-feed system were perfect in the sense that it is free from cross-talk and polarization leakage, then Stokes I and V would be independent of ρ. The two linear polarization Stokes parameters, Q and

[5]Radio astronomers follow the IEEE definitions of polarization while optical astronomers use the opposite convention (i.e., Rybicki and Lightman 1979). For a thorough discussion of this issue and its implications see Robishaw (2008).

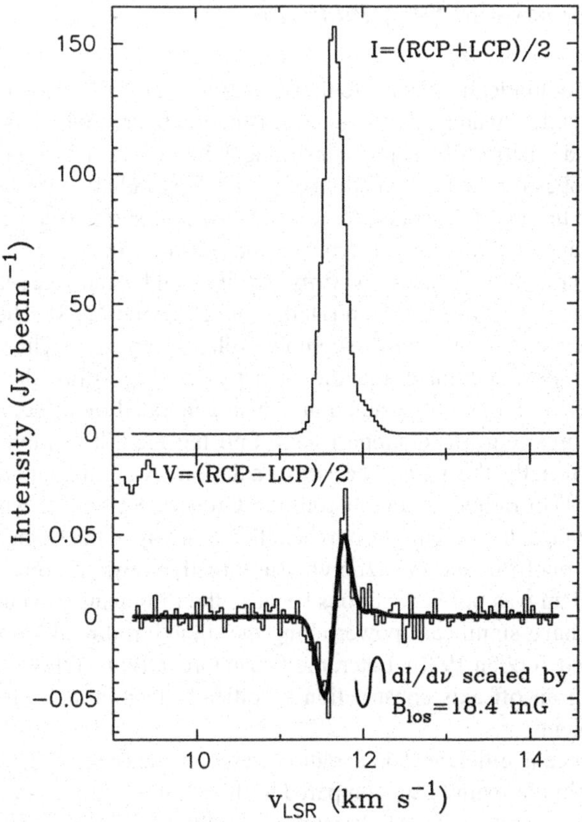

Fig. 4.5 Circularly polarized observations of the 44 GHz Class I CH_3OH maser in OMC-2 showing the Stokes I parameter in the top image. In the bottom image, the data are the Stokes V parameter with a scaled fit of the derivative of the I parameter (dI/dν), which can be used to determine the strength of the magnetic field of the molecular cloud (see Section 2.2.3). Figure from Sarma and Momjian (2011)

U, would then vary sinusoidally with 2ρ. Any departures from this ideal condition defines Mueller matrix of the system (see Robishaw 2008, for further details). Once this radio astronomical Mueller matrix has been obtained for an antenna feed system, it is then multiplied by the matrix of the raw polarization data to produce a calibrated matrix of the XX, YY, XY, and YX cross-correlation products. With this calibrated matrix, a clean V profile can be produced which yields a determination of the strength, and line-of-sight direction of the magnetic field.

4.2.5 Sidelobes and Stray Radiation

A telescope should ideally receive radiation only from the direction towards which it is pointing. Unfortunately, an antenna's diffraction pattern which is the analog of the point spread function of an optical instrument, includes a substantial contribution from the sidelobes for the local maxima of the far-field radiation pattern that are not the main lobe. In radio telescopes, these sidelobes can be many degrees away from the main lobe and pick up radiation from regions distant from the center of the field. This unwanted radiation is known as *stray radiation* and it can be a serious problem when observing the HI line at 21 cm (Heiles and Hoffman 1968). The entire sky is filled with 21 cm emission that has a complex velocity structure. The stray radiation therefore not only contaminates the line intensity but also distorts the line profiles.

Besides the sidelobes, stray radiation is also generated by direct reflections off support structures (e.g., feed-support legs) into the feed. The more blockage an aperture has, usually the worse the stray radiation. A significant exception was the Arecibo 305 m radiotelescope before the Gregorian Upgrade when line feeds were used to collect the radiation from the reflector. Despite the tremendous amount of steel in the platform and the azimuth arm directly above the reflector, the stray radiation problem was not very serious because the illumination pattern of the line feeds did not have significant power along the direction directly below the feeds (see Colgan and Terzian 1986). Later, the 100 m Green Bank Telescope (GBT) was designed using an off-axis construction specifically to minimize the 21 cm stray radiation problem.

The main beam efficiency of a radio telescope measures the fraction of the power pattern in the main beam compared to the whole sky (4π steradian). This is expressed as $\eta_{mb} = \Omega_{mb}/\Omega_a$. All directions not in the main beam are the sidelobes, sometimes referred to as the stray pattern. The stray factor $\eta_{sp} = \Omega_{sp}/\Omega_a$ measures the fraction of the power pattern that is in the sidelobes. The lower this number, the better the telescope as far as stray radiation.

A good antenna design will result in the first sidelobes being about 20 dB down in power from the main lobe but, even at this low level, a stray radiation problem can arise when observing the 21 cm HI line.[6] For example, one can imagine a situation where the main lobe is pointing away from the Galactic plane and picking up an HI signal with an intensity of $\sim$10 K. If some of the sidelobes are projected on the Galactic plane, they can easily contribute a degree in antenna temperature to the spectrum for the given line of sight thereby significantly affecting the result. A clear example of this problem is shown in figures 8 and 10 of Hartmann and Burton (1997). Corrections for stray radiation have been applied to the data from

[6]Most radio observatories will have the power pattern of their antennas available by way of technical memoranda or internal reports. Another correction that is applied to 21 cm data is the confusion effect produced by rotation of the beam on the sky produced by the motion of an alt-azimuth telescope.

the major HI surveys so that this issue is no longer a serious problem when using the online data sets. Although the problem is most severe for the 21 cm line, even high-resolution CO(1-0) observations in extended regions can be affected (e.g., Bensch et al. 2001).

4.2.6 Converting from Antenna Temperature to Beam or Radiation Temperature

A typical radio spectrum plots the antenna temperature of a transition as a function of velocity. The relationship between velocity and frequency is straightforward once the velocity reference is defined (e.g., V_{LSR} or V_{helio}) but the antenna temperature is sometimes a bit mysterious for the novice. It expresses the intensity measured at the telescope as a convenient parameter. Recall from Chap. 2 that in the low frequency limit the intensity of a thermal source is

$$I_\nu = 2kT_B\nu^2/c^2 \quad . \tag{4.8}$$

This defines the brightness temperature, T_B. Besides being linear in the intensity, expressing I_ν in temperature units allows the direct comparison of measurements at different frequencies. However, telescopes do not directly measure the brightness temperature, but a variant called the antenna temperature, or T_A, which is directly related to T_B.

At low radio frequencies (i.e., centimeter wavelengths as opposed to millimeter wavelengths), the power at the output of a telescope antenna when looking at some region of the sky (with coordinates θ, ϕ) with spectral brightness, $I(\nu, \theta, \phi)$ can be expressed as

$$P(\nu) = \frac{1}{2}\int\int A(\theta,\phi)I(\nu,\theta,\phi)d\nu d\Omega \tag{4.9}$$

where $A(\theta, \phi)$ is the effective area of the antenna (see Krauss 1966, for a discussion—basically, the effective area is the power response of the antenna or the ratio of the collecting area of the antenna to its physical area). The factor of 1/2 signifies that only one polarization is being collected by the single receiver (most modern receivers are built to detect both polarizations and are actually two receivers in one).

If one imagines replacing the antenna with a resistor at a temperature equivalent to T_A, then its power output would be identical to equation 4.9 above. We saw in Sect. 4.2.1, for the resistor,

$$P(\nu) = kT_A\Delta\nu \tag{4.10}$$

where $\Delta\nu$ is the bandwidth of the receiver. Thus, the definition of antenna temperature is

$$T_A = \frac{1}{k\Delta\nu}\frac{1}{2}\int\int A(\theta,\phi)I(\nu,\theta,\phi)d\nu d\Omega \quad . \tag{4.11}$$

In real life, the relationship between T_B and T_A is mediated by the antenna efficiency expressed as the beam coupling efficiency $0 \leq \eta_B \leq 1$, such that

$$T_B = T_A/\eta_B \quad . \tag{4.12}$$

The quantity η_B is a function of frequency and parametrizes the efficacy of the antenna at a given frequency. It is important to remember that the antenna temperature is first and foremost a measure of the detected radiation coming from a source. The connection between T_A and a physical temperature depends on whether the object is optically thick and on any background.

If a source is larger than the telescope beam then the antenna temperature measures the intensity of the source. In these cases, T_A expressed in kelvin is the correct way to characterize the spectrum. Once the antenna temperature is corrected for η_B, the value of T_B should be the same from telescope to telescope. Other conditions besides the source being larger than the beam are that there is no resolved structure in the beam or stray radiation issues. If, however, the source is smaller than the beam, then the measured T_A is diluted by a factor that is the ratio between the solid angle of the source to the solid angle of the main beam. *Point sources are thus measured in units of janskys rather than kelvin.* Different radiotelescopes will yield different antenna temperatures for a point source (because of the beam dilution), but ideally all should yield the same flux in Jy. The *sensitivity* or "gain" of a radio telescope is a figure of merit that is the ratio of how many kelvins per jansky an antenna produces when observing a point source. For example, the Upgraded 305-m Arecibo telescope has a sensitivity at 1.4 GHz of 11 K/Jy whereas the 100-m GBT has a sensitivity of 2 K/Jy at the same frequency. Thus, a 1 Jy point source at Arecibo has an antenna temperature of 11 K while the same point source at the GBT will have an antenna temperature of 2 K. In this book, almost all of the structures we discuss are larger than the beam of typical radiotelescopes (at least at mm wavelengths), so that T_A is the relevant measure of their intensity.

This discussion changes somewhat at millimeter wavelengths. For temperatures typical of the molecular ISM, the approximation $h\nu \ll kT$ breaks down so we are no longer in the Rayleigh-Jeans limit. Nevertheless the relationship between the intensity and temperature established in Eq. (4.8) is now just a definition, and so still applies. The radiative transfer equation for optically thick sources now must employ the full Planck function for the source function instead of the Rayleigh-Jeans limit; i.e., $J_\nu(T_B) = J_\nu(T_{ex}) - J_\nu(T_{bg})$, where $J_\nu = B_\nu(T)$:

$$J(\nu, T) = \frac{h\nu}{k}\frac{1}{exp(h\nu/kT) - 1} \quad . \tag{4.13}$$

The quantity $J_\nu(T_B)$ is now referred to as the *radiation temperature* instead of the brightness temperature and it becomes the standard measure of intensity in millimeter-wave spectroscopy. If the source is optically thin, the relevant equation is

$$J_\nu(T_B) = [J_\nu(T_{ex}) - J_\nu(T_{bg})](1 - e^{-\tau_\nu}) \quad . \tag{4.14}$$

Millimeter-wave measurements of T_A have to be corrected using various factors to yield T_R more so than for centimeter radiation. The classic reference here is Kutner and Ulich (1981). A modified radiation temperature, T_R^*, is often used at mm-wave observatories. This is the antenna temperature corrected for ohmic losses, atmospheric losses, spillover, and scattering. This quantity does not correct for the actual coupling of the antenna diffraction pattern to the source. As in the case for centimeter-wave observations, the physically relevant quantity, T_R, can be obtained via a frequency-dependent coupling factor, η_c, so that

$$T_R = T_R^* / \eta_c \quad . \tag{4.15}$$

In recent times it has become more fashionable to report millimeter spectra as T_{mb} vs. velocity, where T_{mb} is the antenna temperature as measured by the main diffraction beam of the telescope. Conversion to T_R follows as in equation 4.14 except that the coupling factor is called η_{mb}.

4.2.7 Interference

Virtually all radio observations are plagued by artificially-generated radio interference. This is not just a problem for radio astronomy (referred to as RAS, Radio Astronomy Service, by the national and international regulatory agencies) but also for scientists involved in EESS (Earth Exploratory Science Services), who use satellites to study the Earth. The telemetry between satellite and terrestrial stations is sometimes compromised by terrestrial radio frequency interference (RFI). Figure 4.6 shows an HI spectrum from the 305-m Arecibo telescope marred by RFI. It is important to note that the 21 centimeter line is in a protected band by international agreement, designated for RAS and EESS passive services (and also for passive SRS—Space Research Service). Despite the band's fairly extensive range (1400–1427 MHz), leakage from harmonics and other sources can sometimes ruin a spectrum far worse than what is shown in Fig. 4.6. This situation is only worsening with time as pressure from the private sector to share protected regions of the electromagnetic spectrum increases.[7]

[7] A discussion of the various issues involved can be found in *Spectrum Management for Science in the 21st Century*, Publication of the National Research Council of the National Academies, The National Academies Press (2010).

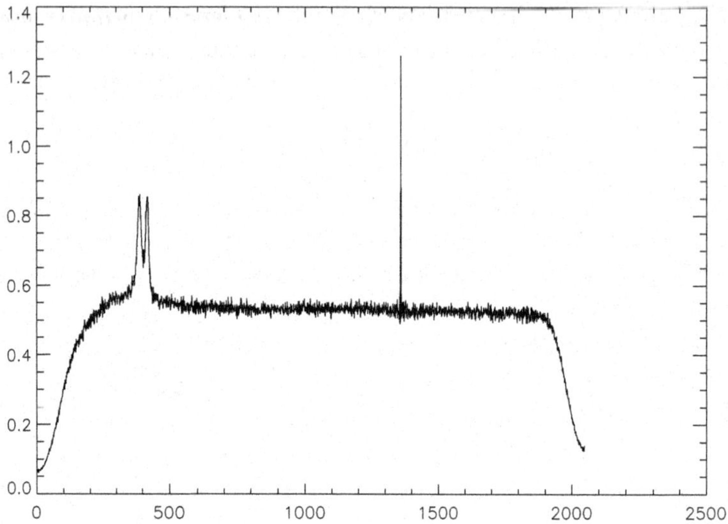

Fig. 4.6 Spectral bandpass similar to Fig. 4.2 showing the effects of interference on an HI spectrum. The y-axis is in autocorrelator units and the x-axis is the channel number

There are three types of radio interference as defined by the Federal Communications Commission (FCC): spurious emission, out-of-band emission, and emissions in adjacent channels. Spurious emission is produced by a transmitter that is emitting strongly at frequencies outside of its assigned band. Even without malicious intent, a transmitter can generate harmonics outside its permitted band of operation. Out-of-band radio emission leaks into neighboring frequencies by the modulation process. These emissions are virtually unavoidable and are the result of limitations in the transmitter electronics. Emissions in adjacent channels are distinguished from out-of-band emission because they are a consequence of limitation in the receiver rather than the transmitter. Regulatory agencies prohibit even unintentional RFI if it rises above a certain level in a protected band. Unfortunately, outside of a protected band, there is no recourse to the problems produced by RFI. In Appendix A, we briefly discuss the structure of the regulatory institutions for deciding which parts of the electromagnetic spectrum are protected and what uses are assigned to these bands.

Despite the many official organizations and the elaborate infrastructure for presenting a scientific case for allocating certain bands for astronomical research, the pressure commercially to use any available portion of the electromagnetic spectrum is tremendous. It is likely that the future of radio astronomy will lie in the hands of RFI excision technology or, in a more distant, optimistic, future, the far side of the Moon.

4.2.8 Mapping Techniques

The speed with which one could obtain a radio spectrum has been so increased that mapping relatively large regions (10^4–10^5 beams on a side) is now possible. At first, each "pixel" of a map was obtained by pointing the telescope, integrating while tracking the position, producing a spectrum, and storing that spectrum in a 3-dimensional data cube (position, position, velocity). This "point and shoot" technique is very inefficient. Large scale mapping made a tremendous leap with the development of the "on-the-fly" technique (OTF—see Magnani et al. 2000). Here, the telescope is driven in some direction and data are collected continuously with spectra produced by dumping the data over a certain time interval into a register, an advance made possible by the vast expansion of computing power in the late 1980s. A common off position is typically used for each row or column of the map. Depending on the dump frequency with respect to the slew speed, the telescope beam may be elongated in the direction of motion. If the dump time is chosen judiciously, then the resulting spectra are nearly identical to those from the point-and-shoot method. However, given the desire to map large areas and the ensuing rapid slew rates, the integration time per spectrum is usually very low, so that several passes are needed for a map with reasonable sensitivity. As the telescope moves, gain changes can distort real features and produce artefacts on the map. This often manifests itself in elongated features in the direction of the scan. A way to mitigate this effect is the basket-weave technique (Emerson and Gräve 1988; Winkel et al. 2012). Here the data are obtained by varying the scanning directions with each pass. Orthogonality of scan directions is often used although this is not a necessary restriction. The scan directions intersect at various crossing points and systematic baseline drifts may be identified and removed.

We discussed in Sect. 4.2.2 the on-off technique and some of the issues involved in finding a suitable off position, free of emission. When observing HI at 21 cm, there is no off position free of emission. Even the so-called "Lockman Hole" has N(HI) of at least $4.5 \times 10^{19}\,\mathrm{cm}^{-2}$ (Lockman et al. 1986) at its lowest level. Consequently, Galactic HI maps must be made in frequency-switched mode. Extragalactic HI observations, where the 21 cm line is redshifted to lower frequencies, of course do not suffer from this restriction.

Almost all mapping of regions in a molecular spectral line are done these days in OTF mode (e.g. Mangum et al. 2000). Usually, enough passes over a region are made to produce a map that rivals conventional point and shoot maps as far as sensitivity. However, baseline stability and gain variations are often issues which affect each spectrum. If one is interested in a low-level extended emission, then the point-and-shoot method allows for high-sensitivity maps with the highest fidelity. When studying features at the limit of a telescope's capability, point-and-shoot should not be forgotten.

4.3 Analysis of Radio Spectroscopic Data

Once a region has been mapped, what next? The most important cloud parameter is the mass. There are several techniques for determining the mass of an interstellar cloud. For example, if the cloud is molecular, the column density of a given transition can be used to determine the mass, if the structure is completely mapped, using the straightforward equation

$$M = N(X_i)C(H_2, X_i)\Omega d^2 \mu m_H \tag{4.16}$$

where N(X_i) is the column density of the species X_i in question, $C(H_2, X_i)$ is the inverse of the abundance of the given species with respect to H_2, Ω is the solid angle subtended by the cloud (obtained from the map), d is the distance to the structure, μ is the molecular weight of the material (it usually includes helium), and m_H is the mass of a hydrogen atom. These types of mass estimates depend critically on the distance and on the conversion factor from the X_i species to H_2. Determining both quantities can be problematic. Another way to determine the mass involves the Virial Theorem described in detail, below.

In addition to the mass, analyses of the velocity field can yield a wealth of information on the kinematics and even the dynamics of the gas. This is something that imaging (e.g., in the infrared) just cannot provide. We describe some of the more common techniques for analyzing the velocity field of a given structure below.

4.3.1 *Virial Masses from Spectral Maps*

There are many derivations of the Virial Theorem (e.g., Goldstein 1951; Collins 1978) . One thing many astronomers forget is that the Virial Theorem does not assume virial equilibrium. In its original form, the Virial Theorem is a statement of the behavior of the second derivative of the moment of inertia under the influence of supportive and compressive forces in a system of particles:

$$\frac{1}{2}\frac{d^2 I}{dt^2} = 2(\mathcal{K} + \mathcal{M}) + \mathcal{U} \quad . \tag{4.17}$$

where I is the Moment of Inertia of the system of particles, $\mathcal{K}$ is the kinetic energy, $\mathcal{M}$ is the magnetic energy, and $\mathcal{U}$ is the potential energy. In molecular clouds, $\mathcal{U}$ is usually only the gravitational potential energy which for a spherical, uniform density cloud is $-\frac{3}{5}GM^2/R$. Unlike equation 4.16, the virial method is based on dynamics. The assumption of virial equilibrium means the particles are bound by the internal and external forces that act on them. If the results of Solomon et al. (1987, see Sect. 8.2.3) really indicate that the clouds are in virial equilibrium, then the compressive force from gravity is balanced by the supportive forces from the thermal and nonthermal gas motions and the magnetic fields, and I is constant. Because it is so difficult to measure, the magnetic field is almost always ignored and twice the kinetic energy is set equal to the negative of the gravitational potential energy so that

$$\mu m_H \sqrt{3}\sigma_{1D}^2 = \frac{3}{5} GM^2/R \tag{4.18}$$

where σ_{1D} is a measure of the dispersion of the radial motions of the gas in the cloud as determined from spectral line data (almost always CO). Assuming the velocity field is isotropic, the 3-dimensional velocity dispersion is just $\sqrt{3}\sigma_{1D}$. If the cloud distance is known and the cloud is fully mapped, R follows immediately and the *virial* mass of the cloud may be determined. The cloud mass can be readily converted to an average column density and, by dividing by the average W_{CO} value, X_{CO} can be determined; a technique which is further discussed in Sect. 8.2.3.

We note that in the most general form of the Virial Theorem there is a surface pressure term, $3\int PdV$, where dV is the volume element in addition to the components described above. Its role in the stability of molecular clouds is still under debate. Keto and Myers (1986) introduced it as a way to keep the small, translucent, high-latitude molecular clouds bound (see discussion in Chap. 11) and McKee and Zweibel (1992) discuss its implications for molecular clouds in general.

4.3.2 A Brief Mention of Statistical Tools

4.3.2.1 Principal Component Analysis

The analytical theory was presented by Heyer and Brunt (2004), and Brunt and Heyer (2013). In general, the idea behind Principal Component Analysis (PCA) is simple. Imagine a sample of measurements (observations). A *known set* of processes *might* contribute to a phenomenon and these can be listed. In the general sense, the *relatively few* physical variables that might be involved renders the problem tractable in the sense that the correlations between the measurements have a relatively low "dimension". Integral measures, such as equivalent width, centroid velocity, and emission measure fit easily into the purview of this method. Think of the attributes of a molecular cloud, e.g., mass, radius, column density, and some attribute, such as the velocity dispersion, and you see that any one of the physical properties is insufficient to explain the observation but, together in some combination, they might give a scaling law. If $\ln(\sigma)$ is a linear combination of the individual possible contributions (recalling that the radius may depend on the mass) then a linear law for $\sigma(M, R, N_{H_2}) \sim M^{e_1} R^{e_2} N_{H_2}^{e_3}$ may better represent the data than any one attribute alone. The idea behind the analysis of turbulent velocity fields using PCA is similar except that the line profile becomes the data and the contribution of some velocity field to every channel of the profile become the set of attributes.

This is a decomposition, not an aggregation application of the technique. While for a small set of parameters this can easily be interpreted, perhaps even recovered from theory, when applied to turbulence it produces a set of "maps" that are the spatial distribution of each of the contributing dynamical fields that then must be re-interpreted as true dynamics (e.g. solenoidal or incompressible fields, potential or harmonic fields).

The algorithm *Clumpfind* (Williams et al. 1994) is popular among the radio astronomy community for identifying clumps in radio maps of various spectral tracers. The spectral data are assembled into a 3-D cube (position, position, velocity), and the algorithm searches for local peaks of emission, following them down to lower intensity levels. The result of running *Clumpfind* on a datacube is a decomposition of the cube into a set of clumps in which the emission is concentrated. This is similar to PCA in that they both attempt to identify coherent structures in an intensity and velocity field by minimizing the velocity interval over which a density structure is spread. Although arithmetically different in having no minimization principle, the results from both methods should be similar. An improvement might be achieved by including a Markov Chain Monte Carlo simulation (MCMC - Diaconis 2009) for the search procedure to remove the bias and account for superpositions of structures along the line of sight.

4.3.2.2 Statistical Analysis of the Velocity Fluctuations

In any channel map, the intensity fluctuations per channel are insufficient to provide relevant density and dynamical diagnostics. This is because the profile may change with position inside a cloud boundary. Instead, integral measures—in this case moments of the profile—can be formed using the full intensity distribution at a given location. The limit imposed by the beam spacing determines the minimum scale in space on which the velocities can be resolved, and the assumption is that the emission does not vary in profile so rapidly that within a single beam (or half-beam) separation the change renders the moment useless. This last point is not trivial since the scale on which dissipation is presumed to occur in a turbulent cascade is almost certainly below the limit of the observational resolution. In forming the moments, it is best to use the intensity distribution without making an assumption of the form:

$$< v(\mathbf{x}) >= \int \Phi_v(\mathbf{x}) v dv \tag{4.19}$$

recognizing that this is the mean including any large scale flow that may be present. Any higher moment is then taken relative to this mean:

$$< (\delta v)^n >= \int \Phi_v (v- < v >)^n dv \tag{4.20}$$

This is the deviation of the profile from the mean *within the beam* (e.g., Dickman and Kleiner 1985). For a Gaussian process with no correlation, all odd moments should vanish so within a single profile one can suspect that the presence of a skew is an indication that turbulence is responsible for the dynamics. However, there are other possibilities that cannot be excluded at this first pass observational stage. Systematic flows, since only the line of sight projected velocity is measured by the profile, can produce a deviation from a homogeneous structure. Only a change is angular sampling can rule this out. The same thing is true for a single channel in the data cube produced when mapping a cloud.

Turbulence is not a discrete process. So when mapping a cloud, you are taking samples of a continuous velocity and density field in a grid. with some fixed spacing and resolution, as determined by the point spread function of the telescope. To make matters worse, this is discrete in all three dimensions since the velocity is also discretely sampled (depending on the bandwidth and number of channels available). But the individual channels are an integral along the line of sight as a continuum. These are weighted by the individual emissivities of structures, depending on density, abundance of the tracer, and temperature. Thus, even if the medium is uniformly Gaussian overall, individual deviations that are unrelated to turbulent motions may locally produce a deviation that appears to be the same as a departure from isotropic motions. These, if small enough, will decorrelate over some spatial scale. But this should serve as a caution: discrete features always result from any turbulent flow when sampled discretely. We will discuss turbulence at length in Chap. 11.

The autocorrelation of the velocity field is obtained from the velocity centroids. It measures the degree of correlation between two points or, said differently, measures how the mean velocity of two points diverge with displacement. Again, in a cosmic source, only one component of the velocity is available so this is not the same as the function for which the Fourier transform yields the power spectrum. It is defined as:

$$C(\mathbf{x}) = \int d\mathbf{x} < v(\mathbf{x}+r) > \cdot < v(\mathbf{x}) > \tag{4.21}$$

The structure function is similar to the probability distribution function in that it is the mean value of different powers of the velocity fluctuations taken over the whole map. Although it is not a correlation function, its definition shows that it is related to $C(\mathbf{r})$,

$$S_p(\mathbf{x}) =< (\delta v(\mathbf{x}+r) - \delta v(\mathbf{x}))^p > \tag{4.22}$$

for an arbitrary moment p. This difference contains the mean velocity at each point in the sample so it is again essential that any systematic large scale motions be removed. Note that there is a fundamental problem when studying cosmic sources that is avoided in the laboratory: you do not know what the mean velocity is.

4.3.2.3 Probability Distribution Function

In effect, a probability distribution function (*pdf*) is the same as a line profile if the turbulent field is homogeneous. That is, looking along any line of sight, the sample contains the integrated contributions of all radial velocity fluctuations. The frequency with which any deviation occurs with respect to the main flow is the same as the histogram of the fluctuations. In a Gaussian process, for which the individual fluctuations are uncorrelated and have a maximum probability of zero and fixed dispersion, the *pdf* should be Gaussian. In contrast, with both correlations and a power law distribution of the fluctuations, those at large value are more frequently seen than expected from a Gaussian. hence, the profiles have long tails. The same holds for the *pdf*. The difference is that the line itself may also contain mean flows that distort the profile weighted along the line of sight by the line emissivity and integrated over the beam of the telescope. We will discuss these issues further in Chap. 11.

References

Bensch, F., Panis, J.-F., Stutzki, J., Heithausen, A., and Falgarone, E. 2001, A&A, 365, 275
Born, M. and Wolf, E. 1959, Principles of Optics, Pergamon Press
Brunt, C.M. and Heyer, M.H. 2013, MNRAS, 433, 117
Chandrasekhar, S. 1950, *Radiative Transfer*, Oxford University Press
Colgan, W.J. and Terzian, Y. 1986, Radio Science, 21, 761
Collins, G.W., II 1978, *The virial theorem in astrophysics*, (Pachart Publishing House: Tucson, AZ)
Condon, J.J. and Ransom, S.M. 2016, *Essential Radio Astronomy*, (Princeton: Princeton U. Press)
Diaconis, P. 2009, Bull. Amer. Math Soc., 46, 179
Dickman, R.L. and Kleiner, S.C. 1985, ApJ, 295, 497
Emerson, D.T. and Gräve, R. 1988, A&A, 190. 353
Goldstein, H. 1951, *Classical Mechanics*, Addison-Wesley
Hartmann, D. and Burton, W.B. 1997, *Atlas of Galactic Neutral Hydrogen*, Cambridge University Press
Hartmann, D., Magnani, L., and Thaddeus, P. 1998, ApJ, 492, 205
Heiles, C. and Hoffman, W. 1968, AJ, 73, 412
Heiles, C 1998, Astro. Lett. and Communications, 37, 85
Heyer, M.H. and Brunt, C.M. 2004, ApJ, 615, 45
Keto, E.R. and Myers, P.C. 1986, ApJ, 304, 466
Krauss, J.D. 1966, *Radio Astronomy*, McGraw-Hill
Kutner, M.L. and Ulich, B.L. 1981, ApJ, 250, 341
Lockman, F.J., Jahoda, K., and McCammon, D. 1986, ApJ, 302, 432
Magnani, L., Blitz, L., and Mundy, L. 1985, ApJ, 295, 402
Magnani, L., Hartmann, D., Holcomb, S.L., Smith, L.E., and Thaddeus, P. 2000, ApJ, 535, 167
Mangum, J.G., Emerson, D.T., and Greison, E.W. 2000, in *Imaging at Radio through Submillimeter Wavelengths*, ASP Conference Series, Vol. 217, 179
McKee, C.F. and Zweibel, E.G. 1992, ApJ, 399, 551
Radhakrishnan, V., Murray, J.D., Lockhart, P., and Whittle, R.P.J. 1972, ApJS, 24, 15
Robishaw, T. 2008, PhD Thesis, Univ. California, Berkeley
Rybicki, G.B. and Lightman, A.P. 1979, *Radiative Processes in Astrophysics*, Wiley-VCH

Sarma, A.P. and Momjian, E. 2011, ApJ, 730, 5
Schlegel, D.J., Finkbeiner, D.P., and Davis, M. 1998, ApJ, 500, 525
Solomon, S.R.R. et al. 1985, J. Atmos. Sci., 42, 1072
Solomon, P.M., Rivolo, A.R., Barret, J., and Yahil, A. 1987, ApJ, 319, 730
van de Hulst, H.C. 1957, *Light scattering by small particle*, John Wiley and Sons
Williams, J.P., de Geus, E.J., and Blitz, L. 1994, ApJ, 428, 693
Wilson, T.L., Rohlfs, K., and Hüttemeister, S. 2010, *Tools of Radio Astronomy*, 5th ed., Springer-Verlag
Winkel, B., Flöer, L, and Kraus, A. 2012, A&A, 547, 119

Chapter 5
Observing the Diffuse ISM: The Space Missions

In a new adventure of discovery no one can foretell what will be found, and it is probably safe to predict that the most important new discovery that will be made with flying telescopes will be quite unexpected and unforeseen.

—Lyman Spitzer, Jr. 1961,
Bulletin of the Atomic Scientists (May 1961), Vol. 17, No. 5, 194.

Abstract This chapter describes how space based observatories have influenced the study of the diffuse ISM by their opening of the full electromagnetic spectrum to imaging and spectroscopic study. From the radio to gamma-rays, satellites have contributed substantially heavily to the field. We review some of the more significant missions and describe some of their special capabilities for studying the diffuse ISM since many of these satellites have produced significant archival databases.

5.1 Introduction

During the last third of the 20th century, astronomy underwent a third revolution in instrumentation, as great as the development of the telescope and spectroscope for astronomical observations had been. The opening of the electromagnetic spectrum finally allowed a complete view of astrophysical objects, not biased by the optical and radio windows. As UV, infrared, X-ray, and finally gamma-ray astronomy developed instruments and techniques to study the sky, the nature of astrophysical research changed—just compare an issue of *The Astrophysical Journal* from 1966 to one from 2016. Of course, part of the problem in developing the instrumentation necessary to explore the portions of the electromagnetic spectrum outside of the optical and radio regimes is the opacity of the Earth's atmosphere. This was finally surmounted by space-based instrumentation following the development of rocketry in the 1950s and 1960s.

L. Magnani, S.N. Shore, *A Dirty Window*, Astrophysics and Space Science Library 442, DOI 10.1007/978-3-662-54350-4_5

This third instrumental revolution in astronomy had, and has, a profound impact on our view of the diffuse ISM. We will survey some of the most salient results in this area in the last four decades, with the understanding that the next four decades will be even more exciting.

5.2 The Radio Regime

Radio astronomy is traditionally done via ground based telescopes and arrays but there have been a few space missions involving radio antennas and instrumentation. At the lowest frequencies, RADIOASTRON (aka Spektr-R) is a joint ground-space VLBI project led by the Astro Space Center of the Lebedev Physical Institute of the Russian Academy of Sciences. It features a 10 m antenna on board an Earth-orbiting satellite which can be coupled with ground-based radio telescopes to produce interferometric baselines of up to 390,000 km. It operates at wavelengths of 1.35–6, 18, and 92 cm. and is used to study the structure of active galactic nuclei, pulsars, and masers. With its 10 m antenna, it is the largest orbiting radio telescope.

Single-dish radio telescopes are, in general, not suitable for space-based platforms because of size considerations and the associated angular resolution issues, but, recently, satellites such as COBE, WMAP and *Planck* featured telescopes that straddled the mm-submm region of the electromagnetic spectrum. In particular, the *Planck* mission[1] had sufficient resolution that it could be considered a mm-wave orbiting telescope and, as such, we discuss some of its initial results relevant to the diffuse ISM.

5.2.1 *The Planck Observatory*

The *Planck* Observatory, an ESA mission with significant NASA involvement, was launched in an Earth-Sun L2 orbit in May 2009 to survey the sky for both continuum and spectral lines in the range 30–857 GHz with angular resolution from 33′ to 5′. *Planck's* primary goal was to study the CMB and, as such, it was the third generation mission to do so after the COBE and WMAP satellites. It functioned from May 2009 to January 2012 with the Low Frequency Instrument (LFI) observing at wavelengths greater than 4 mm. Its 100 GHz channel can be used to trace CO(1-0) emission (115.27 GHz), but it is "contaminated" by ^{13}CO (1-0) emission (110.20 GHz) and CMB emission. Similarly, the CO(2-1) line contributes to the 217 GHz channel, and

[1] http://www.esa.int/Planck.

the CO(3-2) line to the 353 GHz channel. This last channel is sometimes called the "dust" channel because it picks up a lot of thermal emission from cold dust. The CO emission in the 100 GHz channel had to be calibrated so that the ^{13}CO and CMB contributions could be removed (Planck Collaboration, Planck 2013 Results, 2014, Paper XIII; Planck Collaboration, Planck 2013 Results XIII 2014). This was done by comparing its emission data to CO datasets from the Harvard-Smithsonian CfA 1.2 m, the Nanten 4 m, and the FCRAO 14 m radiotelescopes, and the CMB data from COBE/FIRAS.

The Planck Collaboration discussed the relationship between molecular gas and dust emission in several early papers (Planck Collaboration, Planck Early Results 2011, Papers XXI, XXII, XXIII, XXIV, XXV) (Planck Collaboration, Planck Early Results XXI 2011; Planck Collaboration, Planck Early Results XXII 2011; Planck Collaboration, Planck Early Results XXIII 2011; Planck Collaboration, Planck Early Results XXIV 2011; Planck Collaboration, Planck Early Results XXV 2011) and mapped the entire sky in the lower CO rotational transitions (Planck Collaboration, Planck 2013 Results, 2014, Paper XIII) (Planck Collaboration, Planck 2013 Results XIII 2014). Figure 5.1 shows an early release of the *Planck* 100 GHz data for the whole sky showing primarily CO(1-0) emission, along with that from ^{13}CO(1-0) and the CMB. Although the resolution is only 10′, this is the only complete all-sky map in any molecular tracer and, as such, provides the

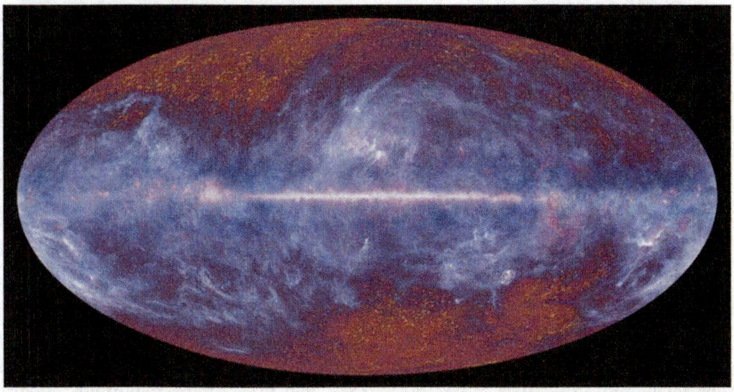

Fig. 5.1 *Planck* whole-sky survey at 100 GHz after one year of data collection. The 100 GHz channel features primarily the CO(1-0) emission from molecular clouds (blue/white regions), but the ^{13}CO(1-0) line is also present. In addition, temperature variations from the CMB are visible at the highest latitudes. Compare to Fig. 7.3 which removes the ^{13}CO and CMB contributions. The image is from the ESA/Planck collaboration

most complete view of the molecular Milky Way to date. The Harvard-Smithsonian Center for Astrophysics molecular map of the sky has 8′ resolution and the telescope is continuing to map the sky accessible from Cambridge, Massachusetts (see www.cfa.harvard.edu/mmw/).

While the sensitivity of the *Planck* maps is adequate to study the molecular emission from GMCs and the larger clouds, the more diffuse and translucent molecular gas is not as well defined. Figures 5.2, 5.3, 5.4, 5.5, 5.6, 5.7 show a translucent and a diffuse molecular cloud (see Chap. 8 for definitions) as mapped in CO by conventional ground-based mm-wave telescopes, by the *Planck* CO channel (100 GHz) and the *Planck* dust channel (353 GHz). Without ^{13}CO and CMB removal, the 100 GHz channel barely shows the outline of the clouds, but at 353 GHz the dust emission detected by *Planck* (e.g., Fig. 5.3) tracks both the IRAS 100 μm (Fig. 5.2 right) and the CO(1-0) emission (Fig. 5.2 left). With its comprehensive frequency and sky coverage, the CO maps from *Planck* are an invaluable contribution to our knowledge of the large-scale molecular gas distribution in the Galaxy.

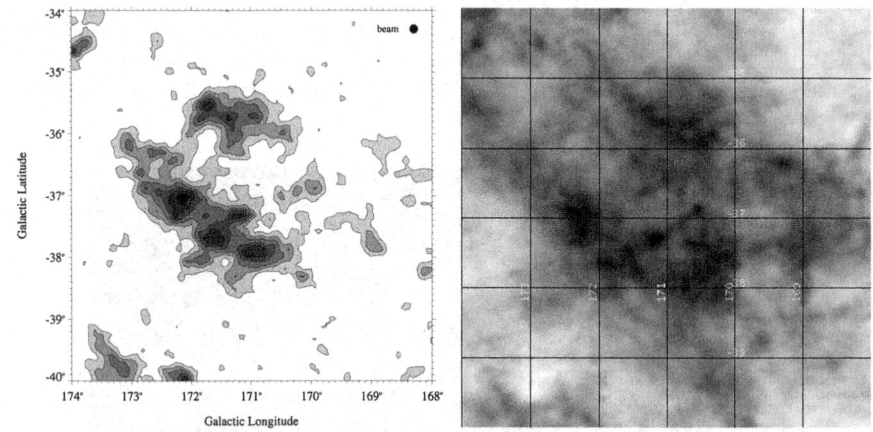

Fig. 5.2 CO and dust comparison for region centered on the high-latitude cloud MBM 16. On the left, CO(1-0) contour map made with 1.2 m mm-wave telescope of the Harvard-Smithsonian Center for Astrophysics. The velocity range of the map is $-10\,\mathrm{km\ s^{-1}} \leq v_{LSR} \leq 15\,\mathrm{km\ s^{-1}}$ and contour levels are 15%, 30%, 45%, 60%, 75%, and 90% of the maximum integrated CO antenna temperature of 6.5 K km s^{-1}. The map is fully-sampled and the resolution is 8.4′. For more details, see Magnani et al. (2003). On the right, a similarly sized map centered on MBM16, showing emission from the IRIS 100 μm data that primarily traces thermal dust. The gray scale is a log scale and ranges from 5.60 to 26.48 MJy/ster. The image was made with the Skyview Virtual Observatory

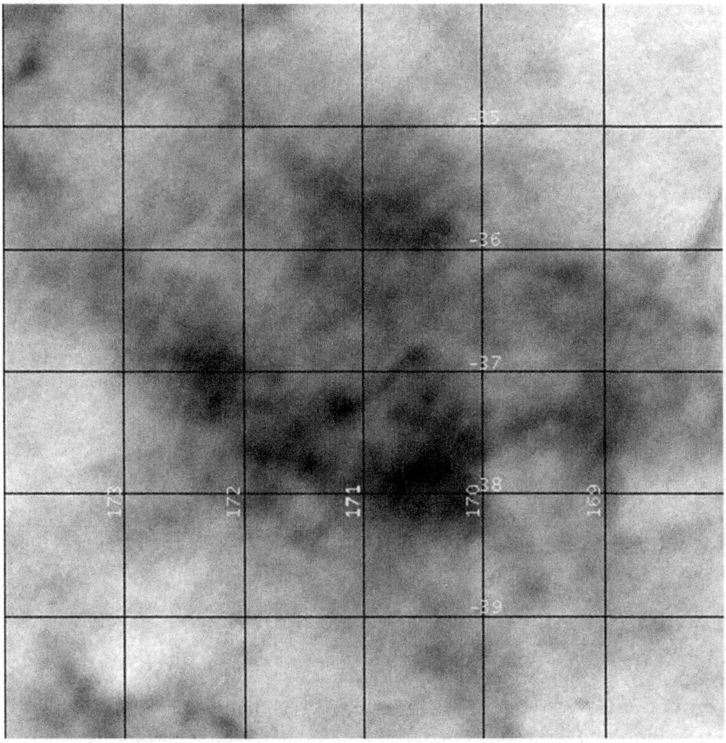

Fig. 5.3 Same region as in Fig. 5.2, centered on MBM16, showing emission from the *Planck* 353 GHz channel which primarily traces dust. The gray scale is linear and ranges from 1.92×10^{-3} K to 12.25×10^{-3} K. The image was made with the Skyview Virtual Observatory

In addition to the CO maps, the three highest frequency *Planck* channels (857, 545, and 353 GHz) can be used to cover the peak thermal emission from dust regions colder than 14 K, and Planck Collaboration, Planck Early Results 2011, Paper XXIII Planck Collaboration, Planck Early Results XXIII 2011 produced the first all-sky survey of Galactic cold clumps. The cold sources they detected included a range of objects from low-mass dense cores to GMCs, so their use of the word "clump" is far more encompassing than what is normally considered a clump in the ISM. There is little question, however, that these sources are associated with molecular gas. A detailed analysis of the clump distribution with the CO(1-0) intensity map just described showed that 95% of the clumps are associated with CO structures (see also Planck Collaboration, Planck 2015 Results, Paper XXVIII Planck Collaboration, Planck 2015 Results XXVIII).

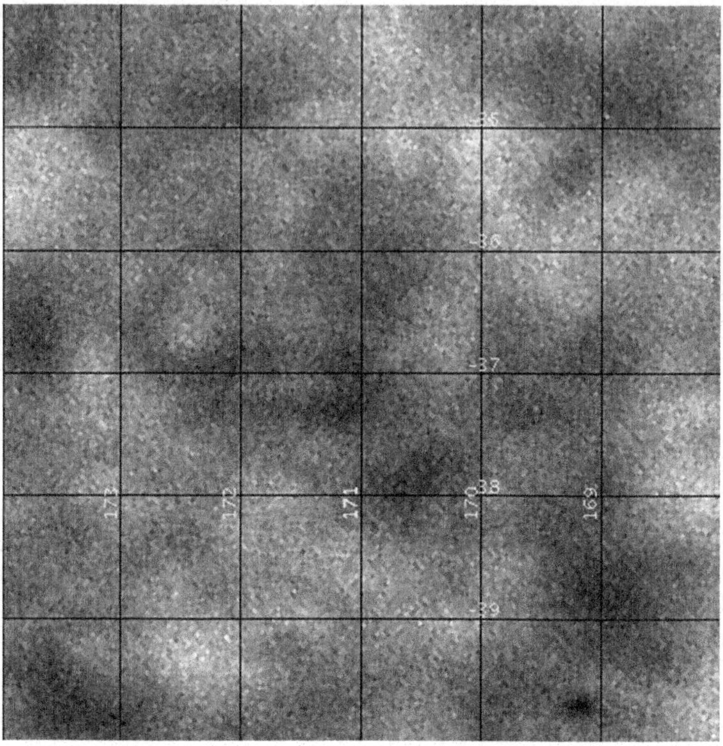

Fig. 5.4 Same region as in Figs. 5.2 and 5.3, centered on MBM16, showing emission from the *Planck* 100 GHz channel which traces emission from the CO(1-0) transition, the CMB, and ^{13}CO(1-0). Notice the contrast between this image and the CO(1-0) map in Figure 5.2, highlighting the effect of the broad band contamination relative to high-spectral resolution studies. The gray scale is linear and ranges from -3.47×10^{-4} K to 7.71×10^{-4} K. The image was made with the Skyview Virtual Observatory

With the ability to trace both CO and dust components, the *Planck* team addressed the issue of "dark" molecular gas (see Sect. 8.4) and proposed that it can correspond up to as much as 28% of the atomic mass and contribute even more molecular mass (118%) in the solar neighborhood than currently determined from conventional CO(1-0) observations (Planck Collaboration, Planck Early Results 2011, Paper XIX)(Planck Collaboration, Planck Early Results XIX 2011). These results were preliminary and are more than likely overestimates—especially for the molecular case. We revisit the issue in Chap. 8.

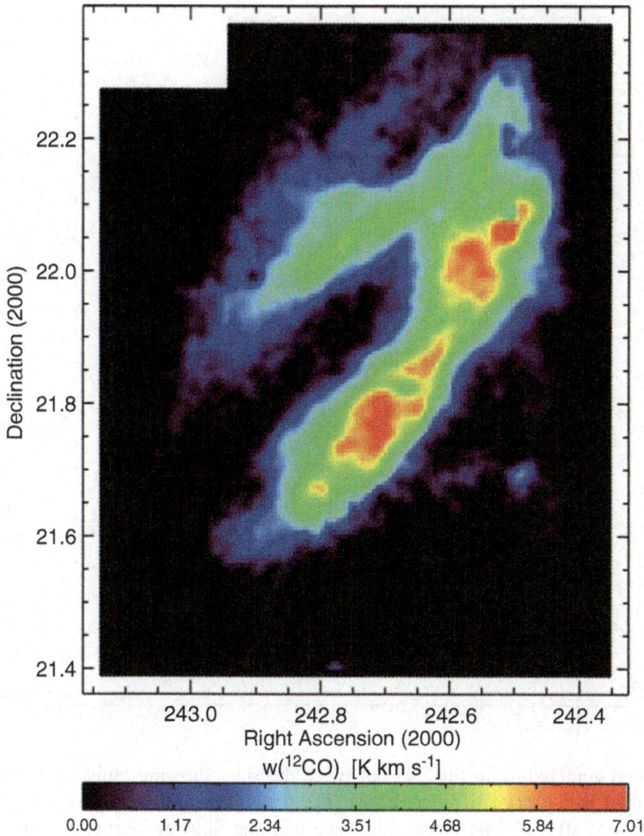

Fig. 5.5 CO(1-0) map in color of the core region of MBM 40 (Chastain 2005). The units are velocity-integrated antenna temperature with the color scale at the *bottom*. The CO data were taken with the FCRAO 15-m telescope and have a resolution of 45″ and are fully-sampled. The 1-σ rms of the antenna temperature is only 0.7 K so there is much more low-level CO emission in this region. See discussion by Cotten and Magnani (2013)

5.2.2 *WMAP*

The Wilkerson Microwave Anisotropy Probe was launched in June 2001 to make fundamental cosmological measurements of the CMB. It made all-sky maps at 22, 30, 40, 60, and 90 GHz at resolutions of 0.93, 0.68, 0.53, 0.35 and <0.23 degrees, respectively (see Fig. 1.5 for the all-sky map at 40 GHz). In the process

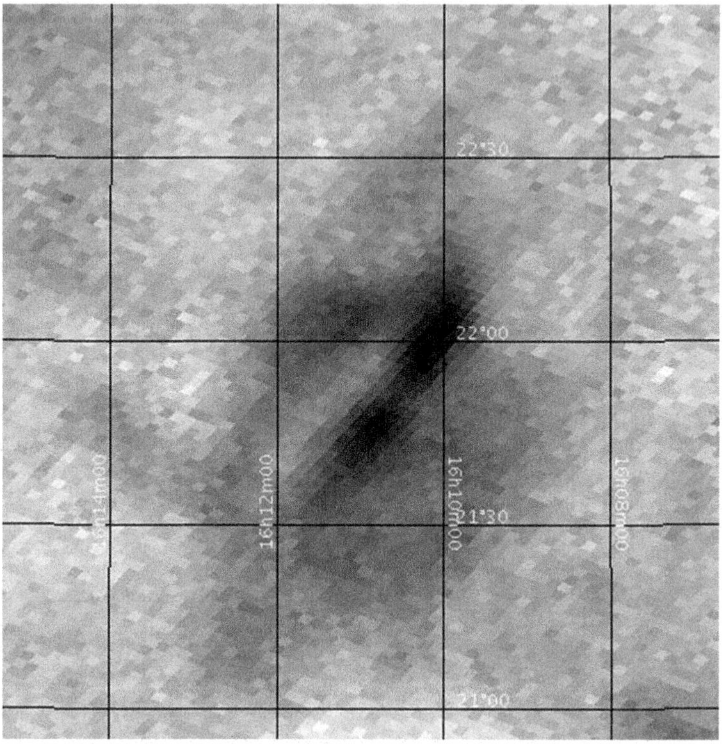

Fig. 5.6 Region centered as in Fig. 5.5 but twice the size, showing emission from the *Planck* 353 GHz channel which primarily traces dust. The gray scale is linear and ranges from 4.94×10^{-4} K to 4.924×10^{-3} K. The image was made with the Skyview Virtual Observatory. Notice how the broadband contamination of ^{13}CO and CMB effectively obliterate the cloud

of obtaining the data so valuable for the CMB, the satellite also picked up Galactic synchrotron and free-free radiation, and some thermal radiation from interstellar dust. A fourth Galactic component, Anomalous Microwave Emission (AME), thought to arise from spinning dust is still somewhat controversial as regards its production mechanism (Sect. 3.5.3). Regardless of its provenance, AME does exist over the frequencies covered by WMAP. This was clear when the Galactic emission from the three established components was removed and there was still a correlation with the dust component associated with the infrared cirrus (Kogut et al. 1996; Leitch et al. 1997). This anomalous emission is discussed further in Sect. 6.6.3 .

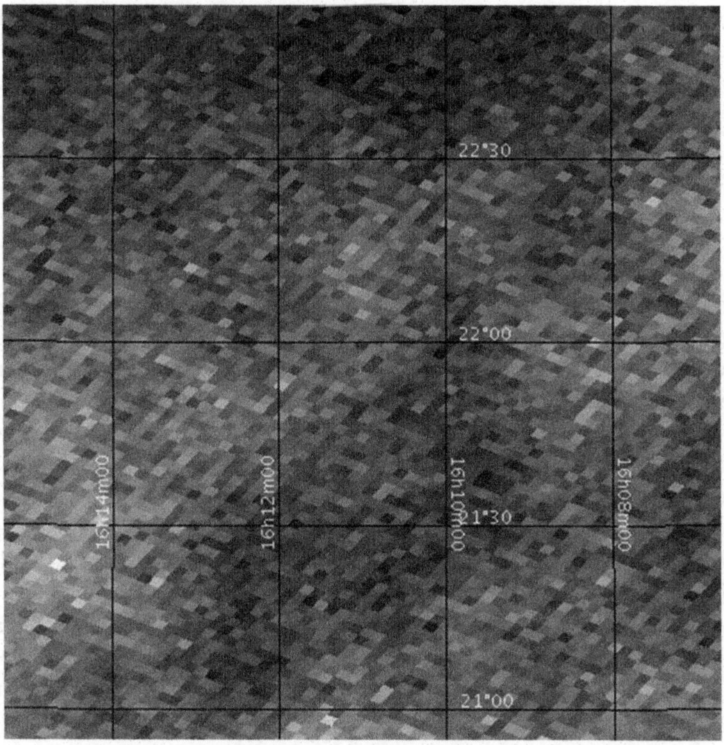

Fig. 5.7 Region centered as in Fig. 5.5 but twice the size, showing emission from the *Planck* 100 GHz channel which traces CO(1-0), the CMB, and ^{13}CO(1-0). The gray scale is linear and ranges from -1.76×10^{-4} K to 4.66×10^{-4} K. Because the CO(1-0) emission in the densest region of MBM 40 is greater than in MBM 16, and the 100 and 353 GHz emission lower, it is likely the gas-to-dust ratio in MBM 40 is higher than in MBM 16 (compare Figs. 5.2–5.4 with Figs. 5.5–5.7). Notice how the broadband contamination of ^{13}CO and CMB effectively obliterate the cloud. The image was made with the Skyview Virtual Observatory

5.3 Probing the Infrared Sky

With the development of infrared detectors and their deployment on satellites providing large-scale mapping of the sky in near and far infrared, Galactic dust can no longer hide anywhere and its properties can be determined with unparalleled precision. It is safe to say that we are just at the beginning of the returns from the opening of the infrared window. The dust and gas in the ISM are well-mixed (Spitzer 1978) with a typical gas to dust ratio, by mass, of about 100. Despite representing only about 1% of the mass of the ISM, the dust is easy to detect (once infrared-detection technology was developed) and gives remarkable views of the structure and distribution of the diffuse ISM. In the next section we briefly review some of the more important infrared space missions for tracing dust in the diffuse ISM.

5.3.1 IRAS

The Infrared Astronomy Satellite (IRAS) was launched on January 25, 1983. This date can be taken as the dawn of a new way to study the ISM. The IRAS mission was initiated in the mid-1970s as a joint project by the United States, the Netherlands, and the United Kingdom. Its principal goal was to produce an unbiased, uniform survey of the entire sky at four infrared wavelengths centered at 12, 25, 60, and 100 μm. The satellite ceased operations in November 1983 when the liquid helium that cooled the telescope had evaporated away. By the end of the mission, it was clear that the IRAS survey was a great success with more than 96% sky coverage [for more details, see Neugebauer et al. (1984), and the IRAS Explanatory Supplement (1988)]. The principal data products of the project were the compilation of a catalog of infrared point sources, a catalog of extended sources less than 8′ in extent, and an atlas of calibrated images of nearly all of the infrared sky. The resolution of these images was 2′ at 100 μm, but, later, reprocessed versions of the original images which resolved several troublesome calibration issues had spatial resolution in the 4′–6′ range.

The impact of the results on the astronomical community was tremendous. Rumors of exciting new findings circulated throughout the summer of 1983, but the proprietary nature of the data prevented external researchers from examining any of the images. The initial, spectacular results of the mission were presented in the March 2004 issue of *The Astrophysical Journal Letters* (Vol. 278). Among the discoveries were distant, powerful infrared galaxies fueled by prodigious rates of star formation, a shell of dust around α Lyrae, surprisingly strong infrared emission from Arp 220, a new comet, and an extended emission component visible at high Galactic latitude primarily at 60 and 100 μm. The extended emission was dubbed "infrared cirrus" because of the ragged nature of the structures which—at some level—covered most of the high-latitude sky.

Although the IRAS discovery of the infrared cirrus (Low et al. 1984) was momentous, the lack of velocity information left open *where* it was located. While most of the authors of the paper announcing the discovery would have favored a Galactic origin, caution forced the inclusion of the following statement:

This dust emission could, perhaps, be associated with HI at anomalous velocities, with small amounts of molecular rather than atomic hydrogen, or with cold material in the outer solar system. Fortunately, the IRAS mission plan of rescanning the sky after an interval of 6 months is well suited to distinguish between dust clouds in orbit around the Sun at various distances and distant interstellar clouds. This aspect of the infrared cirrus remains one of the most enigmatic characteristics of the sky as seen by IRAS and will continue as such until the sky has been examined a second time.

Within two years, the identification of the high-latitude cirrus with a local, Galactic component was firmly established by Weiland et al. (1986). Along with the cirrus, another IRAS discovery of relevance to our discussion was the somewhat surprising amount of infrared emission from the cirrus at 12 and 25 μm. This type of emission was associated with temperature "spikes" as very small grains or PAHs absorb individual, high-energy photons (Sellgren 1984; Boulanger et al. 1985; Puget et al. 1985; Draine and Anderson 1985, Omont 1986, Desért et al. 1986).

Although the infrared data from the IRAS mission is now over three decades old, it is still used productively and offers the simplest and quickest way to look at the dust distribution in the Galaxy (i.e., through the Skyview Virtual Observatory: http://skyview.gsfc.nasa.gov/current/cgi/titlepage.pl).

An improvement to the original data was made by reprocessing the data to better remove the pesky zodiacal light contribution, calibrating and fixing zero levels so as to be compatible with the DIRBE instrument on the COBE satellite[2] (see below), and improving the destriping algorithms. The resulting database is called IRIS and with its 4′ resolution matching the *Planck* high frequency bands, it provides an outstanding view of the dust emission from the Galaxy (Miville-Deschênes and Lagache 2005). Figures 5.2 (right) and 5.8 show IRIS 100 μm data for the translucent cloud MBM 16 and the diffuse molecular cloud MBM 40, respectively. The figures should be compared to Fig. 5.3 and Fig. 5.6 which are the *Planck* 353 GHz (which is equivalent to 849 μm) data for the regions in question.

5.3.2 COBE and ISO

The COsmic Background Explorer (COBE) was launched by NASA on November 18, 1989 primarily to study the CMB in the infrared. It provided the best measurements of the CMB spectrum of that time and revealed the fluctuations in temperature at the 10^{-5} level (which garnered George Smoot and John Mather the Nobel Prize in physics in 2006). Galactic dust emission was also detected by the instruments on board the satellite with better calibration than IRAS, although with lower angular resolution. In particular, the Diffuse InfraRed Background Experiment (DIRBE) and the Far InfraRed Absolute Spectrophotometer (FIRAS) instruments carried forward the legacy of IRAS and led to several important developments.

DIRBE was a photometer with ten bands covering the range from 1.25 to 240 μm at a somewhat poor resolution of 40′ (Silverberg et al. 1993). The FIRAS instrument was a polarising Michelson interferometer with an angular resolution of 7° and a

[2]The IRAS mission also served as the pathfinder for COBE, in the sense that analysis of the IRAS data led to models of the Galactic dust component.

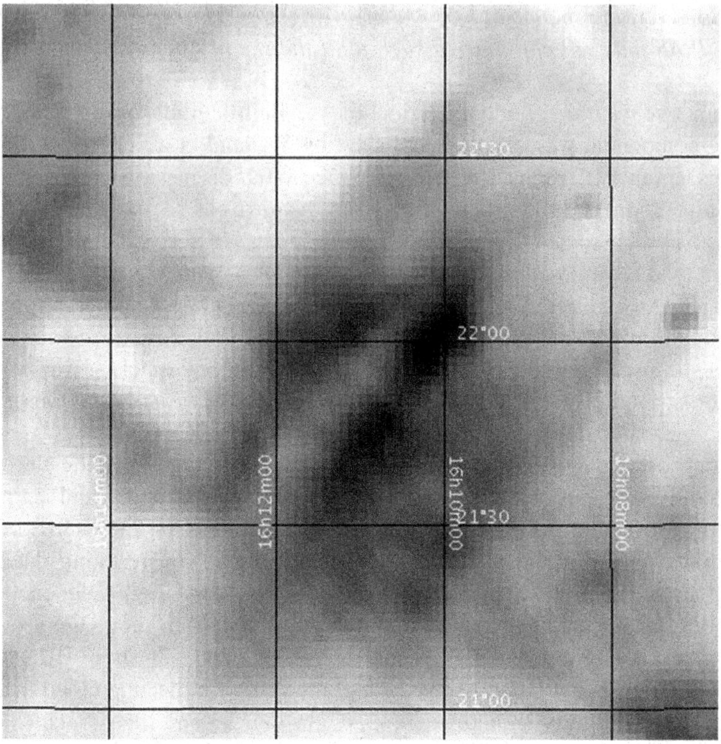

Fig. 5.8 Region centered as in Fig. 5.5 but twice the size (2°× 2°), showing emission from the IRIS 100 μm data which primarily traces thermal dust. The image is centered at (RA, Dec) = (16^h 11^m 00^s, 21° 55' 00") in 2000.0 coordinates. The gray scale is a log scale and ranges from 3.21 to 11.76 MJy/ster. The image was made with the Skyview Virtual Observatory

fixed spectral resolution in two separate bands of 0.57 cm^{-1} (Fixsen et al. 1994). The high-frequency band (from 20 to 96 cm^{-1}) was also used to study the Galactic thermal emission. Although the resolution of this instrument was somewhat poor (0.7 degrees), its careful calibration was very useful in reprocessing IRAS data. For example, the Schlegel et al. (1998) dust maps were based on the DIRBE calibrations.

The Infrared Space Observatory (ISO) was an infrared satellite observatory built by the European Space Agency (ESA) and launched in November of 1995. Its cryogenic reserves lasted until April of 1998 and the instruments on board allowed for observations in the 2.5–240 μm wavelength range (Kessler et al. 1996) and consisted of two spectrometers (SWS and LWS), a camera (ISOCAM), and an imaging photo-polarimeter (ISOPHOT). The spatial resolutions ranged from 1.5″ to 90″ and represented a distinct improvement over the previous missions, although an all-sky survey was precluded. Instead, some 30,000 individual imaging, photometric, spectroscopic, and polarimetric observations of various objects were made. The ISO archive can be accessed at http://iso.esac.esa.int/ida/.

5.3.3 *Spitzer, Herschel, and WISE*

The *Spitzer* Space Telescope (Werner et al. 2004) is the fourth and final of the NASA Great Observatories. It was launched in August 2003 and is still operating. Although primarily designed to study protoplanetary and planetary debris disks, brown dwarfs, and ultra-luminous infrared galaxies, the wavelength range of the instruments is such that interstellar dust can be readily studied (3.6, 4.5, 5.8, 8, 24, 70, 160 μm). Although the liquid helium ran out in May 2009, the mission continues with the two shortest wavelength modules of the IRAC camera (3.6 μm and 4.5 μm still operating and taking data as part of the Spitzer Warm Mission) with a field of view of $5' \times 5'$ and an angular resolution of $1.9''$. While it was cryogenically cooled, IRAC also operated at 5.8 μm and 8 μm.

The telescope consists of an 85 cm mirror and instrumentation to observe the cosmos from 3.6 to 180 μm. Besides the IRAC, Spitzer was also equipped with a Multiband Imaging Photometer (MIPS) which produced images at 24, 70, and 160 μm. The size of the images varied from $5' \times 5'$ at 24 μm to $0.5' \times 5' \mu$m at 70 μm. In addition to the imaging cameras the Infrared Spectrograph (IRS) was capable of both high- and low-resolution spectroscopy at mid-IR wavelengths (5–40 μm). This range makes the data particularly useful for studying PAHs in the ISM.

Of particular interest for the diffuse interstellar medium is the *Spitzer*/GLIMPSE survey of the Milky Way (Churchwell et al. 2004). Three large surveys were undertaken. The first fully mapped the Galactic plane at 3.6, 4.5, 5.8, and 8.0 μm for $\ell = 10° - 65°$ and $|b| \leq 1°$ with a spatial resolution of less than $2''$. The second survey fully imaged the inner 20° of the Galactic plane, and the third surveyed the central bar by extending the latitude coverage to $|b| \leq 3°$.

The *Herschel* Space Observatory was an ESA project and consisted of a single 3.5 m telescope with instrumentation that covered the far-infrared and submillimeter bands from 55 to 672 μm. The *Herschel* mission lasted from May 2009 to April 2013. At the time, it was the largest mirror ever deployed in space. The telescope was placed in an Earth-Sun L2 orbit with imaging instruments that included the Photodetecting Array Camera and Spectrometer (PACS) covering wavelengths from 55 to 210 μm and the Spectral and Photometric Imaging Receiver (SPIRE) covering the wavelength range from 194 to 672 μm. Both instruments also served as low-resolution spectrometers, but the high-resolution spectral work was done by the Heterodyne Instrument for the Far Infrared (HIFI). *Herschel* obtained observations of sub-mm transitions of molecules, and the PRIMAS project took advantage of this to survey transitions of the ions CH^+, SH^+, OH^+, OH_2^+, and OH_3^+; and of the neutrals OH, H_2, NH, NH_2, NH_3, and HF. An entire issue of the journal *Astronomy and Astrophysics* was devoted to its initial results.[3] With its great resolution and sensitivity, *Herschel* provided a fantastic instrument for studying individual objects in the infrared.

[3]*Astronomy & Astrophysics*, Volume 518, July–August 2010.

The *Wide-field Infrared Survey Explorer* (WISE) was launched in December 2009, placed in hibernation (by turning off its transmitter in February 2011), and re-activated in August 2013. In the first phase of its mission, it functioned as an infrared telescope (40 cm diameter) conducting an all-sky survey at 3.4, 4.6, 12 and 22 μm. Once its coolant was depleted, it spent it remaining time searching for near-Earth objects. In its second phase mission, it is recommissioned (NEOWISE) to search for near-Earth asteroids. From its original mission, each image covers fields $47'$ in size with angular resolution of $6''$. Although the resolution is worse than that of *Spitzer* at 3.6 and 4.5 μm, its all-sky survey provides an excellent way to study the dust distribution in the diffuse ISM (see Fig. 5.9).

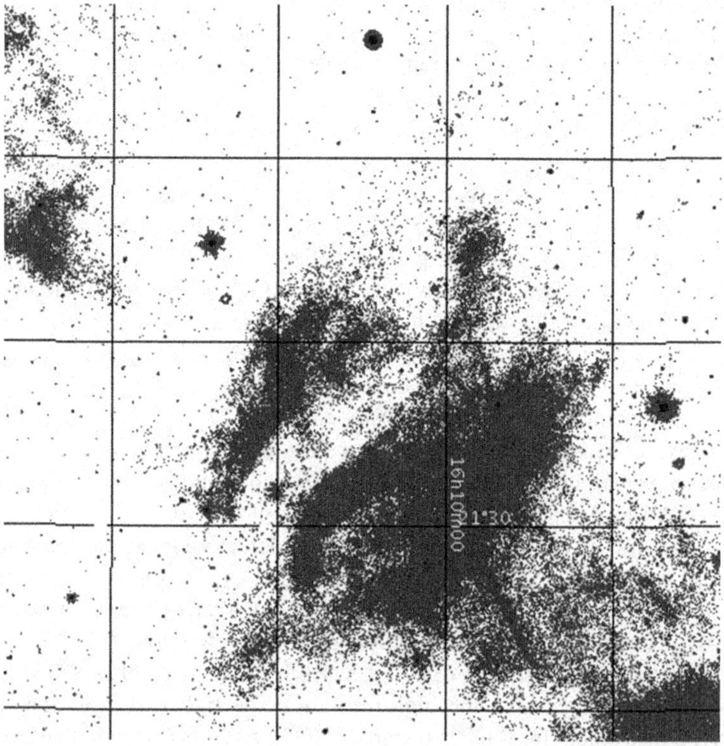

Fig. 5.9 Region centered as in Fig. 5.5 but twice the size (2°× 2°), showing emission from the WISE 12 μm band. The image is centered at (RA, Dec) = (16^h 11^m 00^s, 21° 55′ 00″) in 2000.0 coordinates. The gray scale is a histogram equalization scale to highlight low-level extended emission and ranges from 420 to 3690 Digital Numbers (DN) units. The image was made with the Skyview Virtual Observatory

5.4 Spectral Tracers of Diffuse Molecular Gas in the Far-Infrared

In Chap. 3 we looked at various spectral tracers of diffuse molecular gas in the radio portion of the spectrum. The lines we highlighted were molecular, rotational or hyperfine transitions. In the infrared, the atomic fine structure lines of neutral carbon, CI, and singly-ionized carbon, CII, are the most important because of their visibility in the infrared from space-born platforms in the 1990s.

Along with CO, C^+, is the principal coolant in translucent and diffuse molecular clouds. Thus, lines from this species should be readily detectable in the diffuse, low-density molecular component of the ISM. Moreover, the $^2P_{3/2}$-$^2P_{1/2}$ transition of singly-ionized carbon at 157.7 μm is the dominant coolant in the CNM (see Sect. 1.5.4). In PDRs, C^+ is most abundant at the interface with the atomic ISM and thus the outermost envelope of any cold atomic or molecular cloud will show emission from this coolant. The 158 μm emission as observed by FIRAS aboard the COBE satellite shows strong correlation with neutral atomic hydrogen (Bennett et al. 1994). However, in dense molecular gas the [CII] emission is expected to drop-off as the carbon transforms to C^0 and CO and the UV heating of the grains decreases since the radiation fails to penetrate the more opaque clouds. Even in diffuse and translucent clouds, the [CII] intensity decreases relative to the surrounding diffuse atomic gas. Models indicate that the decrease in 158 μm is produced by attenuation and softening of the ISRF and not from a changes in the dust (Ingalls et al. 2002). In addition, observations of this transition can be used to obtain the median density and pressure in molecular regions.

Two fine structure transitions of CI are the $^3P_1$-$^3P_0$ line at 492 GHz and the $^3P_2$-$^3P_1$ line at 809 GHz. The former transition has been detected in the high-latitude molecular cloud, MBM 12, from ground-based observations by Ingalls et al. (1994—see also, Ingalls 1999). The ratio of C/CO in translucent clouds is expected to be around unity, significantly higher than toward dense Galactic PDRs. Although the column densities of neutral carbon are high, it is relatively difficult to carry out these observations from the ground and CO remains by far the favored tracer of these regions.

The *Stratospheric Observatory for Infrared Astronomy* (SOFIA) currently (2017) offers the best way to obtain infrared spectra and images from the near to the far infrared.[4] Although not a space mission, the observatory consists of a 2.5 meter Bent Cassegrain that is housed in a Boeing 747SP. The plane can fly at altitude of up to 13.7 km where it is above almost all the water vapor in the Earth's atmosphere. The project is a joint venture between NASA and the German Aerospace Center (DLR) and is a successor to the *Kuiper Airborne Observatory* (KAO).

[4] see https://www.nasa.gov/mission_pages/SOFIA/.

5.5 Probing the Ultraviolet Sky

Observations of absorption lines from electronic transitions in the UV portion of the spectrum are analyzed in the same way as optical absorption observations described in the Chap. 3; the key difference is that the observations must be conducted from earth orbit because of the opacity of the Earth's atmosphere to UV radiation. In the ISM, the widespread presence of atomic hydrogen effectively terminates UV observations below the Lyman limit (912 $\AA$).

The first observation of H_2 were from sounding rocket flights (Carruthers 1970), and the UV transitions of CO were detected immediately after (Smith and Stecher 1971). These discoveries paved the way for the *Copernicus* satellite (operated from 1972 to 1980) which produced most of the early information on the abundance and excitation of H_2 in the diffuse ISM (Spitzer et al. 1973). A key project for *Copernicus* was the survey of atomic and molecular hydrogen towards 109 stars leading to the two classic papers by Savage et al. (1977) and Bohlin et al. (1978) (see §2.3.1).

For the diffuse ISM, the absorption lines are produced as molecules in the ground electronic state absorb UV photons and transition to excited electronic states. Unfortunately, the dust in the ISM is an excellent absorber and scatterer of these photons—with increasing efficiency toward the shorter UV wavelengths, so that observations are possible only in regions with less than a magnitude of visual extinction (the corresponding extinction in the UV is anywhere from a few times to 15 times higher depending on the wavelength). Nevertheless, a lot of interesting work was done in the near UV from the ground. Notable UV observations of the diffuse ISM include observations of the OH line at λ3078Å by Crutcher and Watson (1976), NH, CH, and OH for the line of sight to ζ Per, all below 3400 Å (Chaffee and Lutz 1977), five OH transitions out of the $X^2\Pi_{3/2}$ ground state between 3070 Å and 3080 Å (Felenbok and Roueff 1996), FUSE observations of the line of sight to HD 34078 (Boissé et al. 2005), and OH observations towards 10 sightlines by Weselak et al. (2009).

The H_2 ground state consists of both electrons in the lowest $1s\sigma$ orbital and it is denoted by $^1\Sigma^+$. In the ultraviolet, the Lyman absorption lines from the ground state to the higher $^1\Sigma$ states were observed by space telescopes such as *Copernicus* in the 912 Å-1215 Å range against background early-type stars. But if one wishes to observe lines in emission, the lowest rotational levels in the $v = 0$ state are the only ones that are populated under the physical conditions of the diffuse ISM. However, because of selection rules, the transitions between odd and even J levels are forbidden. Thus, H_2 has para and ortho levels corresponding to even and odd J levels with transitions allowed only within each set of levels. The selection rules allow only those transitions with $\Delta J = 2$ so that the lowest allowed transition is the para J = 2-0 transition at 28 μm. Unfortunately, the J = 2 level is 500 K above the ground level and thus is not very populated in cold molecular clouds, although some observations of this transition have been made in shocked gas regions.

Notable UV telescopes after *Copernicus* were the *International Ultraviolet Explorer* (IUE—Kondo 1987), the Hopkins Ultraviolet Telescope (Blair et al. 1996), ORFEUS (Barnstedt et al. 2000), and the *Far Ultraviolet Spectroscopic Explorer* (FUSE—Moos et al. 2000) . In addition to these satellites, we can add the Cosmic Origins Spectrograph (COS) on the *Hubble Space Telescope* (HST) since it allows spectroscopy of astronomical objects below $Ly\alpha$.

5.5.1 IUE and HST

For longevity and productivity, the *International Ultraviolet Explorer* and the *Hubble Space Telescope* rank as the most successful missions to date. The IUE satellite, launched in Nov. 1977 (operational from January 1978, through September 1996) was, like *Copernicus*, an exclusively ultraviolet spectroscopic satellite. It consisted of a 45 cm telescope with two absolutely calibrated solar blind spectrographs (1150–2000 Å, 1900–3200 Å) that used two apertures, 10×20 arcsec^2 and 10×10 arcsec^2, and two resolutions (about 300 and 10000). Launched in Nov. 1977 into a geosynchronous orbit, it produced over 10^5 spectra in its lifetime covering sources (Galactic and extragalactic) distributed over most of the sky (see Fig. 5.10). For diffuse ISM studies, it was singularly important for the enormous number of sight lines it probed, in many cases as a collateral product of the primary project. The only limitation was its SIT vidicons detectors (these, unlike CCDs, could physically saturate and were nonlinear at their upper operational range). The archive of all spectra is available through the MAST facility at Space Telescope Science Institute.[5]

HST was launched in Apr. 1990, the first of the NASA Great Observatories program, and is still operating. Placed in a low Earth orbit by the Space Shuttle, it was designed to be serviced by periodic astronaut visits with reboots for its orbit and change-outs for instrument updates. The low orbit, however, reduces its efficiency because of Earth occultations. But that notwithstanding, the complement of high resolution spectrographs carried in the last two decades makes the satellite archive a major resource for interstellar medium studies. The Goddard High Resolution Spectrograph (GHRS) was the first generation high resolution (up to 100,000) single order spectrograph. Its limitation was its detector, an image intensifier coupled to a linear silicon diode array, so only a limited part of the spectrum could be observed at any time. The Space Telescope Imaging Spectrograph (STIS), a second generation instrument on HST, vastly extended capabilities for ultraviolet spectroscopy by using two dimensional detectors (CCDs and Multianode Microchannel Arrays, MAMA) along with echelle gratings extending to resolutions of 100,000 from 1150–3200 Å and medium resolution gratings, R$\approx$30,000.[6]

[5]http://archive.stsci.edu.

[6]The UV spectra shown in Chap. 3 for the nova T Pyx were obtained with STIS.

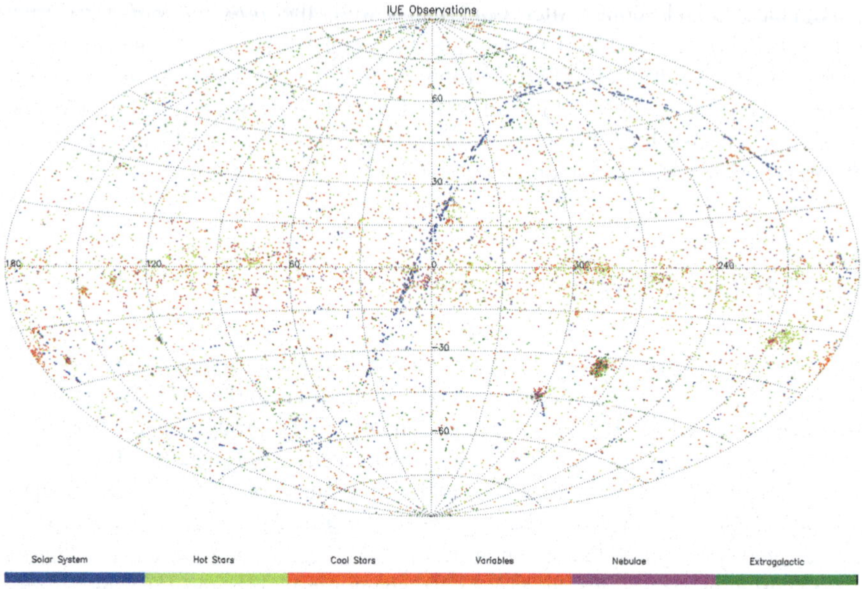

Fig. 5.10 Distribution of lines of sight observed by IUE over its lifetime. The spectrophotometry ranged from high (0.1–0.3 Å) to low (6–7 Å) resolution between 1150 Å and 3200 Å . More than 100,000 UV spectra were obtained with the IUE satellite from January 26, 1978 to September 30, 1996. Image from MAST

5.5.2 FUSE

The FUSE satellite was designed to extend the range of ultraviolet spectroscopy to wavelengths shorter than those detectable by the Hubble Space Telescope (Moos et al. 2000; Sahnow et al. 2000). The satellite was launched in 1999 and remained in operation until 2008. The FUSE instrument channeled the light into 4 channels each of which was comprised of a grating, spectrograph and FUV detector. The four channels overlapped but covered the range from 917–1188 $\AA$. This allowed for high-resolution (R $\sim$ 20,000) FUV spectra of hot point sources. A graphical display of the observed lines of sight is shown in figure 5.11. Of particular interest to the diffuse ISM was the presence of several bands of H_2 and CO within FUSE's observable channels. Observations of these lines along the line of sight to various QSOs, Seyferts, BL Lacs, and Galactic Halo stars provided direct, key, information about these molecules in the diffuse ISM.

In particular as far as diffuse molecular gas is concerned, Snow et al. (2000) detected H_2 in the translucent cloud along the line of sight to HD 73882. This initial work was followed up by a more extensive survey towards translucent clouds by Rachford et al. (2002) which primarily studied the correlations between N(H_2) and N(CH), N(CH^+), N(CN), and N(CO). They also noted that many "translucent" lines of sight were really a superposition of several diffuse clouds which combined to produce A_V values greater than 1 magnitude. These lines of

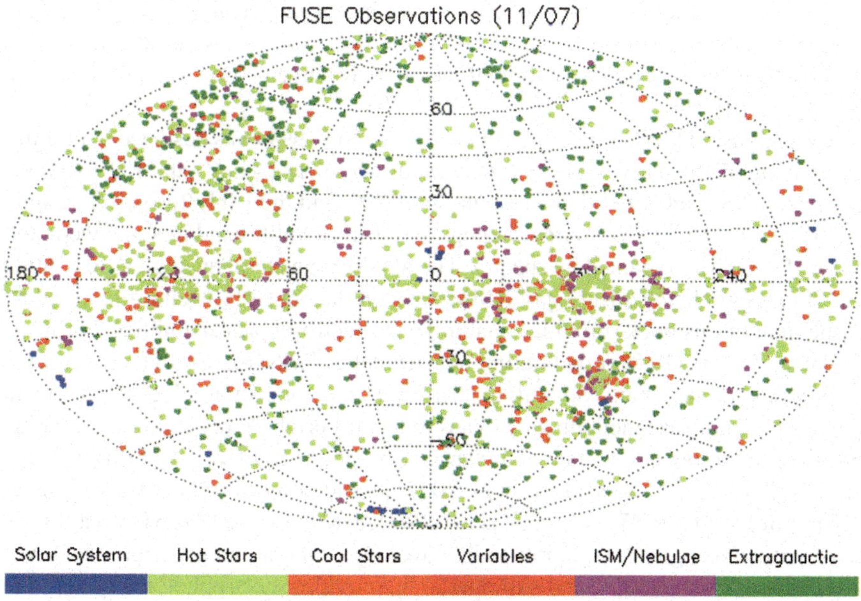

Fig. 5.11 Distribution of lines of sight observed by FUSE over its lifetime (June 24, 1999–October 18, 2007). The Aitoff projection in Galactic coordinates shows nearly 3000 separate astronomical targets that resulted in over 6000 observations. See text for more details. Image from MAST

sight, though "translucent" in the sense that they had substantial extinction, did not represent translucent *clouds*. In this respect, maps of the CO(1-0) line are the best method for identifying these versus a superposition of diffuse clouds along a line of sight (though see discussion in Chap. 7). Richter et al. (2003) surveyed interstellar absorption lines of H_2 in the Lyman and Werner bands toward 56 mostly extragalactic sources. They found $N(H_2)$ values of 10^{14}–$10^{17}\,cm^{-2}$ in 14 intermediate-velocity clouds (IVCs) in the lower Galactic halo implying that CNM in these clouds is widespread. This result reinforces the idea that virtually any line of sight in the Galaxy has molecules along it, but usually not in sufficient quantities to be deemed a "molecular cloud".

5.6 Probing the X-Ray Sky

X-ray emission is generally associated with compact objects. However, at low X-ray energies (about 1/4 keV) a diffuse background from the HIM is present over most of the sky. As discussed in Chap. 1, the heating for this gas is from supernovae and stellar wind bubbles. At higher energies (above 1/2 keV), this soft background is still visible but it is not as prominent. The isotropy of the radiation at this energy implies an extragalactic origin. Finally, above 1 keV, the diffuse background is likely a superposition of emission from distant extragalactic objects (typically quasars and

AGNs). At the lower energies, nearby dense objects such as molecular clouds can absorb the X-ray emission and produce "shadows" which can be used to study both the absorbing objects and the nature of the more distant emission (e.g., Burrows and Mendenhall 1991).

Like IRAS, ROSAT (operating between 1990 and 1999) was designed to produce an imaging all-sky survey in the soft X-ray range (0.1–2.4 keV) and EUV (0.006–0.2 keV). The PSPC provided a resolution of 20″, and the High Resolution Imager (HRI) achieved 5″. Point source catalogs and diffuse multi-band imaging were not, unlike IRAS, accompanied by spectra. The all-sky maps have an angular resolution of 12′ to highlight the diffuse gas. XMM-*Newton* (launched 1999 and still operating at the time of this book) provides imaging and spectroscopic capabilities. For interstellar studies, the Reflection Grating Spectrometer (RGS) (0.35–2.5 keV) achieves resolutions greater than 100, sufficient to measure ionization edges of abundant species in the diffuse medium when viewed toward bright X-ray sources, providing limits on high ionization state abundances (the presence of very hot diffuse gas). *Suzaku* (2005–2015, see Mitsuda et al. 2007) provided spectrophotometric imaging capabilities with the XIS (Koyama et al. 2007), extending the ROSAT capability in sensitivity and angular resolution while covering a similar energy range. The Swift satellite launched in 2004, and still operating, provides spectrophotometry with the X-ray telescope (XRT) spanning the energy range 0.2–10 keV with a resolution is ≈ 40 at 6 keV. This resolution is sufficient to distinguish strong emission lines from point sources and continuum edges, as with XMM-*Newton*. The lower energy range is also especially useful as a probe of the line of sight neutral hydrogen seen against bright sources. Comparing the derived absorption optical depths with 21 cm measurements is especially useful since the continuum absorption is independent of excitation conditions. The *Chandra* x-ray observatory (Tananbaum et al. 2014), launched in 1999 and still operating (2017), is the third of the Great Observatories series. It operates in the 0.09-10 keV range with an effective area of 400 cm^2 with imaging (with better than 1″ resolution) and grating spectroscopy. Because the field of view is small, 31 by 31 $arcmin^2$, ISM studies are best performed on *extragalactic* sources (e.g., imaging of spiral and elliptical galaxies) or targeted study of individual lines of sight (absorption spectroscopy) or broadband imaging of supernova remnants within the Galaxy.

5.7 Probing the Gamma-Ray Sky

Although important milestones in detecting gamma rays from space were achieved with the detectors flown on the NASA SAS-2 and the European COS-B satellites launched in 1972 and 1975, respectively,[7] gamma-ray astronomy really took off with the advent of NASA's *Compton* Gamma-Ray Observatory (CGRO) launched

[7]The first firm detection of Galactic gamma rays was with the experiment flown on the OSO-3 satellite in 1967.

in 1991, the second of the NASA Great Observatories. The failure of a gyroscope on the spacecraft in 2000 caused NASA to deliberately de-orbit the telescope which then crashed into the Pacific Ocean. A review of the discoveries and accomplishments of the mission was given by Gehrels and Shrader (2001). The four instruments on board (BATSE, OSSE, COMPTEL, and EGRET) covered the energy range from 20 keV to 30 GeV with unprecedened resolution and sensitivity. An earlier Franco-Soviet mission, *Granat*, had covered the energy range from 3 keV to 1.3 MeV. Gamma-ray emission from the ISM is produced by cosmic rays interacting with the gas component of the Galaxy (see below); the principal physical processes involved are neutral-pion decay, bremsstrahlung, and inverse Compton scattering. The key instrument on the Compton for wide-field imaging was EGRET; and a whole sky map made with EGRET at $>$ 100 MeV is shown in Fig. 1.3.

In 2008, the *Fermi* Gamma-Ray Space Telescope was launched with new generation instruments: The Gamma-Ray Burst Monitor (GBM) is sensitive to X-rays and gamma-rays in the 8 keV to 40 MeV range and is designed to identify gamma-ray point sources. Although smaller than BATSE, it has the same sensitivity. The imaging instrument is the Large Area Telescope (LAT) which measures the arrival direction, energy, and time of individual gamma rays with energy in the range of 20 Mev to 300 GeV. The LAT's field of view surveys about 20% of the sky at any given time, and scans continuously, covering the whole sky every three hours (two orbits around the Earth). As the images are stacked, ever more sensitive gamma-ray maps of the sky are made. Most gamma-ray sources can be located to within $10'$ and the sensitivity is an order of magnitude better than EGRET. Whole-sky maps of the gamma-ray sky above 1 GeV and 10 GeV can be found at the website of the Fermi LAT collaboration (https://www-glast.stanford.edu/). The Milky Way is prominent in the whole sky maps, a result of the interaction of hydrogen nucleons with cosmic rays. In addition to point sources, the diffuse gamma-ray background comes from the Galaxy and, extragalactically, from unresolved quasars. We describe below how the diffuse galactic gamma-ray background can be used to trace the distribution of H_2.

5.7.1 How Gamma Rays Trace Molecular Gas

G.G. Fazio reviewed in 1967 the principal ways to produce gamma-rays from the Galaxy (Fazio 1967). As far as the diffuse ISM is concerned, cosmic ray nucleons interact with interstellar hydrogen and helium nucleons resulting in gamma-rays with the decay of π^o mesons, cosmic ray electrons interact with the ionized portion of the interstellar gas via bremsstrahlung, and with the interstellar radiation field through the inverse Compton process. In the presence of a magnetic field, cosmic ray electrons can also produce synchrotron radiation. At high energies, the most important emission mechanisms are pion-decay, inverse Compton, and bremsstrahlung. There is a contribution to the diffuse gamma-ray spectrum by an extragalactic component, but it is typically down by a factor of several from the Galactic component in the energy range 20 MeV–20 GeV.

Because most of the high-energy ($\geq$ 50 MeV) interactions occur with interstellar gas, it was quickly realized that studying the diffuse gamma-ray emission at these energies was a way to trace the distribution of Galactic atomic and molecular hydrogen (the helium is assumed to be distributed equally in both phases). The contribution to the diffuse gamma-ray emission from the inverse Compton effect has to be modeled and removed correctly for this technique to work out, but Bloemen (1985) showed how this could be done and concluded that the contribution of inverse Compton gamma rays to the observed gamma-ray diffuse emission was negligible down to 10 MeV. The spectral distributions for pion-decay and electron-bremsstrahlung peak at different energies and so in the 300 MeV–5 GeV range pion-decays dominate.

Given the above considerations, the Galactic diffuse gamma-ray emission can be used to trace directly the column density of molecular hydrogen along a given line of sight. The basic idea is that once the contributions from inverse Compton emission and instrumental effects are removed, the diffuse gamma-ray emission map traces the column density of hydrogen nucleons along a given directions. If the contribution from atomic nucleons is removed (by determining N(HI) from 21 cm data, for example), the remaining gamma ray emission must be from hydrogen in molecular form. The LAT instrument on Fermi should be able to image at least the larger high-latitude molecular clouds and, since they are local objects, would provide information on the local cosmic ray flux. Moreover, observations in varying energy bands can be used with 21 cm data and CO velocity-integrated main-beam antenna temperature data (W_{CO}) to determine the calibration of the CO-H_2 conversion factor. This is discussed further in Sect. 8.2.4.

The high latitude molecular clouds seem like ideal candidates for studies using the technique described above. They are localized and their lines of sight are not confused by foreground and background objects. However, Torres et al. (2005) argued that this population did not have sufficient γ-ray emission to be detected by the EGRET instrument on the CGRO. However, the LAT instrument on the *Fermi* should be able to detect the larger clouds, providing a new way to characterize their H_2 content (see, e.g., Abrahams and Paglione 2015).

References

Abrahams, R.D. and Paglione, T.A.D. 2015, ApJ, 805, 50
Barnstedt, J., Gringel, W., Kappelmann, N., and Grewing, M. 2000, A&AS, 143, 193
Bennett, C.L., et al. 1994, ApJ, 434, 587
Bloemen, J.B.G.M. 1985, A&A, 145, 391
Bohlin, R.C., Savage, B.D., and Drake, J.F. 1978, ApJ, 224, 132
Blair, W.P, Long, K.S., and Raymond, J.C. 1996, ApJ, 468, 871
Boissé, P. et al. 2005, A&A, 429, 509
Boulanger, F., Baud, B., and van Albada, G.D. 1985, A&A, 144, 9
Burrows, D.N. and Mendenhall, J.A. 1991, *Nature*, 351, 629
Carruthers, G. 1970, ApJ, 161, L81
Chaffee, F.H. and Lutz, B.L. 1977, ApJ 213, 349

Chastain, R.J. 2005, PhD Thesis, University of Georgia
Churchwell, E. et al. 2004, PASP, 121, 213
Cotten, D.L. and Magnani, L. 2013, MNRAS, 436, 1152
Crutcher, R.M. and Watson, W.D. 1976, ApJ, 203, L123
Desért, F.X., Boulanger, F., and Shore, S.N. 1986, A&A, 160, 295
Draine, B.T. and Anderson, N. 1985, ApJ, 292, 494
Fazio, G.G. 1967. ARAA, 5, 481
Felenbok, P. and Roueff, E. 1996, ApJ, 465, 57
Fixsen, D.J. et al. 1994, ApJ, 420, 445
Gehrels, N. and Shrader, C.R. 2001, AIPC, 587, 3
IRAS Catalogs and Atlases: Explanatory Supplement 1988, ed. C.A. Beichman, G. Neugebauer, H.J. Habing, P.E. Clegg, and T.J. Chester, Washington, D.C.:GPO
Ingalls, J.G., Bania, T.M., and Jackson, J.M. 1994, ApJ, 431, L139
Ingalls, J.G. 1999, Ph.D. Thesis, Boston University
Ingalls, J.G., Reach, W.T., and Bania, T.M. 2002, 579, 289
Kessler, M.F. et al. 1996, A&A, 315, 27
Kogut, A. et al. 1996, ApJ, 464, L5
Kondo, Y. 1987, *Exploring the Universe with the IUE Satellite*, Astrophysics and Space Science Library, Springer
Koyama, K. et al. 2007, PASJ, 59, 23
Leitch, E.M., Readhead, A.C.S., Pearson, T.J., and Myers, S.T. 1997, ApJ, 486, L23
Low, F.J. et al. 1984, ApJ, 278, L19
Magnani, L. et al. 2003, ApJ, 586, 1111
Mitsuda, K. et al. 2007, PASJ, 59, S1
Miville-Deschênes, M.-A. and Lagache, G. 2005, ApJS, 157, 302
Moos, H.W. et al. 2000, ApJ, 538, L1
Neugebauer, G. et al. 1984, ApJ, 278, L1
Omont, A. 1986, A&A, 164, 159
Planck Collaboration, Planck Early Results XIX, 2011, A&A, 536, A19
Planck Collaboration, Planck Early Results XXI, 2011, A&A, 536, A21
Planck Collaboration, Planck Early Results XXII, 2011, A&A, 536, A22
Planck Collaboration, Planck Early Results XXIII, 2011, A&A, 536, A23
Planck Collaboration, Planck Early Results XXIV, 2011, A&A, 536, A24
Planck Collaboration, Planck Early Results XXV, 2011, A&A, 536, A25
Planck Collaboration, Planck 2013 Results XIII, 2014, A&A, 571, A13
Planck Collaboration, Planck 2015 Results XXVIII, 2014, arXiv:1502.01599
Puget, J.L., Leger, A., and Boulanger, F. 1985, A&A, 142, 19
Rachford, B.L. et al. 2002, ApJ, 577, 221
Richter, P., Wakker, B.P., Savage, B.D., and Semabach, K.R. 2003, ApJ, 586, 230
Sahnow, D.J. et al. 2000, ApJ, 538, L7
Savage, B.D., Bohlin, R.C., Drake, J.F., and Budich, W. 1977, ApJ, 216, 291
Schlegel, D.J., Finkbeiner, D.P., and Davis, M. 1998, ApJ, 500, 525
Sellgren, K. 1984, ApJ, 277, 623
Silverberg, R.F et al. 1993, SPIE, 2019, 180
Smith, A.M. and Stecher, T.P. 1971, ApJ, 164, L43
Snow, T.P., et al. 2000, ApJ, 538, L65
Spitzer, L, Drake, J.F., Jenkins, E.B., Morton, D.C., Rogerson, J.B., and York, D.C. 1973, ApJ, L116
Spitzer, L. 1978, *Physical Processes in the Interstellar Medium*, New York: Wiley-Interscience
Tananbaum, H., Weisskopf, M., Tucker, W., Wilkes, B., and Edmond, P. 2014, RPPhy, 77, 6902
Torres, D.F., Dame, T.M., and Digel, S.W. 2005, ApJ, 621, L29
Weiland, J.P. et al. 1986, ApJ, 306, 101
Werner, M.W. et al. 2004, ApJS, 154, 1
Weselak, T., Galazutdinov, G., Beletsky, Y., and Krelowski, J. 2009, A&A, 499, 783

Chapter 6
Observing in the Dark: The Dust-Gas Connection

Watch my dust.

—Babe Ruth

For dust thou art, and unto dust shalt thou return.—Genesis 3:19

Abstract Along with the gas there is a ubiquitous solid component of the diffuse medium, loosely referred to as *dust*. Its presence was first inferred by distance dependent chromatic extinction (reddening) of starlight. But with satellite observations in the ultraviolet and far infrared, its thermal properties, spatial distribution, and physical makeup have been inferred. In this chapter we discuss the measurement of dust and its relation to the diffuse atomic and molecular gas.

6.1 Introduction

Besides gas and cosmic rays, the ISM contains a significant and readily observable population of small particles whose sizes are considerably larger than molecules. These solids, collectively referred to as *dust* or *grains*, cannot be formed *ab initio* from the diffuse gas since the densities are far too low to grow macroscopic aggregates. But they are unmistakably present as ices so at least some part of these grains has accumulated in much denser environments with subsequent dispersal. These tell-tale spectral signatures are, however, all found in other than visible wavelengths. The broad absorption feature at around 2175 Å is characteristic of amorphous carbon that otherwise shows only continuous extinction in the optical and infrared. Indications of water, ammonia, and CO_2 ices are, instead, found in the 3–10 μm portion of the infrared, while silicates produce a feature observed at 10 μm. These bands are all broad, otherwise structureless features. The two principal bands, at 0.2 μm and 10 μm, have widths of order $\Delta\lambda/\lambda \approx 0.1$ while the ice bands in the mid- and near-infrared are narrower by a factor of a few. A view of the historical development of our general understanding of this component of the interstellar gas is provided by the set of *Annual Reviews* overviews over the last 50 years: Lynds and Wickramasinghe (1968), Savage and Mathis (1979), Mathis (1990), and Draine (2003).

L. Magnani, S.N. Shore, *A Dirty Window*, Astrophysics and Space Science Library 442, DOI 10.1007/978-3-662-54350-4_6

The gas to dust ratio in the Milky Way has canonically been taken as about 100 from early studies of neutral hydrogen clouds and HII regions (see Spitzer 1978). More recent studies, e.g., Draine et al. (2007) confirm this value based on observations and more sophisticated dust models: $M_{dust}/M_H \approx 0.01$ and $M_{dust}/M_{gas} = [M_{dust}/M_H]/1.36$ accounting for helium. However, the importance of dust in the ISM is not simply its mass fraction. In addition to scattering starlight, dust grains absorb between 30–50% of the optical and ultraviolet light emitted by stars, and eventually re-radiate this in the far-infrared. Dust thus plays a large role in the overall bolometric luminosity of a spiral galaxy. More important, perhaps, is the role of grains in astrochemistry and the production of complex molecules.

With the spectacular advances in infrared and millimeter technology in the last 20 years, measuring the dust content and properties of the cold diffuse ISM has never been easier. The constancy of the dust-to-gas ratio has actually reversed its observational role: the dust is now often used as a surrogate for tracing gas. Early works, such as Savage et al. (1977) and Bohlin et al. (1978), firmly established the relationship between E(B-V) and $N(H_{total})$ [which is the column density of hydrogen nucleons: N(HI) + $2N(H_2)$]. We will explore in this chapter some of the ways that this relationship can be exploited to identify diffuse molecular gas and determine its properties. While we will assume that the gas-to-dust ratio in the Milky Way is 100, it is always worthwhile to remember that there is a large scatter in the observed values. In Spitzer's classic book *Physical Processes in the Interstellar Medium* (1978) he quotes values for the dust-to-gas ratio ranging from 1/20 to 1/700. There is no question that the dust can be used to find the gas, but as far as getting the gas mass by observing dust, your mileage may vary.

6.2 Extinction and Color Excess

Dust grains absorb, re-emit, scatter, and reflect light, and also radiate without incident electromagnetic radiation. In this chapter we will touch only on those basic considerations that are important for studying diffuse molecular gas in the ISM. We begin by discussing how dust absorbs primarily UV and optical light (scattering is usually included in the extinction since it removes light from the observed line of sight). Extinction by dust depends on wavelength.

In the visible and UV, the problem of passing photons through the dusty interstellar medium is rather simple. The grain density is so low that the light is produced only by a distant source whose intensity, I_ν, is known independently from its spectrum (or inferred after inverting the transfer process). There is no diffuse radiation to complicate the problem, unlike a planetary or stellar atmosphere where the source function from scattering is J_ν, the mean intensity. Even if multiple scattering is required, the scatterers are so few in number density along a line of sight that their individual contributions to the total optical depth add linearly and the problem is one of pure extinction. The infrared is a different matter. In this spectral interval the grains contribute to the photon pool because they have an intrinsic thermal emission. But the grain opacity at long wavelengths is sufficiently small relative to the visible and ultraviolet, and the line of sight column densities low enough, that any extinction is handled similarly to that at shorter wavelengths. Only in dense clouds does the necessity of treating the full transfer problem arise. Diffraction effects complicate matters relative to atoms and molecules because they are wavelength—size—dependent, whatever the intrinsic opacity of grains, and dominate the extinction. On the other hand, this difficulty affects only the treatment of the opacity, not the general solution of the transfer problem.

Exactly as we saw in Chap. 2 for a simple obscuring medium with no internal sources, consider a ray of light passing through a medium with extinction coefficient, $\kappa_{\nu,d}$. The ray will be reduced in intensity from its value at the source, $I_{\nu,0}$, by a quantity dI_ν, according to the pathlength, dl, in the obscuring medium, $dI_\nu = -\kappa_{\nu,d} I_{\nu,0} dN_d$, where N_d is the total dust column density. The optical depth, τ_ν, is defined in the usual way, $d\tau_\nu = \kappa_{\nu,d} n_d dl$ for number density n_d so integrating along the line of sight gives

$$I_\nu = I_{\nu,0} e^{-\tau_\nu} \quad . \tag{6.1}$$

For an unobscured source of light, the difference between the apparent and absolute brightness, in any band, expressed in magnitudes, is called the *distance modulus*, $m - M$, and is defined as

$$m - M \equiv -2.5 \log \frac{F(d)}{F(10pc)} \tag{6.2}$$

where $F(d)$ is the flux density or the flux (integrating over the point spread function of the detector-telescope combination for a point source at distance d). To normalize the magnitude system, the quantity $F(10pc)$ is the flux of the source when placed at a distance of 10 pc from the observer in the absence of obscuration. This is an arbitrary, but conventional, choice for absolute magnitude, as is the zero point of the magnitude system [see Sterken and Manfroid (1992) and Bessell (2005)]. With extinction the flux is reduced by a factor of $\exp(-\tau)$ so the distance modulus becomes

$$m - M = 5 \log d - 5 - 2.5 \log(e^{-\tau}). \tag{6.3}$$

As we noted, the extinction depends on wavelength so since $-2.5 \log[\exp(-\tau)] = 1.086\tau$, and calling the extinction at some wavelength A_λ, eq. (6.3) becomes

$$m_\lambda - M_\lambda = 5 \log d - 5 \; + \; A_\lambda \quad . \tag{6.4}$$

Here A_λ is measured in magnitudes and is produced by the whole intervening medium between the source and the observer. Thus, by the formal definition of magnitudes, dust optical depth unity is equivalent to 1.086 magnitudes of extinction.[1] If the volume number density of spherical grains is n_D, then the column number density of grains within a column of length L is $N_D = n_D L$. If the grains have an extinction cross-section, C_{ext}, then

$$\tau_\lambda = N_d C_{ext} \ . \tag{6.5}$$

so extinction is then directly proportional to the *total* dust column density along the line of sight. We will relate C_{ext} to some basic grain properties in Sect. 6.2.1. Note that this is rather different from the line transfer problem since here the dust absorption does not saturate.

The wavelength dependence of the extinction is quantified by the *color excess* (CE), or "reddening". The scattering and absorption contributions to the dust opacity are not distinguished in the definition, which is purely photometric and usually expressed in standard photometric systems (e.g. Johnson UBV). As we will discuss below, blue light is more extinguished than red light so the color excess, defined in magnitudes is

$$E(B-V) = (B-V) - (B-V)_o \quad , \tag{6.6}$$

where $(B-V)_0$ is the intrinsic color. Studies of nearby stars have shown that the ratio of the visual extinction to the color excess, called R, is roughly constant in the diffuse ISM

$$R = \frac{A_V}{E(B-V)} \approx 3.1 \quad . \tag{6.7}$$

Although this value holds for the vast majority of sightlines in the Galaxy, there are striking exceptions and R has a *range* of about 50% (e.g. Fitzpatrick and Massa 2007). The difficulty in relating this quantity to grain properties comes from knowing the intrinsic colors of the source. Those rely on spectroscopic calibrations of the photometric systems, based on classification, or on a model for the flux distribution of the source. For early type stars (classes O, B, and A) this is relatively straightforward since their visual continua are rather simple. It becomes more uncertain for late type stars, such as spectral types K or M because it also depends on the metallicity when the energy distribution is dominated by photospheric absorption lines.

The weak wavelength dependence of extinction in the Johnson UBV system makes reddening vectors follow an almost universal slope, $E(U-B)/E(B-V) = 0.72$ (see Fig. 6.1). In the UBV plane, however, there is a middle portion (roughly corresponding to A to G stars) that is not monotonic. For earlier type stars there

---

[1] For example, A_V is the obscuration in magnitudes in the Johnson V filter with a bandpass of about 800 Å and an effective wavelength of 5500 Å . A_B is the extinction in the B filter with a similar bandpass and an effective wavelength of 4450 Å.

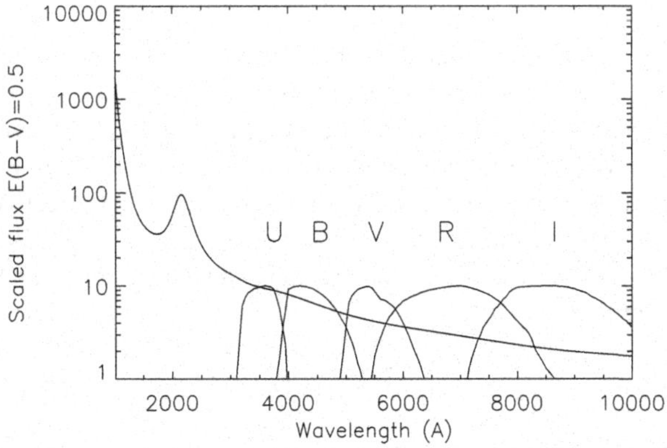

Fig. 6.1 The Galactic extinction law with superimposed filter profiles of the Johnson-Cousins photometric system. This figure shows the correction factor in flux, not magnitudes, for E(B-V)=0.5 and R=3.1. Note the strong extinction feature at 2175 Å, attributed to amorphous carbon, and the steep rise in the UV, attributed to small grains (see text for details)

is a combination that renders the colors independent of differential extinction. This is a single parameter, $Q = (U - B) - [E(U - B)/E(B - V)](B - V)$ from which the intrinsic colors (based on the optical properties of diffuse dust) can be derived (Bessell 2005). Similar indices can be formed for the Strömgren system (Sterken and Manfroid 1992). Because both extinction and color excess are proportional to the dust column density, they can be used to determine the gas-to-dust ratio, or to find regions of molecular gas (see Sect. 6.7).

6.2.1 A Bit of Grain Optics

The optical depth is a parameter that conflates a large number of grain properties. In this section we describe some of the basic relationships between τ_λ and physical grain properties.[2] We begin by considering a spherical grain of radius a. The

[2]The theory of scattering of electromagnetic waves from extended, isolated dielectric solids with a variety of shapes—spheres, spheroids, and cylinders—was first developed by Mie (1908) soon after the foundational work on electromagnetic theory by Maxwell, Helmholtz, and Hertz. It is masterfully presented by van de Hulst (1957), Kerker (1969), and Bohren and Huffman (1983) in their books on light scattering by small particles. The applications are very broad. The analogies between the engineering problem of antennas and the beginnings of radar meteorology and radar detection, involving nonspherical dielectrics and conductors led to parallel developments of the theory between engineering and astrophysics. A prime example of this synergy is the review by Greenberg (1968). Much of the work has proceeded in parallel between requirements of atmospheric scientists and astrophysicists. The radar work concentrates on understand the scattering properties of ices. The astronomers need this too but also concern themselves with the absorption and re-emission issues. The reader is urged to consult specific journals, such as *Journal*

geometric cross section is obviously given by $\sigma_{gr} = \pi a^2$ but it is the effective cross section that the grain presents to an incoming electromagnetic wave that is the more relevant quantity. This is parametrized by the dimensionless extinction efficiency of the grain, Q_{ext}, which depends on a, the wavelength of the light interacting with the grain, λ, and the index of refraction of the grain, n, which consists of both a real and an imaginary part: $n = n_r - in_i$. If n_r is large, the grain is an effective scatterer of light, if n_i is large, then it is an effective absorber.

With these considerations, the optical depth produced by dust grains along a line of sight can be written

$$\tau_\lambda = N_{gr}\sigma_{gr}Q_{ext} \tag{6.8}$$

where N_{gr} is the dust grain column density. Q_{ext} can be broken up into a scattering and an absorption part:

$$Q_{ext} = Q_{scat} + Q_{abs} \quad . \tag{6.9}$$

There is more than one way to extinguish light toward an observer. A scatterer, whatever its geometry, has an effective radius that depends on its composition but also has an irreducible diffractive contribution. Absorption is different, it is a loss mechanism for the light beam and a heating mechanism for the target. It cannot be treated as a purely geometric phenomenon. But the two aspects, scattering and absorption, are connected by the word "effective". The geometric diffractive effects are modified by the frequency (time) dependent interaction of the electromagnetic field with the material. Consider a quasi-static field, **E**, directed along the symmetry axis of a sphere with dielectric constant ϵ. The solution satisfying the boundary conditions for an external field that varies as $E_{ext} = E\cos\theta$ and an internal field D that must be regular at the center of the sphere. The external field $D_{ext} = \epsilon_0 E_{ext}$. Solving for the condition that there is no free surface charge (just an induced dipole), the polarization of a sphere of radius a is

$$P = \frac{4\pi}{3}a^3\frac{(\epsilon/\epsilon_0) - 1}{(\epsilon/\epsilon_0) + 2} \quad , \tag{6.10}$$

well known in the static limit as the Clausius-Mossotti equation [see Stratton (1941), Panofsky and Phillips (1962), Jackson (1962)]. But this should also hold for very slowly varying fields, at the low frequency limit or for wavelengths that are very large compared to the size of the particle. To pass to a measurable optical property, note that $\epsilon/\epsilon_0 = n^2$. The refractive index is a complex number because the imaginary part encodes the absorption (dissipation).

---

of Quantitative Spectroscopy and Radiative Transfer, *The Journal of the Atmospheric Sciences*, and *Journal of the Optical Society of America* for both ideas and the latest computational developments.

Full scattering (absorption) can be computed only by including all geometric constraints on the fields. This is the basis of Mie scattering theory (see, e.g., van de Hulst 1957). The scattering and absorption coefficients are not independent, which greatly reduces the complexity of the extinction computation. The connection comes through the Kramers-Kronig relations (we discuss this further in Appendix C). Strongly scattering particles can be subdivided into two categories, depending on their radius relative to the wavelength of the light, $x \equiv 2\pi a/\lambda = ka$. The geometric radius a is defined either by the particle's area or volume. It is, however, not as simple as it might seem to arrive at this value.[3] Real grains are not symmetric, geometrically ideal, filled things. Having different origins and histories, individual grains are more probably somewhere between fluff and billiard balls. Even aside from the size distribution of the solid constituent of the ISM, which we discuss below (Sect. 6.6.4), grain structure depends on the formation process in complicated, not completely understood, ways. A grain is an aggregate structure. Crystalline solids have definite symmetries and, therefore, specific bands in which they resonate. This renders the comparison between transmitted light and laboratory samples difficult. For a filled uniform sphere, for instance, A is numerically proportional to the square root of the area. But for a fluffy, possibly fractal aggregate, it may have a different dimensional relation to the area, a property called *fractal* (see Mandelbrot 1977; Barnsley 1993, e.g.).

The long-wavelength limit is known as Rayleigh scattering. For a classically approximated dielectric sphere with no absorption, Q_{scat} is

$$Q_{scat} = \frac{8}{3}x^4 \left|\frac{n^2-1}{n^2+2}\right| \tag{6.11}$$

Since for this case there is no absorption and the frequency is assumed to be far from the resonance, n is real and $Q_{scat} \propto \lambda^{-4}$. The scattering is strongly polarized since the phase function is that of a dipole, $(1+\cos^2\theta)$, for an angle θ between the source and observer's line of sight, so the degree of polarization is $p = \sin^2\theta/(1+\cos^2\theta)$, which gives the strongest effect at $\theta = \pi/2$. You will recognize this as the explanation for blue skies (Chandrasekhar 1950; see also Shore 2002). The short wavelength case corresponds to Mie scattering. The behavior of Q_{scat} is now an oscillatory function of x and approaches 2 for x very large. Finally, if $Q_{abs} > Q_{scat}$, the absorption term is the imaginary part of the index of refraction and

$$Q_{abs} = -4x\, Im\left(\frac{n^2-1}{n^2+2}\right) \tag{6.12}$$

so that $Q_{abs} \propto \lambda^{-1}$. This rather nicely approximates the behavior of interstellar extinction in the UV to near-infrared regions of the spectrum.

---

[3]While for atomic absorption the elementary cross section is σ_e, the Thomson cross section, there is no such scaling value for a grain other than a shape parameter for solids.

6.3 The Correlation Between Gas and Dust

Gas and dust in the ISM are usually well-mixed. This can be seen immediately by comparing the HI distribution of the 21 cm line with extinction (Lilley 1955, Heiles 1967), or by examining the CO(1-0) line intensity and infrared emission from dust grains (e.g., figure 9 of Dame et al. 1987). However, the first quantitative measurements of the relation of atomic and molecular hydrogen with E(B-V) to the stars in question were by Savage et al. 1977, and Bohlin et al. 1978. These authors used ultraviolet absorption-line spectroscopy of reddened stars within 1 kpc of the Sun to establish a direct relationship between the column density of hydrogen nucleons [N_H = N(HI) + 2N(H_2)] and the measured color excess to the stars in question. Their empirical relationship is

$$N_H/E(B-V) = 5.8 \times 10^{21} \quad cm^{-2}\, mag^{-1} \tag{6.13}$$

or, rewriting using equation 6.7:

$$N_H/A_V = 1.9 \times 10^{21} \quad cm^{-2}\, mag^{-1}\ . \tag{6.14}$$

The typical scatter about the mean line defined by equations 6.13 and 6.14 is ∼30% for E(B-V) > 0.10 mag. However, recent work by Liszt (2014*a*) indicates that equation 6.13 may not be the most accurate relationship between hydrogen and extinction if only the atomic hydrogen is considered. He derived

$$N(HI) = 8.3 \times 10^{21} E(B-V) \quad cm^{-2} \tag{6.15}$$

for regions with E(B-V) ≤ 0.1 and at $|b| \geq 20°$. In a subsequent paper, Liszt (2014*b*) noted that an inflection in the N(HI)/E(B-V) relation occurring at E(B-V) ≥ 0.08 mag is caused by the formation of significant quantities of H_2. Thus, the traditional reddening for the onset of H_2 and the detection of weak emission of molecular lines is at a reddening of ∼ 0.1 mag and not the more conventional threshold of 0.3 mag (equivalent to A_V ∼ 1 mag).

The above relationships are the standard way to derive a gas column density from an estimate of E(B-V) or A_V. However, even ignoring variations in the gas-to-dust ratio (which can be quite sizeable; see Spitzer 1978), one has in hand only N_H; to break this quantity down into its constituent atomic and molecular components is far from trivial. We will examine this problem in detail in Chap. 8.

6.4 Diffuse Reflection Nebulae

The scattering of visible light from dusty clouds located behind bright stars produces the well-known *reflection nebulae*, examples of which are the nebulosity around the Pleiades (Fig. 6.2), IC 2118 (the Witch Head nebula), and M78. Because of the wavelength dependence of the scattering cross section, reflection nebulae tend to be bluish. The proximity of bright stars to the dust which produces the reflection can also give rise to reddish emission from the heated gas [i.e., M20 (the Trifid nebula)] or photoluminescence of some of the carbonaceous grains. This latter emission is sometimes known as Extended Red Emission (ERE) and consists of a featureless emission band between 6100–8200 Å (Szomoru and Guhathakurta 1998). Thus, reflection nebulae can also be seen in the red even if there are no nearby bright stars. Reflection nebulae were cataloged by Lynds (1965) and the vast majority of regions with $|b| \geq 20°$ are seen in both the blue and the red POSS plates. There are many more of these Lynds Bright Nebulae (LBN) at high Galactic latitude than there are Lynds dark nebulae [that catalog is Lynds (1962)].[4] Both van den Bergh (1966) and Sandage (1976) remarked that some of the bright nebulae at high latitudes could be dust clouds reflecting the integrated star light of the Galactic plane below them. An example of this can be seen in Figs. 3.9 and 3.10.

6.5 Diffuse Galactic Ultraviolet Emission

Diffuse Galactic emission is expected because grain extinction in the ultraviolet has a scattering component. The observation is complicated by Solar System (zodiacal) dust scattering of sunlight and fluorescent emission from molecular hydrogen, but a surface brightness at 1350–2850 Å of 250–550 photons cm^{-2} sr^{-1} s^{-1}Å^{-1}, has been reported by Sujatha et al. 2009 using the GALEX satellite. The object they studied, known as the Sandage nebulosity, is the high-latitude molecular cloud MBM 30. Scattering by grains in the diffuse medium surrounding bright early type stars has also been reported by Murthy and Henry (2011). They detect halos up to 5° from the stars at 1521 Å and 2320 Å. This is small angle scattering compared to the Sandage nebulosity, but the derived dust optical properties are similar with a phase function asymmetry > 0.5. A similar detection was reported in Ophiuchus by Lee et al. (2008) and Hamden et al. (2013) have mapped the diffuse emission, proposing that there is a detectable scattering component. Murthy (2016) performed Monte Carlo scattering simulations of this emission, finding that it can be explained with a relatively small number of early type stars (about 1000) within less than 600 pc of the Sun based on the Hipparcos sample. The best fitting models have a reflectivity of between 0.3 and 0.6 (the grain albedo) and an asymmetry of 0.6. Consequently, the question is not whether such scattering exists but what this says about the grain properties in the UV and how it relates to heating. Although this is a comparatively minor contributor to the overall energy budget of the gas in the clouds, the observations constrain dust properties and radiative transfer modeling that may be important for the future observations of early stages of galactic evolution when higher redshift systems are studied in the near future with JWST. Scattering

[4]The dark nebulae are denoted L followed by a number or, sometimes, LDN followed by a number.

has also been detected around X-ray sources, for small angle scattering [see van de Hulst (1957), Martin (1978), and Bohren and Huffman (1983)]. Hoffmann and Draine (2016) noted that photoelectric absorption affects the extinction of background starlight within the grains and must be included in any determination of interstellar abundances from X-ray diffuse gas absorption edges.

6.6 Diffuse Galactic Infrared Emission

The dust grains that produce the obscuration and reddening described above are bathed in the interstellar ultraviolet radiation field produced by the ambient stars. The grains then re-emit the radiation in the ten—several hundred micron range producing a pervasive galactic background of diffuse infrared radiation. Thermal continuum emission comes from normal-sized grains ($\geq$ 100 Å or 0.01 μm) in thermal equilibrium with the local radiation field produced by stars. Typical dust temperatures range from 15–40 K and the resulting emission dominates at wavelengths greater than 30 μm. A component from very small grains ($\sim$ 5–50 Å in size) was identified by IRAS at 12 and 25 μm and arises from the non-thermal heating of these grains by UV photons [first mentioned by Sellgren (1984) at shorter IR wavelengths]. These grains are not in thermal equilibrium and have temperatures ranging from hundreds to thousands of K.

The first whole-sky infrared survey was made in the early 1980s by IRAS (Sect. 5.3.1). The satellite surveyed the sky at 12, 25, 60, and 100 μm and revealed the presence of extensive infrared radiation from the Galaxy (Hauser et al. 1984, see figure 1.2). In addition to the expected infrared emission from dust along the Galactic plane, a surprising amount of low-intensity 60 and 100 μm radiation was detected at high Galactic latitudes and, because of its wispy morphology, was dubbed the infrared "cirrus" (Low et al. 1984—see §5.3.1).

Later satellite missions such as COBE and WMAP produced better-calibrated, longer wavelength surveys (although not at the resolution of IRAS—see Sect. 5.3.2). These led to dust maps combining IRAS data, calibration, and temperature-correction using the Diffuse Infrared Background Experiment (DIRBE) on the COBE satellite (Schlegel et al. 1998—see §6.7.1) This latter survey can play a key role in identifying high-latitude molecular gas, as we will explore in Chap. 8.

If the grains are diffusely illuminated by a composite interstellar radiation field, they can reach thermal equilibrium. Since they are very large aggregates compared to, say, PAHs, the number of possible modes over which the energy is internally distributed after absorption is large enough to approximate a solid in thermal equilibrium and radiative balance. Therefore, the emissivity will follow a Planck distribution, $B_\nu(T)$ at some temperature T_d. The emissivity, j_ν, is obtained at each frequency from the Kirchhoff law,

$$j_\nu = \kappa_\nu B_\nu(T_d) \quad . \tag{6.16}$$

In radiative balance, the condition is that the bolometric luminosity of the grain balances the total absorbed flux. The grain properties enter here through κ_ν according to

$$\int_0^\infty F_{\lambda,DIRF} Q_{\lambda,abs}(a) d\lambda = \int_0^\infty Q_{\lambda,abs} B_\lambda(T) d\lambda \tag{6.17}$$

where T is the solution of this radiative balance. Notice that the particular optical model for the grain is characterized by the factor a/λ and for Mie scattering this is approximately a power law, $Q_{\lambda,abs} \sim (a/\lambda)^n$, so calling the left hand side of equation 6.16 Γ_{rad}, we obtain

$$T \sim \Gamma^{1/(4+n)} \tag{6.18}$$

which is proportional to the efficiency of absorption in the ultraviolet and visible. The emission is weighted to the infrared so two different segments of the extinction curve contribute to this balance. The 2175 Å feature of carbonaceous grains, coupled to the absence of such resonances in the infrared, lead to higher temperatures (efficient absorption, inefficient emission). In contrast, silicates lack the UV resonance but have strong features in the infrared at 10 μm and 20 μm (inefficient heating, efficient cooling) so they reach lower equilibrium temperatures. These two species of grains have similar visible band scattering properties (aside from the size distribution) but the temperature difference in the diffuse medium is a tracer of the grain properties in the far infrared.

The small grains and PAHs differ in that they never emit in equilibrium (see Tielens 2009). Photon absorption is sporadic and there are few modes for redistribution. Consequently, these particles behave more like very large molecules than solids and the emissivity is not a Planck function. The process is sporadic fluorescence, as discussed in Chap. 2: UV DIRF photons are absorbed, resulting in a cascading down-conversion of the energy through a large number of lines. The characteristic bands seen in the near and middle infrared, between 1 and 20 μm, attest to this in a variety of candidate species. The bands seen in this region are generic but not from solid state resonances, unlike the UV and IR broad bands of carbonaceous and silicate grains. Although the emitted spectrum can be described by a color temperature, this is not from a system in thermal equilibrium and is not the same as a kinetic or excitation temperature.

6.6.1 Light Echoes, Halos, and Reflection Nebulae

A star seen through dust in the diffuse medium behaves much like the Sun seen through a thin mist. Aerosols scatter light whether they are grains or droplets and many of the effects seen in the atmosphere have important analogs in the interstellar medium. One familiar example is the blue misty, filamentary emission

Fig. 6.2 The reflection nebulosity from the Pleiades (M45). Image credit: NASA

around the stars in the Pleiades (see Fig. 6.2). The dust lies both in front of and behind the cluster, but reconstructions seem to be more consistent with reflection from foreground, inclined sheets (Gibson and Nordsieck 2003). The reflection is, however, from steady sources and the individual contributors can, in theory, be identified from their spectra in the reflected light. Sudden, short lived events are, however, richer in information. For instance, just as lightning at large distance, sometimes called *heat lightning* when seen in summer thunderstorms, is a reflection from clouds along the line of sight, a nova or supernova produces a burst of light that scatters toward the observer along many lines of sight. Since the event is a time-limited pulse, the reflection is seen along directions of constant travel time to the observer and, therefore, will surround the source but be delayed. These are called *light echoes*. The first such event was registered at the start of modern astrophysics when rings appeared around the site of the classical nova GK Per 1901 (Couderc 1939). The explanation for these features, that seemed to expand at the speed of light (so they couldn't be dynamical) set the paradigm for interpretation of the rings around SN 1987A in the LMC (see Fig. 6.3). Supernova 1987A in the Large Magellanic Cloud is the most studied such event, being detectable for more than 5 years after the event in the visible and UV. Matching the observed spectra of the supernova with those obtained from the echo showed the time delay quite clearly (Crotts et al. 1995). Rest et al. (2008) found halos around two historical supernovae, Tycho 1572 and Cas A (around 1670) for which subsequent observations obtained the spectrum of both supernovae (SN 1572, Krause et al. 2008*a*, and Cas A, Krause et al. 2008*b*). These were observed spectroscopically more than three *centuries* after the event. While the surface brightness is very low, unresolved SN environments would produce substantial contributions in the late time observations. A distant example is SN 2014J in M82 (Yang et al. 2015), and others, e.g. SN 1991T, SN 1998bu, and SN 1995E have been associated with echoes.

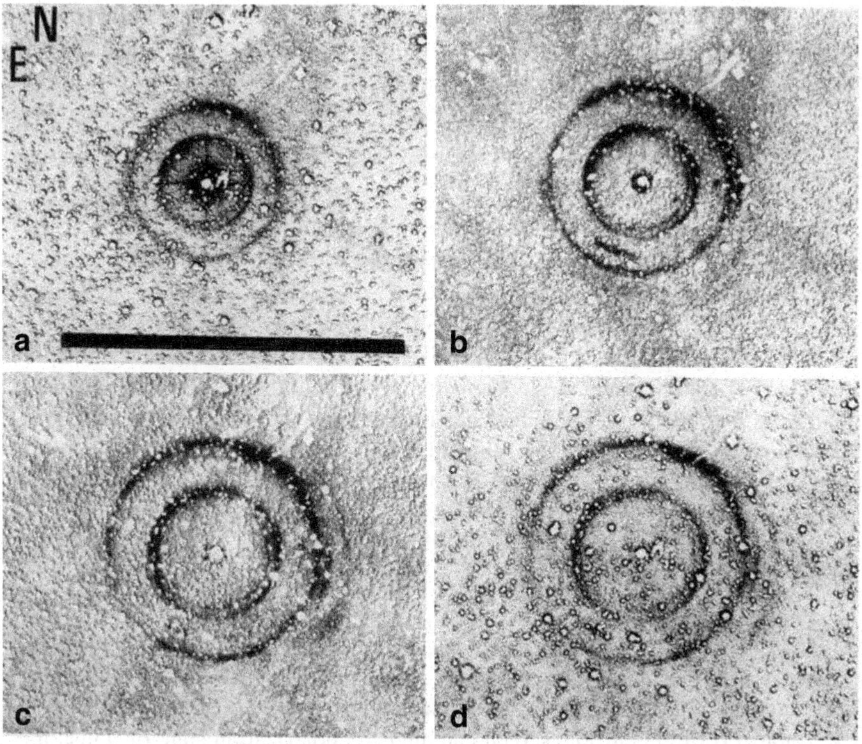

Fig. 6.3 Light echoes from SN 1987A. Panels (**a**)–(**d**) are 5′ fields taken 420, 626, 826, and 886 days after the event and are produced by two sheets of material in front of the supernova. Image from paper by Spyromilio et al. 1995

The properties that produce extinction make the dust an efficient scatterer, the key process in echo formation. Thus, an optical model for the grains obtained from visible light measurements toward stars can be used directly to treat echoes since it is derived from the diffuse gas, and not from a circumstellar environment. The grains also produce polarization independent of their alignment properties since the light is scattered at potentially large angles. Note that the photometry is altered by the optical properties of the grains, being bluer than that transmitted, but will pass through essentially the same line of sight as that of the event that produced the echo in the first place.

Halos are not, however, limited to visible wavelengths. Observations of bright Galactic X-ray sources shows that these too are surrounded by halos that, when the sources are observed in outburst, show the same echo phenomenon. The difficulty in treating the light propagation, however, comes from the near transparency of grains at high energies. They have ionization edges, depending on their heavy element composition, but they also are geometrically quite large relative to the wavelength of the light and, consequently, normal Mie scattering theory fails. An alternative,

Rayleigh-Gans theory, applies when the index of refraction is nearly unity [see e.g., Chap. 7 of van de Hulst (1957)]. It takes a modified form for the Clausius-Mosatti relation for the polarizability

$$C_{sca} \sim \frac{8\pi}{3} k^4 |n - 1|(1 + \cos^2 \theta) \tag{6.19}$$

and the Kramers-Kronig relation is applied to the absorption resonance spectrum to obtain the transmission. The grain scattering has the same angular dependence as Rayleigh scattering from infinitesimal dipoles—mainly forward scattering (where θ is the angle relative to the line of sight)—so the halos are quite narrow and strongly dependent on wavelength. However, this all assumes no resonances in the absorption coefficient. For more realistic grains, there are continuum edges and these must be included in the modeling. Recent studies indicate that for lower energies, below 10 keV, grain optics may still require Mie, rather than Rayleigh-Gans, modeling (Smith and Dwek 1998; Corrales and Paerels 2015). Multiple scattering is also possible, depending on the line of sight extinction.

6.6.2 *Polarization*

Any scattering process produces a polarized signature because of its coherence but only some sort of anisotropic screen can alter the polarization state of transmitted light. The evidence that grains are large aggregates is their polarization. This depends on grain composition and its environment. We will not go into the details of the alignment mechanisms. For comprehensive overviews see, e.g., Draine (2003) and Andersson et al. (2015). But the very presence of the effect in *transmitted* light indicates a significant ordering of the grain axes along sight lines in the diffuse gas. Far infrared measurements, from WMAP and more recently from *Planck*, highlight the connection. In transmission, the polarization is perpendicular to the symmetry axis of the grain, for emission it is parallel. Therefore, an important diagnostic of grain geometry is the comparison toward background OB stars, seen in transmission, between the linear polarizations in the optical and far infrared-submillimeter spectral intervals.

Wavelength dependent optical polarization in the diffuse gas obeys a broadly applicable empirical relation, called the *Serkowski law* (Serkowski 1973):

$$\frac{p}{p_{max}} = \exp[-K \ln^2 \left(\frac{\lambda_{max}}{\lambda}\right)] \tag{6.20}$$

where $p_/p_{max}$ is the peak polarization at peak wavelength λ_{max} for any line of sight. This is in the visible part of the spectrum *only*, no comparable law is available for other spectral ranges. The fit is only a representation, requiring three parameters when K is a constant (Whittet 1996). This representation has, however, been extended to include a functional form for $K = c_0\lambda_{max} + c_1$ where the c_j values are constants.

In the infrared, the situation is somewhat different. The polarization law is simpler, $p/p_{max} \sim (\lambda/\lambda_{max})^{\beta}$ where $\beta \approx 1.5 - 2$ (Martin and Whittet 1990; Whittet 2004). In the far infrared, dust thermally and nonspherical grains emit polarized light when aligned in ambient magnetic fields, whatever the mechanism. The *Planck* results are particularly important. The High Frequency Instrument (HFI) uses paired bolometers at four frequencies from 100 GHz to 353 GHz to obtain the two Stokes parameters from which the polarization and position angle can be measured. Actually, what is measured is the polarized flux, not the percentage (hence the units are *Jy*) and the position angle (Planck Collaboration, Planck 2015, Paper XXI). For this FIR radiation the dust is optically thin so the light is unmodified in its passage through the diffuse medium and the polarization angle is the sum of the intensity weighted contributions along the viewing direction. The optical (transmission) results are heavily weighted toward regions of relatively low column densities while those in the FIR and mm are mainly contributed by the high column density, warmer dust. Emphasizing that only aspherical grains will produce any polarization in transmission, PAHs and very small grains, if nearly spherical, will only produce a polarized signature from scattering (reflection), not transmission at visible and near infrared wavelengths. In transmission, the DIBs—if associated with small grains, fullerenes, and perhaps PAHs—show no signature (Cox et al. 2011). Therefore, the correlations between selective (reddening) and total extinction and the polarization parameters (p_{max} and λ_{max}) are rather weak for $A_V > 10$ while, instead, there is a steep drop in p_{max} with increasing extinction for $A_V \leq 5$ (Whittet 2004).

The problem of a local measure, such as grain-induced polarization, being combined with an integral property, such as line of sight extinction, is like that of rotation measure (see Chap. 2). The observational challenge is illustrated in Fig. 6.4 that shows the *upper envelope* of p_{max} relative to the total extinction A_V. Including translucent clouds, the scatter is large but the upper envelope of the scatter plot is a linear relation for the two quantities but this breaks down for sufficiently large extinctions.

6.6.3 Anomalous Microwave Emission and Spinning Grains

A prediction of dust modeling is that in the diffuse medium the grains are charged. But even neutral grains can have electric dipole moments if they are sufficiently irregular. Then if the neutral grains are rotating they can radiate at low frequencies as spinning dipoles with an emissivity varying as $D_g^2\omega^4$, according to the Larmor law, where D_g is the grain dipole moment and ω is the dust rotation frequency

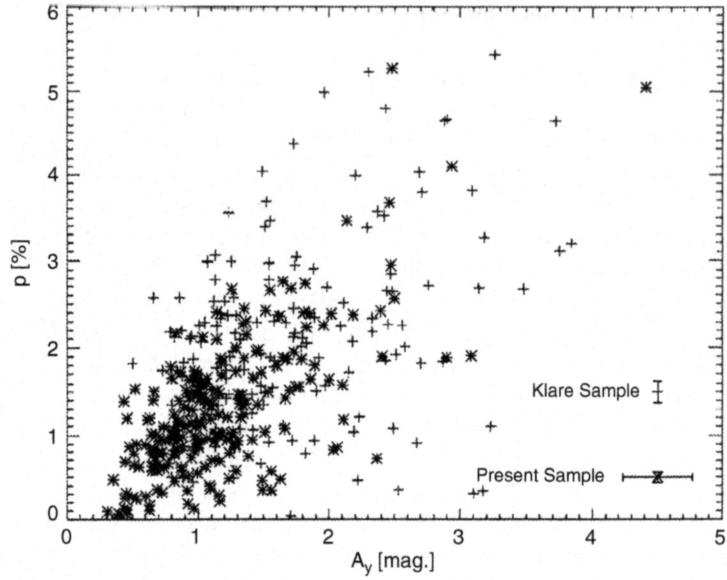

Fig. 6.4 Plot of the polarization as a function of visual extinction for data from the Southern Coalsack. *Asterisks* represent data from Andersson and Potter (2005) and *plus signs* from Klare et al. (1972). Representative error bars are shown in the lower right hand corner. Figure from Andersson and Potter (2005)

(e.g. Draine and Lazarian 1998). A prediction of the model is that the dipole emission is mainly from small grains, similar to PAHs and fullerenes in size. This implies a strong correlation of this low frequency radiation with regions showing fluorescent emission by stochastic, nonequilibrium heating of the small grains. The most recent results from the *Planck* satellite have greatly clarified the Galactic picture (Planck Collaboration, Planck Intermediate Results, Paper XV 2014; Planck 2015 results, Paper XXV 2015) The AME peaks in the 20–35 GHz spectral interval. Individual regions have, however, been found to have deviant spectra relative to the bulk sample. The HII regions, for instance NGC 1499, the California Nebula (associated with the O star ξ Per) show peaks as high as around 50 GHz. Tibbs et al. (2013) found that the AME in G159.6-18.5, powered by the OB star HD 278942, is correlated with IRAS 24 μm PAH-related emission on scales of $2'$, while Génova-Santos et al. (2011) find only a weak upper limit ($< 6\%$ at 12–18 GHz). Harper et al. (2015) detected the emission with the GBT at 13.7 GHz in the cloud LDN 1622 with a rising spectrum toward higher frequencies. The association of anomalous emission with weaker extinction (lower column densities) and the lack of polarization further supports the identification of its responsible agent as small grains.

6.6.4 Grain Size Distributions

A vexing problem in any population is its intrinsic distribution of fundamental properties, especially when the sample is heterogeneous. For grains, polarization measurements show that there is a range of sizes (one single *mean* grain size seems insufficient). Mathis et al. (1977) proposed a single power law distribution for the radii of spherical grains

$$\frac{dn(a)}{da} \sim a^{-3.5} \tag{6.21}$$

in the size range 50–2500 Å. Weingartner and Draine (2001) used, instead, a more complicated functional parametric fit and separated the carbonaceous and silicate components. For the silicates, they found a monotonic distribution with $dn/da > 0$ and an upper size cutoff at around 0.3 μm and virtually no grains smaller than around 10^{-3} μm. Carbonaceous grains were best modeled with several components, with peaks at around 5×10^{-4} μm, 5×10^{-3} μm, and a monotonic larger grain population with a maximum size of around 1 μm (see Fig. 6.5).

Determining the size distribution is complicated by the heterogeneity of the grain population. Regardless of the environment several distinct groups are present that in different combinations, produce the observed extinction curve. The ultraviolet is dominated by small grains and some combination of an amorphous phase of carbon, possibly including something resembling graphite (Weingartner and Draine 2001; Clayton et al. 2003) that contribute the 2175 Å resonance and the FUV rise. The infrared is dominated by large grains and silicates that produce the broad features at 10 μm and 18 μm and the relatively weak wavelength dependence longward of the B band. Several methods have been introduced to address the inherent complexity of separating the effects of composition and size, e.g. maximum entropy (Clayton et al. 2003, 2015). Principal component analysis (Sect. 4.3.2) has not yet been applied but this seems a natural statistical tool for decomposing the grain population.

A related problem to that posed by the size distribution is that for the grain masses. As for polarization and grain optics, the gas to dust ratio and the mass distribution both depend on assumptions about the dust structure. If the particles are large, filled spherical volumes then $m(a) \sim a^3$ so one distribution immediately supplies the other. But if grain properties depend on size and composition, whether they are fractal or irregular aggregates, then passing from the sizes (which come from the optics) to mass fractions is more conjectural. Something midway between filled and fractal, a sort of compromise structure is afforded by *porous* grains (e.g. Voshchinnikov et al. 2006). Unlike fractals, for which the density depends on size, this model uses holes within the aggregate to produce an effective dielectric constant and radius, like a dense irregular volume with vacancies. Porosity as high as 0.9 has been suggested for grains of a wide range of compositions (filled has a porosity of zero) with similar areas as those inferred from extinction but much lower masses and possibly different metal depletions from the diffuse medium.

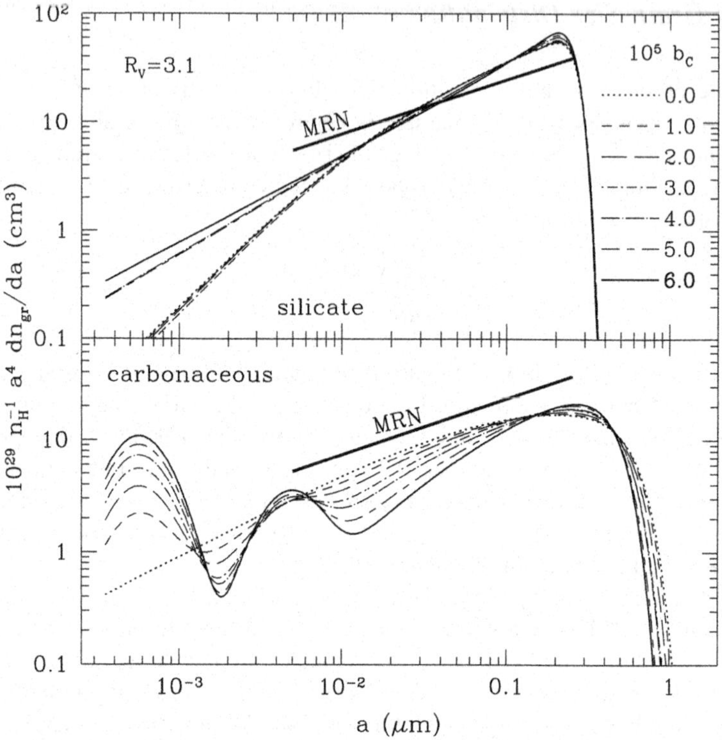

Fig. 6.5 Grain size distribution for $R_V = 3.1$ for silicate and carbonaceous. The parameter b_C is the total C abundance per hydrogen nucleus. The *heavy, solid lines* are the Mathis et al. (1977) distribution as a comparison. Figure from Weingartner and Draine 2001

6.7 Tracing Molecular Gas Using Dust Maps

If the gas and dust components of the ISM are well-mixed and the infrared emissivities of the dust associated with atomic and molecular hydrogen are the same, then it is straightforward to use dust tracers to find molecular gas: subtract the contribution to the dust opacity or column density due to HI (the latter easily determined with 21 cm observations) and the residual emission is likely due to molecular gas (if heating of the dust has been correctly accounted for). This technique has been used for a long time to find molecular gas, particularly at high galactic latitudes: de Vries et al. (1987); Heiles et al. (1988); Désert et al. (1988); Blitz et al. (1990); Reach et al. (1994); and Meyerdierks and Heithausen (1996). More recently, studies based on the IRAS and COBE data culminated in the catalog of infrared-excess clouds (i.e., clouds that had more infrared emission than the HI along the line of sight would have predicted) by Reach et al. (1998) and the Schlegel et al. (1998) dust maps which will be discussed below. Both works have led to many new detections of molecular clouds (e.g., Onishi et al. 2001; Magnani and Smith 2010).

The all-sky survey by the *Planck* satellite (see Sect. 5.2.1) represents a qualitative leap in this area of astrophysics because of its outstanding ability to trace dust emission and determine opacities. In particular, Planck Collaboration, Planck Early Results 2011, Paper XIX , combined *Planck* and IRAS data to determine the all-sky temperature and optical depth distributions of the dust component of the Galaxy at $|b| \geq 5°$. The results were compared to existing HI and CO data to constrain the dark gas in the Solar neighborhood (see further discussion in Sect. 8.4), but their map of the excess column density derived from 857 GHz data (their figure 8) is an outstanding indicator of diffuse local molecular gas. Nearby molecular clouds (e.g., the Taurus, RCrA, and Chamaeleon complexes) stand out as cold dust environments with dust temperatures $\approx$ 13 K.

6.7.1 The Schlegel, Finkbeiner, and Davis (SFD) Dust Maps

Before the *Planck* satellite, the best infrared database for identifying molecular gas was the Schlegel et al. (1998) E(B-V) data for the whole sky (we will refer to this database as the SFD dust maps). They combined the IRAS/ISSA maps without the zodiacal foreground and point sources, and then used the COBE/DIRBE data to produce a temperature-corrected 100 μm intensity map of the sky that is free of striping and accurately calibrated to the photometry of the DIRBE instrument. They calibrated their maps, whose flux is proportional to the dust column density, to E(B-V) by assuming a standard reddening law and using the colors of elliptical galaxies to determine the reddening per flux density of 100 μm emission. Although the color excess maps have a resolution of $\sim 6'$, the dust temperature maps have a resolution of one degree (because of the poorer DIRBE resolution compared to IRAS). Thus, the E(B-V) maps are according to SFD "a proxy to the average extinction along the line of sight, and not to the detailed extinction for fine structure in the interstellar medium." Nevertheless, they have proven to be significantly better than the IRAS 100 μm database for identifying regions with possible molecular content (e.g., Magnani and Smith 2010).

Schlafly and Finkbeiner (2011) examined measurements of dust reddening of stars from the Sloan Digital Sky Survey and suggest that a 14% re-calibration of the SFD dust maps, to lower extinctions, based on their analysis. That paper also provides a table of conversion coefficients from the SFD E(B-V) maps to extinction in 88 different bandpasses for four values of R_V. The recalibrated maps are used by the NASA/IPAC Extragalactic Database Coordinate and Extinction Calculator (http://ned.ipac.caltech.edu/forms/calculator.html) and are an excellent first estimate of the extinction in any given direction. The results are likely to be more reliable for high Galactic latitude sightlines where a single dust temperature can be used to describe the bulk of the dust that is absorbing and/or scattering the background

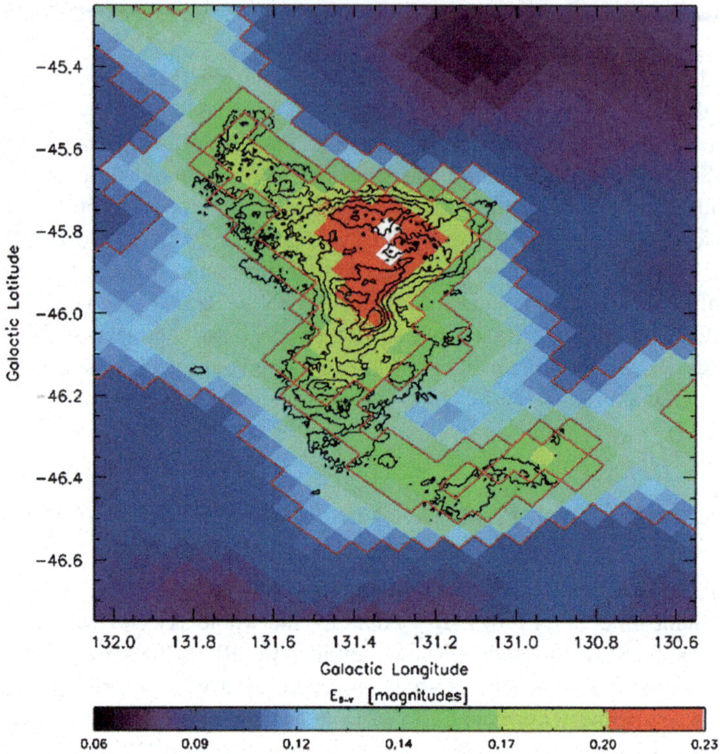

Fig. 6.6 The relationship between E(B-V) in magnitudes (see color scale at bottom of figure) and CO(1-0) emission in MBM 3. The *black contours* represent W_{CO} at levels of 2.5, 4.0, and 5.5 K km s^{-1}. The *red contours* are E(B-V) at levels of 0.10, 0.14. and 0.16 mag. Figure from Shore et al. (2006)

starlight. An example of the efficacy with which the SFD maps track CO(1-0) emission is shown in Fig. 6.6 for MBM 3.

6.8 The E(B-V) Molecular Threshold

That hydrogen nucleons and dust were well-mixed and co-spatial was known from 21 cm line observations and from using the CO(1-0) line as a surrogate for H_2. However, the optical depth of the CO(1-0) line makes estimates of N(H_2) tricky (see Chap. 8). In the diffuse ISM, direct detection of CO(1-0) lines was difficult in the 1970s so the first large-scale quantitative derivation of the relationship between E(B-V) and atomic and molecular hydrogen was accomplished in the UV by Savage et al. (1977) and Bohlin et al. (1978). These authors used the *Copernicus* satellite to derive N(H_2) in the direction of 109 early-type stars. The column density of H_2 was determined from (1,0) Lyman lines and N(HI) came from Lyman-α observations. the resulting relationship (Eq. 6.13) was the standard way to convert from a measurement of E(B-V) to a gas column density. Sometimes, even visual extinction, A_V was converted to N(H), or vice-versa, using the standard ratio of total to selective absorption, R. A similar result was found by Diplas and Savage (1994) using only N(HI). It is notable that in the 1977 paper, Savage et al state that the ratio of $N(H_2)/N(H)$ undergoes a transition from low (less than 0.01) to high (greater than 0.01) for E(B-V) values $\approx$ 0.08 mag, equivalent to $A_V \approx 0.25$ mag. This was not of much importance in the 1970s when millimeter radio receivers were not very good and detection of CO lines was difficult in regions with $A_V < 1$ mag. However, today it appears that regions where there is sufficient column density of CO to be detectable with radio techniques are indeed around E(B-V) $\sim$ 0.1 mag. For instance, Magnani and Smith (2010) and Cotten and Magnani (2013) have detected weak CO(1-0) emission from regions with color excesses as low as 0.08 mag according to the SFD dust maps. This question is very pertinent since several papers have appeared during the last decade citing the possibility that "dark" molecular gas exists. This is an unfortunate choice of adjective because the gas has nothing to do with dark matter, but only signifies that the gas is spectroscopically undetectable (Grenier et al. 2005). We will revisit this issue in Sect. 8.4. For the moment, we focus on the nature of the relationship between E(B-V) and molecular gas; in particular, whether a threshold can be empirically determined from observations of the CO(1-0), CH 3335 MHz, or the OH 18 cm main lines.

The SFD dust maps cover the entire sky and with a resolution of $6'$ are a good match in angular resolution for the CO(1-0) data from the Harvard/CfA 1.2 m millimeter-wave radiotelescope. In Chap. 9, two high-latitude ($|b| \geq 30°$) CO surveys made with this telescope are described in detail, but for now it suffices to note that the entire sky that culminates at least 30° above the horizon from Cambridge, MA (the location of the 1.2 m radiotelescope) was mapped at one-degree true angle spacing. This resulted in nearly 5000 pointings in the southern Galactic sky and nearly 11,000 in the north. Each pointing has an rms of 0.1 K and, with a beam $\sim 8'$, the survey matches fairly well with the SFD E(B-V) data. Cotten (2011) compared the two data sets and we discuss some of the results in light of establishing a threshold in E(B-V).

In Fig. 6.7 (top), we show a histogram of E(B-V) for the northern sky and for that portion of the southern sky observable from Cambridge, MA. Clearly, the bulk of the surveyed points have E(B-V) $\leq$ 0.6 mag; there is just not that much obscuration from dust at high Galactic latitudes. Each of the more than 15,000 points in the top of Fig. 6.7 were observed in the CO(1-0) transition as described above. Figure 6.7 (bottom) shows the detection rate of CO emission. As long as E(B-V) is less than 0.1 mag, the detection rate is negligible. However, when the color excess increases to 0.3 mag (equivalent to $\sim$ 1 mag of visual extinction), the detection rate for CO emission is around 25%. A reasonable question to ask is how would the statistics change if the rms threshold was lowered by a factor of 2 or 10. We make the case in Sect. 8.4.1 that deeper integrations in CO will reveal more CO emission and the prevalence of spectroscopically undetectable molecular gas is overstated.

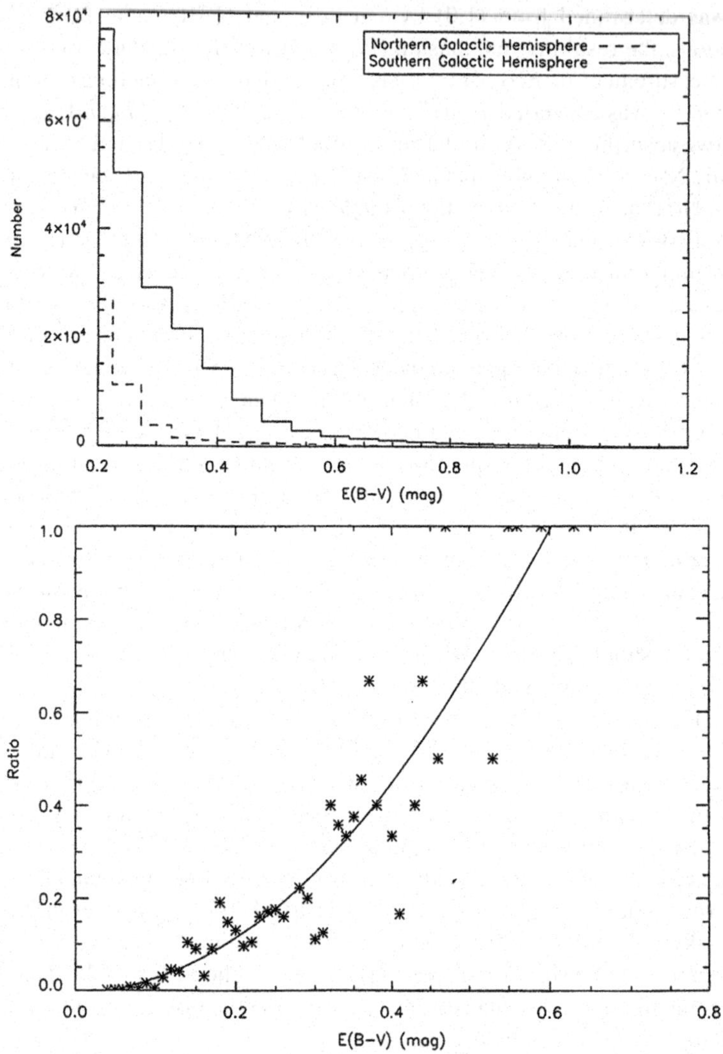

Fig. 6.7 The relationship between W_{CO} and E(B-V) for the high-latitude sky. The top figure is a histogram of E(B-V) as determined from the SFD dust maps for the nearly 11,000 points in the NGH and 5000 point in the SGH. The bottom figure shows the detection rate of CO(1-0) emission as a function of E(B-V) for the points observed in the NGH and SGH during the course of the Georgia-Harvard CfA high-latitude surveys (see Sect. 9.6). Details can be found in Cotten (2011)

In summary, E(B-V) is an excellent tracer of diffuse gas in the ISM and maps of the distribution of E(B-V) can be used to search for molecular gas with the understanding that there is no absolute threshold (at least for E(B-V) $\leq$ 0.6 mag where detection of molecular gas in emission by radio spectroscopic techniques is assured. The conditions in the ISM are varied enough that the depth of the PDR regions at which molecular gas becomes prevalent changes significantly from region to region. The interstellar radiation field is probably most responsible for this and because it is difficult to determine accurately from region to region, the E(B-V) threshold for detecting molecular gas by rotational and hyperfine spectral lines will range between 0.1–0.6 mag.

6.9 Ices and Their Signatures

Grains and ices are not quite the same thing. In the diffuse medium for $A_V \approx 1.6$, there is little evidence for absorption from coatings on the raw grain. Although the dust is the site of molecular formation and chemical processing, only darker clouds appear to accumulate sufficiently thick surface layers to produce a spectral signature seen against background sources (see Herbst 2014; van Dishoeck 2014; Boogert et al. 2015). Photo-desorption, the erosion of mantles when the grains are exposed to the UV of the DIRF, and cosmic ray sputtering appear to be the main agents. Grains in denser environments, such as cold cores, show a rich array of absorption features from different ices, mainly CO_2, water, methane, and ammonia. In a few lines of sight, notably toward star 12 of Cyg OB2, there are weak CO and CO_2 absorption bands that likely arise from denser clumps along the line of sight. Since the bands are far broader than even the DIBs, the profiles contain no dynamical information so any association with specific structures must be somehow inferred from the resolved line profiles in the same direction.

All this is especially important for the treatment of galactic evolution in cosmological simulations. The chemical processing of the diffuse medium affects the radiative cooling properties of the gas. If the heavy metals are depleted onto grain surfaces, and therefore cannot radiate by electron and molecular collisional excitation, the gas is inhibited in its cooling.

Ice absorption (and emission) features have been identified in translucent and star forming clouds. Since the bands are located in the middle infrared, at much higher frequencies than the CMB or Galactic nonthermal emission, the study of such spectral signatures is biased since only objects that have strong continua in the 5–20 μm range are suitable sources. This means we know far more about the dust mantles and compositions toward imbedded protostars than in the diffuse medium (Herbst 2014). But a few things are sufficiently common to be generic statements.

The principal ice related absorption bands are from H_2O (3 μm), NH_3 (3.47 μm), CO (4.76 μm), and CO_2 (15.2 μm) (Boogert et al. 2015) provide a comprehensive review of properties currently known for interstellar ices). A general evaluation is that water ices require $A_V > 1.6$ mag, so these absorption features should be observable in translucent clouds [see, in particular, the study of the sight line toward Cyg OB2 No. 12 by Whittet (2015)]. The ices are an admixture with CO_2, NH_3, and CH_4. Optical and near infrared absorption is weak but not unknown in diffuse lines of sight, possibly signalling the presence of smaller, denser structures in translucents that remain spatially unresolved but that contribute to the absorption lines (including the bands).

6.9.1 *Other Sources of Information: Ices in the Solar System and Laboratory*

When exposed to the combined effects of solar activity, the solar wind, and ultraviolet and X-rays from the corona and chromosphere, ices in the most local environment are heavily processed. For comets, for instance, the question was whether there are effects, other than thermal, that govern their spectrophotometric signatures and surface properties. This goes under the jargon *space weathering*.[5] An example is how the bodies outgas, and if they develop hard crusts from progressive charged particle bombardment. While this seems distant from the diffuse ISM, it is quite relevant for understanding how the grains chemically evolve. Large ice samples, at least compared to interstellar grains, can be placed in accelerators and exposed to a range of particle beams. But they can't be processed with the same spectrum as the cosmic rays of the ISM so the appearances of the surface layers may be different.

6.10 Gas to Dust Ratio from Cosmic Ray-Produced γ-Rays

Cosmic ray protons with energies below 1 GeV also provide an essential calibration of the dust-to-gas ratio. The protons have several interaction channels in their passage through matter. As for the electrons, their main losses are from ionization of the background. However, there is a threshold for π meson production at 1232 MeV. This is especially important for π^0 since these decay through $\pi^0 \rightarrow \gamma\gamma$ (Grenier et al. 2015). These photons were first detected in the MeV range by Compton scattering calorimeters such as Comptel on the *CGRO*. The conversion efficiency is weakly dependent on energy above threshold and the only unknown is the matter column density since these photons are nearly unattenuated by the intervening gas. The production rate, averaged over the proton energy spectrum, is not isotropic since these are particles confined to comparatively small distances from the Galactic plane, but they produce a diffuse γ-ray background proportional to the column density. This is the key to the calibration. Since dust is such a low number density constituent of the medium, the diffuse phase, the diffuse Galactic γ-ray background depends mainly on the gas column density. To be more specific, because of the interaction responsible for the γ-ray production, the Δ-resonance

[5]This term should not be confused with *space weather*, referring to the solar wind and its fluctuations induced by photospheric and coronal activity.

that produces neutral pions, protons are the target nucleus. But this is independent of whether these are bound in molecules or free hydrogen atoms or ions. Thus, with the dust optical depth, or extinction, the dust to gas ratio can be calibrated using a model for the dust optical properties as they depend on the grain composition and radius. This also gives the total dust column density, which cannot be directly obtained from the infrared since the emissivity is temperature dependent. In contrast, the γ-ray flux produced by CR π^0 is independent of the optical coefficients of the grains (Strong et al. 1988; Bloemen 1989). The importance of this calibration is that the extinction is more easily measured, at least differentially, than the total gas column density. The 21 cm hydrogen emission gives the neutral column density but it cannot yield the *total* gas fraction unless lines of sight can be observed in the far ultraviolet to obtain equivalent widths for the H_2 ground state lines.

Although easily outlined in broad terms, this estimate faces considerable difficulties in so inhomogeneous an environment as the ISM. Trapping of the lower energy protons and modulation by propagation effects produces depletion of the energy distribution below a few MeV. At higher energies, there are still problems in understanding the diffusion through smaller scale magnetic structures (on the sub-parsec level).

Tracing dust using γ-ray observations inverts the astrochemical argument. The intensity of the emission, due mainly to π^o decays following their production by cosmic ray protons colliding with interstellar gas (also protons), can be used to obtain the cosmic ray flux within the Galaxy. But even without a precise knowledge of the cosmic ray intensity, the emission itself should be correlated with the baryonic component of the Galaxy. Along any line of sight, the intensity I_γ is linearly proportional top the total proton column density (neutral and ionized atomic hydrogen and H_2). This intensity is then correlated with the dust emissivity maps and extinction measures to give the relation between E(B-V) (or A_V) and $N(H_{total})$.

An estimate of the cosmic ray intensity is supplied by H_3^+ in the diffuse medium since, as we discussed in Chap. 3, this reaction is uniquely driven by charge exchange with incident cosmic ray protons. But the ionization rate, ζ, still varies within the Galaxy and also depending on the environment. Padovani et al. (2009) argue that the penetration and trapping of CR protons in molecular clouds significantly alters the ionization and chemistry through a variety of channels not available in the diffuse medium. The intensity is the energy integrated product of the emission yield and the CR spectrum, so it becomes a single coefficient that depends on the assumed production mechanisms. Note that the cosmic rays responsible for the ionization state of the gas are far lower energy, a few tens of keV to about 1 MeV, than those producing $p(p,p')\pi^0 \rightarrow 2\gamma$, (above 10 MeV) but they are connected by the energy distribution function so they can eventually be cross calibrated. Thus, as a future application of *Fermi* observations of emission from the diffuse gas, the link can be established between the high energy γ emission and the cosmic ray ionization rate in different environments.

The alternate use of the diffuse background is to obtain a measure of the molecular fraction, hence the numerical conversion between H_2 column density and CO emissivity, X_{CO}. As Bolatto et al. (2013) note, other infrared calibrations relating the dust to gas ratio also link to I_γ. However, there is a caveat. Observing the Magellanic Clouds, Abdo et al. (2010) found only a poor link between the γ-ray intensity measured by *Fermi* and the 21 cm traced gas distribution. The variation in the conversion factor is also an indication of the low emissivity molecular gas that may be linked to dust but otherwise is faint in CO.

6.10.1 Extinction from Hydrogen Lyman Continuum Absorption Measurements

There is yet another way to obtain information about the column density of neutral hydrogen, and also heavy elements such as the CNO group, using X-ray observations. Although still in development (e.g., Gatuzz et al. 2016), using *XMM* and *Chandra* X-ray band spectroscopy and *Swift* X-ray and UV photometry work rather well in discriminating continuum absorption intrinsic to the source and that from the gas along the line of sight. The idea is quite straightforward. Absorption above the Lyman limit decreases rather slowly with energy so even in the soft X-ray spectral range the neutral gas has an optical depth $\tau(E) \approx \tau_{LyC}(E/13.6eV)^{-3}$. Studies of explosive events, such as classical novae, show that the minimum N_H obtained from the X-rays equals that inferred from the Lyα and other UV and optical resonance transitions. From spectrophotometry, the absorption edges of heavy elements can be studied independent of the line excitation conditions. Finally, using X-ray sources, mainly supernova remnants and low mass X-ray binaries observed in the continuum, Güver and Özel (2009) derived an alternate relation between atomic hydrogen column density and extinction,

$$N_H = (2.21 \pm 0.09) \times 10^{21} A_V \quad cm^{-2}. \tag{6.22}$$

The results from this method are very close to those discussed in Sect. 6.3.

6.10.2 Time Variability of the Diffuse Gas

Over the last decade, a time interval covered by similarly sensitive spectroscopic monitoring, indications of time variability have emerged for diffuse lines of sight in the ISM. The line of sight toward κ Vel, that has been especially well studied (e.g. Crawford 2002; Smith et al. 2013), displays detectable changes in CH, CaI, and CaII over decade timescales. The Carina nebula has also been observed over a long enough time that small scale changes have been observed (e.g. Danks et al. 2001). The issue of whether this is the motion of the background star relative to the gas, individual filaments moving across the line of sight, or both is still uncertain. But, in one sense, it does not matter; having structures on scales of less than 0.1

pc that are detectable along kiloparsec sight lines is sufficient whatever the relative motion. In another sense, however, it *is* important whether it is the stellar or gas motion that produces the changes. Small scale structure has been known for some time in expanding envelopes around star forming regions (e.g. for the "Orion Cloak" region, surrounding the Ori OB1 association (see Cowie et al. 1979; Shore 1982; Cantiello 2007), and even smaller scales are well known from molecular studies of translucent clouds (e.g. Magnani et al. 1993; Falgarone et al. 1998). The motion of these molecular cloud structures, corresponding to velocities of a few km s^{-1} at most, would produce only long timescale changes unless the medium is far denser and even more filamentary on smaller scales. Instead, if associated with larger dynamical structure, fragments will move across the line of sight at much higher speeds. Since it is enough that time pass for these motions to be detectable, future studies should distinguish the origin and significance of such variability and connect it with fluctuations in radio sources (such as giant radio burst events seen in galactic nuclei and interstellar scintillation).

References

Abdo, A.A. et al. 2010, A&A, 512, A7
Andersson, B.-G and Potter, S.B. 2005, MNRAS, 356, 1088
Andersson, B.-G, Lazarian, A., and Vaillancourt, J.E. 2015, ARAA, 53, 501
Barnsley, M.F. 1993, *Fractals Everywhere*, (Cambridge: Academic Press Professional)
Bessell, M.S. 2005, ARAA, 43, 293
Blitz, L., Bazell, D., and Desert, F.X. 1990, ApJ, 352, L13
Bloemen, H. 1989, ARAA, 27, 469
Bohlin, R.C., Savage, B.D., and Drake, J.F. 1978, ApJ, 224, 132
Bohren, C.F. and Huffman, D.R. 1983, *Absorption and scattering of light by small particles*, (New York: John Wiley & Sons)
Bolatto, A.D., Wolfire, M., and Leroy, A.K. 2013, ARAA, 51, 207
Boogert, A.C.A., Gerakines, P.A., Whittet, D.C.B. 2015, ARAA, 53, 541
Cantiello, M. 2007, Laurea MSc, University of Pisa
Chandrasekhar, S. 1950, *Radiative Transfer*, (London: Oxford University Press)
Clayton, G.C. et al. 2003, ApJ, 592, 947
Clayton, G.C. et al. 2015, ApJ, 815, 14
Corrales, L.R. and Paerels, F. 2015, MNRAS, 453, 1121
Cotten, D. 2011, Ph.D. Thesis, University of Georgia
Cotten, D.L. and Magnani, L. 2013, MNRAS, 436, 1152
Couderc, P. 1939, Annales d'Astrophysique, 2, 271
Cowie, L.L., Songaila, A., and York, D.G. 1979, ApJ, 230, 469
Cox, N.L.J. et al. 2011, A&A, 531, 25
Crawford, I.A. 2002, MNRAS, 334, 33
Crotts, A.P.S., Kunkel, W.E., and Heathcote, S.R. 1995, ApJ, 438, 724
Dame, T.M., et al. 1987, ApJ, 322, 706
Danks, A.C. et al. 2001, ApJ, 547, L155
Désert, F.X., Bazell, D., and Boulanger, F. 1988, ApJ, 333, 353
de Vries, H.W., Heithausen, A., and Thaddeus, P. 1987, ApJ, 319, 723
Diplas, A. and Savage, B.D. 1994, ApJ, 427, 274
Draine, B.T. and Lazarian, A. 1998, ApJ, 508, 157

Draine, B. T. 2003, ARAA, 41, 241
Draine, B.T. et al. 2007, ApJ, 663, 866
Falgarone, E. et al. 1998, A&A, 331, 669
Fitzpatrick, E.L. and Massa, D. 2007, ApJ, 663, 320
Gatuzz, E. et al. 2016, A&A, 588, A11
Génova-Santos, R., Rebolo, R., Rubiño-Martín, J.A., López-Caraballo, C.H., and Hildebrandt, S.R. 2011, ApJ, 743, 67
Gibson, S.J. and Nordsieck, K. H. 2003, ApJ, 589, 347
Greenberg, J.M. 1968, in *Stars and Stellar Systems, Vol. VII: Nebulae and Interstellar Matter*, eds. B. Middlehurst and L. Aller, (Chicago: University of Chicago Press), 221
Grenier, I.A., Black, J.H., and Strong, A.W. 2015, A&A, 53, 199
Grenier, I.A., Casandjian, J.-M., and Terrier, R. 2005, Science, 307, 1292
Güver, T. and Özel, F. 2009, MNRAS, 400, 2050
Hamden, E.T., Schiminovich, D., and Seibert, M. 2013, ApJ, 779, 180
Harper, S.E., Dickinson, C., and Cleary, K. 2015, MNRAS, 453, 3375
Hauser, M.G. et al. 1984, ApJ, 278, L15
Heiles, C. 1967, ApJ, 148, 299
Heiles, C., Reach, W.T., and Koo, B.-C. 1988, ApJ, 332, 313
Herbst, E. 2014, *Faraday Discuss.*, 168, 617
Hoffmann, J. and Draine, B.T. 2016, ApJ, 817, 139
Jackson, J.D. 1962, *Classical Electrodynamics*, (New York: John Wiley & Sons)
Kerker, M. 1969, *The scattering of light and other electromagnetic radiation*, (New York: Academic Press)
Klare, G., Neckel, Th., and Schnur, G. 1972, A&AS, 5, 293
Krause, O. et al. 2008*a*, *Nature*, 56, 617
Krause, O. et al. 2008*b*, *Science*, 320, 1195
Lee, D.-H., et al. 2008, ApJ, 686, 1155
Lilley, A.E. 1955, AJ, 60, 167
Liszt, H. 2014*a*, ApJ, 783, 17
Liszt, H. 2014*b*, ApJ, 780, 10
Low, F.J. et al. 1984, ApJ, 278, L19
Lynds, B.T. 1962, ApJS, 7, 1
Lynds, B.T. 1965, ApJS, 12, 163
Lynds, B.T. and Wickramasinghe, N.C. 1968, ARAA, 6, 215
Magnani, L., LaRosa, T.N., and Shore, S.N. 1993, ApJ, 402, 226
Magnani, L. and Smith, A.J. 2010, ApJ, 722, 1685
Mandelbrot, B.B. 1977, *The Fractal Geometry of Nature*, (San Francisco: Freeman)
Martin, P.G. 1978, *Cosmic dust, its impact on astronomy*, (Oxford: Clarendon Press)
Martin, P.G. and Whittet, D.C.B. 1990, ApJ, 357, 113
Mathis, J.S., Rumpl, W., and Nordsieck, K.H. 1977, ApJ, 217, 425
Mathis, J.S.1990, ARAA, 28, 37
Meyerdierks, H. and Heithausen, A. 1996, A&A, 313, 929
Mie, G. 1908, Annalen der Physik, 330 (3), 377
Murthy, J. and Henry, R.C. 2011, ApJ, 734, 13
Murthy, J. 2016, MNRAS, 459, 1710
Onishi, T. et al. 2001, PASJ, 53, 1017
Padovani, M., Galli, D., and Glassgold, A.E. 2009, A&A, 501, 619)
Panofsky, W.K.H. and Phillips, M. 1962, *Classical Electricity and Magnetism*, (New York: Addison-Wesley)
Planck Collaboration, Planck Early Results XIX, 2011, A&A, 536, A19
Planck Collaboration, Planck Intermediate Results XV, 2014, A&A, 565, A103
Planck Collaboration, Planck 2015 Results XXI, 2015, A&A, 576, A106
Planck Collaboration, Planck 2015 Results XXV, 2015, arxiv:1506.06660
Reach, W.T., Koo, B.-C., and Heiles, C. 1994, ApJ, 429, 672

Reach, W.T., Wall, W.F., and Odegard, N. 1998, ApJ, 507, 507
Rest, A. et al. 2008, ApJ, 681, L81
Sandage, A. 1976, AJ, 81, 954
Savage, B.D., Bohlin, R.C., Drake, J.F., and Budich, W. 1977, ApJ, 216, 291
Savage, B. D. and Mathis, J. S. 1979, ARAA, 17, 73
Schlafly, E.F. and Finkbeiner, D.P. 2011, ApJ, 737, 103
Schlegel, D.J., Finkbeiner, D.P., and Davis, M. 1998, ApJ, 500, 525
Sellgren, K. 1984, ApJ, 277, 623
Serkowski, K. 1973, ApJ, 179, 101
Shore, S.N. 1982, in *Advances in Ultraviolet Astronomy: Four Years of IUE Research*, NASA CP-2238, eds. Y. Kondo, J.M. Mead, and R.D. Chapman, 370
Shore, S.N. 2002, *The Tapestry of Modern Astrophysics*, (Wiley-VCH)
Shore, S.N., LaRosa, T.N., Chastain, R.J., and Magnani, L. 2006, A&A, 457, 197
Smith, R.K. and Dwek, E. 1998, ApJ, 503, 831
Smith, K.T. et al. 2013, MNRAS, 429, 939
Spitzer, L. Jr. 1978, *Physical Processes in the Interstellar Medium*, John Wiley & Sons, New York
Spyromilio, J., Malin, D.F., Allen, D.A., Steer, C.J., and Couch, W.J. 1995, MNRAS, 274, 256
Sterken, C. and Manfroid, J. 1992, *Astronomical Photometry. A Guide*, (Dordrecht: Kluwer)
Strong, A.W. et al. 1988, A&A, 207,
Stratton, J.A. 1941, *Electromagnetic Theory*, (New York: McGraw-Hill)
Sujatha, N.V., et al. 2009, ApJ, 692, 1333
Szomoru, A. and Guhathakurta, P. 1998, ApJ, 494, 93
Tielens, A.G.G.M. 2009, ARAA, 46, 289
Tibbs, C.T. et al. 2013, ApJ, 768, 98
van de Hulst, H.C. 1957, *Light scattering by small particles*, (New York: John Wiley & Sons)
van den Bergh, S. 1966, AJ, 71, 990
van Dishoeck, E. 2014, The Observatory, 134, 9
Voshchinnikov, N.V., Il'in, V.B., Henning, Th., and Dubkova, D.N. 2006, A&A, 445, 167
Weingartner, J.C. and Draine, B.T. 2001, ApJ, 548, 296
Whittet, D.C.B. 1996, *Polarimetry of the interstellar medium*, eds. W.G. Roberge and D.C.B. Whittet, ASPC, 97, 125
Whittet, D.C.B. 2004, ASP Conference Series, 309, 65
Whittet, D.C.B. 2015, ApJ, 811, 110
Yang, Y, et al. 2015, arXiv:1511.02495

Chapter 7
Surveying the Molecular Milky Way

Aren't the clouds beautiful? They look like big balls of cotton... I could just lie here all day, and watch them drift by... If you use your imagination, you can see lots of things in the cloud formations... What do you think you see, Linus?
Charles M. Schulz, The Complete Peanuts, Vol. 5: 1959–1960

Abstract After a brief review of the Galactic molecular surveys, we discuss the distinct types of Galactic clouds. We briefly describe the large GMCs and then turn our attention to the smaller objects, like dark clouds, but with an emphasis on the flotsam of the ISM, the small diffuse and translucent clouds. We also discuss how recent work on diffuse clouds is revising our ideas of this component of the ISM.

7.1 Introduction

Discussing the molecular component of the ISM is difficult because the term "molecular component" means different things to different people. The FUSE observations by Shull and co-workers show that very few sight lines are free of H_2 absorption. It is striking that at least some H_2 is detectable along nearly 90% of lines of sight at high Galactic latitude (e.g., Snow 2005; Wakker 2006). Of course, most of the directions have only trace amounts of H_2, with N(H_2) values of order 10^{14} cm^{-2}, six or more orders of magnitude below the atomic hydrogen column density. So, although molecular gas is present along these sight-lines, we will not consider them part of the molecular component of the ISM. For this chapter we define a *molecular* line of sight as one with an $n(H_2)/n(H_{total})$ ratio of at least a few percent. Gillmon and Shull (2006) used FUSE data to show that half of the sky at $|b| \geq 30°$ has sheetlike cirrus structures with molecular hydrogen fractions between 1%-30%. These clouds have been considered CNM and their molecular content was largely ignored. It is likely that some of this gas has been misidentified as "dark gas"; while not all of it may be traceable with CO(1-0) emission, there are other ways to trace it spectroscopically (see Sect. 8.4). Thus, diffuse molecular clouds can be mostly atomic and yet still be referred to as molecular clouds. Translucent clouds have higher fractions of molecular gas and are not usually considered part of the CNM (however, see Sect. 1.6). The two principal properties shared by diffuse and translucent clouds are turbulence as their principal structuring mechanism and lack

L. Magnani, S.N. Shore, *A Dirty Window*, Astrophysics and Space Science Library 442, DOI 10.1007/978-3-662-54350-4_7

of star formation. In these respects, these objects are very different from the dark molecular clouds or the envelopes of GMCs. Reviews of the molecular component of the ISM that do not focus primarily on the star-forming aspects of the molecular gas include those by Combes (1991) and Heyer and Dame (2015).

The CO(1-0) transition was the first systematic tracer of the molecular gas and so we begin with a discussion of what was revealed by the early CO surveys in the 1970s. We then discuss how the molecular gas is organized as "big clouds" and "little clouds", and the sometimes misleading terminology employed in the taxonomy of the latter objects. Given the excitation characteristics of CO(1-0), it is not suitable for detecting molecular gas where $N(H_2) \lesssim 10^{19}$ cm^{-2} (because the volume density in these regions is almost always significantly less than 10^2 cm^{-3}).[1] Nevertheless, most of our knowledge of molecular cloud comes from CO observations. We briefly review them in the next sections.

7.2 The CO Surveys Along the Galactic Plane

7.2.1 Early Results

The study of molecular clouds began in earnest with the detection of the lowest CO rotational transition at 115.271 GHz in 1970 by Wilson et al. (1970). Large-scale surveys of CO(1-0) emission are considered to be maps of the molecular gas distribution of the Galaxy with most of this gas contained in discrete, gravitationally bound entities knowns as molecular clouds (a counterpoint to this view is discussed at the end of this chapter). Most of our knowledge of molecular clouds has come from many extensive surveys conducted in the 1970s and 80s. Because the molecular gas is associated with the thin disk, all early CO surveys were made at relatively low Galactic latitudes. The early CO surveys (Burton et al. 1975; Scoville and Solomon 1975; Cohen and Thaddeus 1977; Burton and Gordon 1978; Stark 1979; Robinson et al. 1984; Israel et al. 1984; and Sanders et al. 1984) focused almost exclusively on latitudes within a few degrees of $\ell = 0°$. It was not until relatively recently that wide-latitude, fairly complete surveys of the Galactic molecular distribution became available (Dame et al. 1987, 2001). Interior to the Sun, the distribution of molecular gas defines a thin plane. However, beyond 12 kpc the midpoints of the molecular distribution deviate significantly from $\ell = 0°$ and mapping the molecular outer Galaxy was more problematic. Specific, targeted surveys (using IRAS sources to track the molecular gas) were carried out by Wouterloot and Brand (1989) and Wouterloot et al. (1990).

[1]It is this diffuse molecular component that has recently caused much confusion about its quantity and importance in the ISM. The rise in popularity of the idea of a "dark" molecular component (in its original definition, a component of molecular gas that was spectroscopically undetectable, and, later, one that could not be traced by the CO(1-0) transition) will be examined in the next chapter.

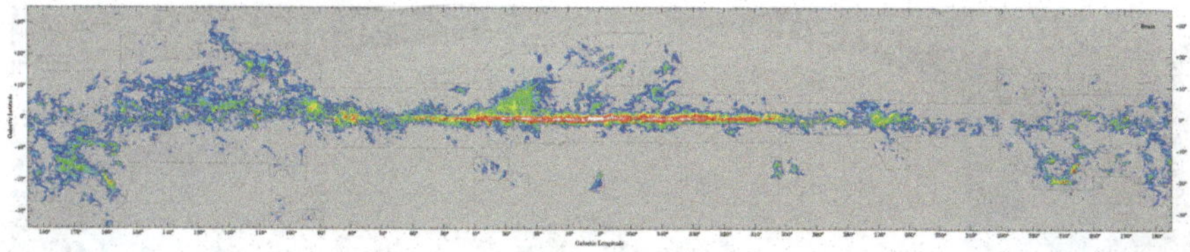

Fig. 7.1 The Dame et al. (2001) CO(1-0) map of the Galaxy made with the 1.2 m millimeter-wave telescopes at Cambridge, Massachusetts and Cerro Tololo, Chile. The color scale represents the log of the CO(1-0) velocity-integrated line emission from ~ 0 to 2.4. The angular resolution is $9'$ over most of the map, but lower ($15'$ or $30'$) over some regions. A large version of this map is available at https:///www.cfa.harvard.edu/mmw/MilkyWayinMolClouds.html Figure courtesy of Tom Dame

The principal results from the early CO(1-0) surveys are the following: (1) Surveys in the CO(1-0) line are effectively surveys of H_2. (2) The bulk of the molecular gas is distributed in the inner Galaxy along the Galactic plane. The mass surface density peaks at the Galactic Center (GC) and then decreases rapidly between 1–3 kpc from the GC. There is a strong concentration between 3–6 kpc known as the "Molecular Ring",[2] and then a gradual falloff to the Solar Circle and beyond. See, for example, the radial CO distribution in Clemens et al. (1988). (3) Individual contiguous objects exist and are more clearly delineated when velocity information is used in addition to the two-dimensional spatial data. The bulk of the molecular gas along the plane is tied up in large entities known as Giant Molecular Clouds (GMCs). These objects are tens of pc in size and 10^3-10^6 $M_\odot$ in total mass. Their properties make them among the largest entities in the Milky Way. (4) The average gas number density, $n(H_{total})$, inside the Solar Circle is of order 1 cm^{-3}, while the average density in GMCs is about 10^2 cm^{-3}, so that the volume filling factor of GMCs along the Galactic Plane is $\sim 1\%$. (5) Most of the star formation in the Galaxy is in GMCs.

Figure 7.1 shows the Galactic CO(1-0) emission from the most recently published survey by Dame, Hartmann, and Thaddeus based on over 270,000 CO spectra. Several characteristics of the molecular distribution are readily apparent: the most intense CO(1-0) emission is located at the Galactic center and in the inner Galaxy (significantly different from the HI distribution that has 70% of its mass outside the Solar Circle—e.g., Dickey and Lockman 1990). The strong CO emission is confined to low latitude, although regions known to contain dark clouds such as the Taurus-Auriga or ρ Ophiuchi cloud complexes lie at higher latitudes and, consequently, must be fairly nearby.

Even a perfunctory glance at Fig. 7.1 indicates that the molecular emission regions at the higher latitudes have lower surface brightness than regions towards the Galactic center or along the plane in the Inner Galaxy. The higher latitude

[2]These values are for a Solar Circle at 8 kpc; using the originally accepted value of 10 kpc would increase the galactocentric locations by 20%.

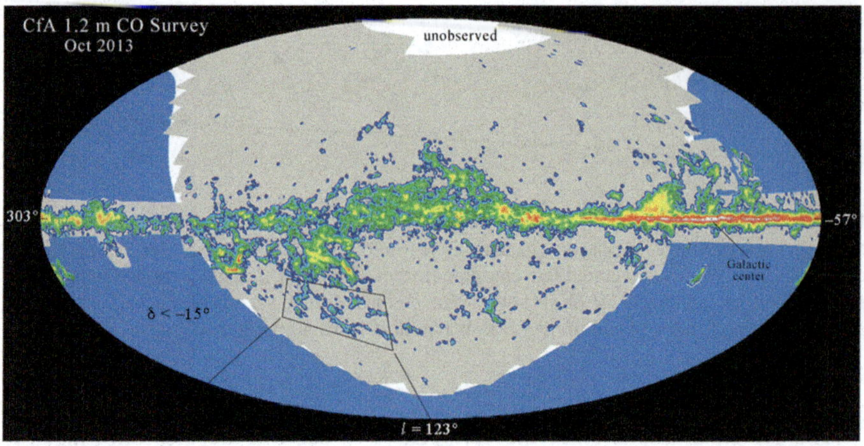

Fig. 7.2 The most recent CO map of the Galaxy from Thaddeus and Dame that extends the map in Fig. 7.1 to higher Galactic latitudes. See https://www.cfa.harvard.edu/mmw/ for more details and progress reports. The Aitoff projection of the CO data is centered at $\ell = 123°$ to show the high-latitude molecular clouds more prominently. Figure courtesy of Tom Dame

CO-emitting regions in the figure (note that the latitude extent of the map ranges from $+32.5°$ to $-32.5°$) are relatively nearby (a few hundred parsecs) and, with the exception of the Orion molecular clouds, are composed of smaller molecular clouds whose basic properties are discussed below.

7.2.2 *Recent Surveys*

Most of the Galactic plane CO surveys were completed by the late 1980s, and only a few have been made in recent years from ground-based mm-telescopes. The emphasis in mm-wave radio astronomy has shifted to studies of individual clouds or star-forming regions, and, most notably, to external galaxies. However, the Thaddeus/Dame mm-wave group at the Harvard-Smithsonian Center for Astrophysics continues their systematic survey of the CO(1-0) line in the Galaxy and as of 2013 their mapped area had extended over a significant portion of the high-latitude sky (see Fig. 7.2).[3]

In the late 90's the Five College Radio Astronomy Observatory's (FCRAO) 14-m mm-wave telescope was used to map a substantial portion of the Outer Galaxy between 102.49°–141.54° in Galactic longitude and $-3.03°$ to $+5.41°$ in latitude. With a spacing of 50″ (about the beam size at 115 GHz), this survey includes nearly

[3] As of the summer of 2016, the Harvard-Smithsonian mm-wave group had completely mapped the Northern Galactic Hemisphere in the CO(1-0) line.

1.7 million spectra and represents perhaps the best combination of coverage and angular resolution over a wide area. Heyer et al. (1998) described the initial results. Voids are present in the molecular distribution possibly from the action of UV radiation fields and stellar winds. Radial profiles of the mass surface density and variations of the midplane and scale height of the molecular distribution in the outer Galaxy are presented completing the picture of the distribution of GMCs begun with the early surveys. Most surprisingly, a great deal of low surface brightness, low mass surface density molecular gas is present in the spiral arms of the Galaxy and in the vicinity of GMCs. Blitz and Williams (1999) refer to this gas as "the chaff" and link it with the smaller, lower extinction molecular clouds discussed below. This raises the question of what fraction of the overall molecular gas is contained in these smaller objects. Regarding the molecular gas, the early surveys implied that 90% of the H_2 mass is tied up in GMCs, but the results from the more sensitive, better sampled FCRAO survey indicate that up to 50% of the CO *intensity* may be contained in this low surface brightness gas surrounding the dense cores of a GMC (Carpenter et al. 1995).

Galactic CO emission peaks in the Molecular Ring at 3–6 kpc, and molecular gas dominates over atomic gas inside the Solar Circle (see figure 6 by Scoville and Sanders 1987). Beyond the Solar Circle, the molecular gas peters out rather rapidly and hydrogen in atomic form overwhelmingly dominates. Despite this, there are molecular clouds, even GMCs, beyond the Perseus arm (at around 11 kpc from the Galactic Center). Surveys by Kutner and Mead (1981),[4] Mead and Kutner (1988), Wouterloot and Brand (1989), Digel et al. (1994), and Heyer et al. (1998), established that molecular gas can be found out to 20 kpc, albeit in the form of isolated clouds. However, beyond a galactocentric radius of 16 kpc, the molecular mass surface density has dropped to 0.095 $M_{\odot}$ pc^{-2} compared to 0.5 $M_{\odot}$ pc^{-2} at the Solar Circle. The *Planck* survey, discussed in Chap. 5, in the 100 μm band offers a complementary view of the molecular gas distribution in the Galaxy, and Fig. 7.3 shows an early release of the *Planck* 100 GHz data for the whole sky showing emission from the CO(1-0) line since ^{13}CO and CMB contributions have been removed.

Early results on the Galactic CO emission detected by Planck were reported by the Planck Collaboration, Planck 2013 results, paper XIII (2014). The CO(1-0), (2-1), and (3-2) rotational transitions contribute significantly to the signal in the Planck 100, 217, and 353 GHz HFI channels. Preliminary extraction of the signals were compared to existing CO maps of the sky and give excellent results. However, it is important to point out that the Planck satellite produced imaging data. The 100 GHz survey is virtually useless for studying GMCs throughout the Galactic plane because of the lack of kinematic information. The *Planck* survey is likely to be more useful at high Galactic latitudes where the molecular clouds are isolated both spatially and in velocity.

[4]Some of the detections of the initial paper were disputed by Solomon et al. (1983).

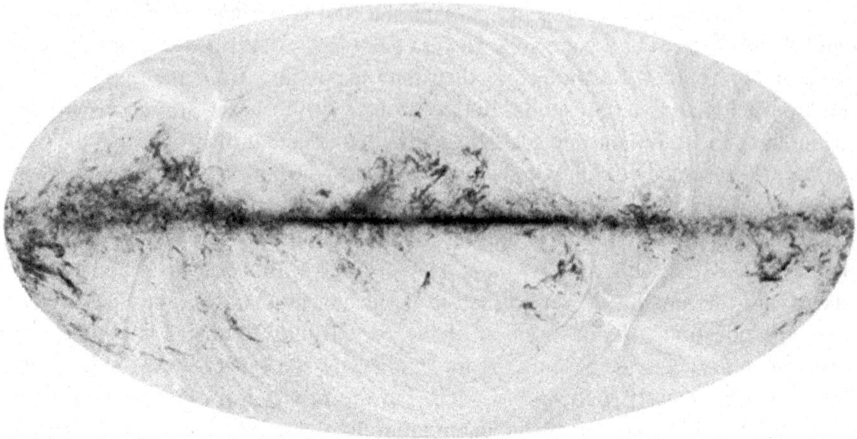

Fig. 7.3 The *Planck* CO(1-0) map of the entire sky based on imaging data in the 100 GHz channel. The contributions from ^{13}CO and the CMB have been removed. The molecular cloud distribution along the Galactic plane arises primarily from GMCs. CO emission away from the plane (i.e., at $|b| \geq 20°$) is relatively nearby, in the solar neighborhood. Compare this map with the CO(1-0) maps made with the Harvard-Smithsonian CfA 1.2 m telescope (see Figs. 7.1 and 7.2). Image credits: ESA/Planck Collaboration

7.3 Giant Molecular Clouds

The building blocks of the CO emission in Figs. 7.1 and 7.3 are enormous concentrations of molecular gas known as Giant Molecular Clouds or GMCs. They are among the largest individual entities in the Galaxy with masses ranging from 10^3 to 10^6 $M_\odot$, dimensions of up to 100 pc, temperatures around 10 K (at least away from star forming regions) and volume averaged densities of order 10^2 cm^{-3} (Blitz 1979; Sanders et al. 1985). By the end of the 1980s, large CO surveys of the Galactic plane had shown that nearly 90% of the H_2 in the Galaxy contains about 5000 GMCs with masses greater than 10^5 $M_\odot$ and of which about 1000 clouds with sizes greater than 50 pc and masses greater than 10^6 $M_\odot$ containing about half of the total H_2 mass (see, however, Sect. 7.4.3). More importantly, for Galactic evolution, virtually all of the most massive stars and most of the lower mass stars form in these structures. GMCs nearly always have clusters of OB stars associated with them and many of their names are identical to the accompanying OB association (e.g., Cep OB3, Ser OB1, Cyg OB1, *etc.*). The connection between GMCs and OB associations is so marked that the lower limit to the mass range for GMCs shows that clouds smaller than 10^3 $M_\odot$ cannot produce OB associations. In light of these considerations, in order to understand the stellar content of a galaxy and the overall life-cycle of stars, GMCs must be studied thoroughly.

The GMCs differ somewhat in their general properties depending on their location in the Galaxy. The overwhelming majority of molecular gas is inside the Solar Circle.[5] In general, local GMCs are the best-studied and their general properties are summarized in several review articles (e.g., Scoville and Sanders 1987; Solomon and Rivolo 1987; Friberg and Hjalmarson; Blitz 1991*a*,*b*; Combes 1991; Williams et al. 1999; Blitz and Williams 1999; Heyer and Dame 2015). GMCs are internally highly structured with the denser clumps occupying a small fraction of their total cloud volume (~10% - Lada 1990). Both the GMC as a whole and its constituent dense clumps are gravitationally bound with masses a factor of ten greater than a Jeans mass (see Chap. 11). The formation of O and B stars wreaks havoc within the GMCs and they do not appear capable of surviving more than a few generations of massive star formation (Franco et al. 1994). Cloud dispersal times are then 10^6–10^7 years after star formation begins.

The broad scenarios for molecular cloud formation have been widely discussed: (1) collisional agglomeration from smaller clouds (e.g. Kwan 1979; Scoville and Hersh 1979; Stark 1979); (2) a gravi-thermal instability in the Galactic plane (e.g. Parker 1966; Mouschovias et al. 1974; Elmegreen 1982a,b); and (3) formation in shock fronts (e.g. Woodward 1976). While much work has been done in this area (see review by Elmegreen 1990; Elmegreen and Palous 2006), a definitive answer still eludes us. The relatively short lifetime of GMCs compared to the Galactic rotation period implies that they are for the most part confined in or near the spiral arms of the Galaxy. This enables astronomers to trace the structure of the Galactic spiral arms using the global distribution of CO emission (Liszt 1984; Dame et al. 1986; Clemens et al. 1988).

Although we know how GMCs are destroyed, the support mechanisms before the onset of star formation for both the large and small molecular clouds are not fully understood. Self-gravity considerations imply that these objects should collapse on a few times the free-fall time scale [$t_{\rm ff} \sim (3G\rho)^{-0.5}$; less than a few times 10^6 years—see Genzel 1991 for a discussion of the energy balance of molecular clouds]. Recent work indicates that turbulence (either externally or internally driven) may be the principal supporting mechanism for molecular clouds (LaRosa et al. 1999, and references therein; Hennebelle and Falgarone 2012).

An important uncertainty for GMCs concerns their lifespan. Early estimates based on the fraction of H_2 compared to total nucleons in the ISM favored very long lifetimes ($\sim 10^9$ years). The idea is based on mass conservation arguments or continuity of the hydrogen mass between the various ISM phases (in this case, atomic, molecular, and ionized). It seems plausible that the mass flux from one phase to another should be balanced (see Scoville and Hersh 1979). Scoville (2014) then writes the mass flux equality as:

$$\dot{M}_{H_2 \rightarrow HI+HII} \, (\equiv M_{H_2}/\tau_{H_2}) = \dot{M}_{HI+HII \rightarrow H_2} \, (\equiv M_{HI+HII}/\tau_{HI+HII}) \tag{7.1}$$

Since in the inner Galaxy, most of the hydrogen nucleons are sequestered in H_2, the lifetime of the molecular gas is significantly longer than the HI, which is thought

[5]The distribution of molecular gas with respect to Galactic radius can be seen in Fig. 8.1.

to be 10^8 yr based on their dynamical time scale or orbital passage between the spiral arms. However, the recognition that dynamical processes were critical to the formation and destruction of these objects led to estimates more in the 10^7–10^8 year range (e.g., Blitz and Shu 1980). By the late 1980s, almost everyone in the field agreed that GMC lifetimes depended on how quickly star formation begins, in the sense that the onset of star formation would destroy the cloud in 10^6–10^7 years. Methods to determine their ages have employed (1) determining the HI/H_2 ratio, (2) ^{13}CO depletion on dust grains, (3) the deuterium fractionation of N_2H^+, and (4) ratio of ortho-H_2D^+ to para-H^2D^+. The first method basically assumes that the formation of GMCs occurs within large atomic regions, so that the H_2/H ratio is small at first and then increases so that by the time star formation begins the clouds is almost entirely molecular. To exploit this idea, Goldsmith and Li (2005) proposed to obtain cloud ages using the abundance of HI Narrow Self-Absorption (HINSA) lines in clouds. These absorption lines are due to the presence of atomic hydrogen within the molecular cloud. The strength of the HINSA features decreases as the cloud ages and becomes progressively more molecular until an equilibrium is reached between the formation of H_2 and its destruction by cosmic rays.

Without question, the study of GMCs dominates observational molecular astrophysics. However, the nearest GMC to the Sun is the Orion molecular cloud system, about half a kiloparsec away. There are molecular clouds closer to the Sun than that, but they are much smaller than GMCs. Although GMCs contain very dense clumps, their envelope and interclump medium share some of the properties of the diffuse molecular gas discussed in this book. We will now focus on the smaller clouds and, in particular, the diffuse and translucent molecular clouds which are clearly part of the gas inventory of the *diffuse* ISM.

7.4 The Smaller Molecular Clouds

Historically, small molecular clouds ($\lesssim 10^3$ $M_\odot$) were divided into diffuse molecular clouds studied primarily by optical and ultraviolet absorption lines against the background continuum of early type stars, and dark clouds, so named because their large column density of dust obscured background starlight.

An important development occurred in the late 1980s, when John Black and Ewine van Dishoeck proposed a quantitative distinction, dividing the small clouds into three broad categories defined by visual extinction: diffuse ($A_V < 1$ mag), translucent (1 mag $\leq A_V \leq 5$ mag), and dark ($A_V > 5$ mag) (Van Dishoeck and Black 1988). This was not a mere taxonomic exercise. These cloud types represent different physical and chemical states of molecular gas. The astrochemistry of diffuse clouds is regulated primarily by photoprocesses and molecular abundances tend to be low. Before 2000, they had been studied mainly by optical astronomers using a limited set of absorption lines. In contrast, dark cloud chemistry is dominated by collisional processes and typically leads to higher abundances of most molecular species. Consequently, the column densities of many species suffice for radio spectroscopic studies. Because the diffuse and dark clouds were studied primarily by different techniques and communities in the late 20th century, the literature on these objects did not overlap significantly (notable exceptions include the work of Steve Federman, Robert Willson, Ken Lang, Harvey Liszt, and Robert Dickman), especially throughout the 1970's and early 80's, when millimeter receiver technology was in its infancy. The transition objects, the translucent clouds, could be studied by both techniques and represent an astrochemical regime where both photoprocesses and collisional reactions are important. Moreover, for the most important tracer of molecular gas, CO, the translucent clouds represent the regime where the CO abundance with respect to H_2 rises precipitously from less than 10^{-6} to $\sim 10^{-4}$ and the carbon content changes from atomic form to mostly CO over an increase of an order of magnitude in $N(H_{total})$ (Van Dishoeck and Black 1988). Another significant difference between dark molecular clouds and the translucent ones is that the former, for the most part, are gravitationally bound while the latter are not. Below, we will take a closer look at each type of object.

7.4.1 Dark Clouds

The small, dark clouds tend to be local objects because at low Galactic latitudes and distances much greater than a kiloparsec their modest CO emission tends to blend with that of the background GMCs. Thus, only the nearer objects or those at high latitudes are seen without confusion. Examples of isolated small clouds are Lynds 134 and Barnard 68 (see Figs. 7.4 and 1.7, respectively), while some dark cloud complexes include the Taurus-Auriga-Perseus dark clouds and the ρ Ophiuchi clouds (see Astronomy Picture of the Day—http://apod.nasa.gov.apod—for 2009 July 8). Isolated clouds tend to be a few parsecs or less in size while dark cloud complexes made up of many dozens of small clouds can stretch for tens of parsecs (e.g., the Taurus-Aurigae dark cloud complex; Ungerechts and Thaddeus 1987). At the smallest extreme are the Bok globules (Bok et al. 1971; Bok 1977) which are compact, spherical entities of size a few tenths of a parsec and mass $\sim 10\ M_{\odot}$. Dark clouds were recently reviewed by Bergin and Tafalla (2007) with emphasis on their star-forming capabilities.

One of the earliest systematic searches for obscured regions or dark nebulae in the Galaxy was conducted by Barnard (1919) and included a catalog of 182 such objects. Interestingly, in that work, Barnard left open the possibility that some of these regions were not dark nebulae but could be produced by "vacancies" in the stellar distribution. Later, Barnard (1927) published a photographic study of dark nebulae distributed along the Galactic plane; only 3 of 50 photographic plates were at Galactic latitudes (old Galactic coordinates) $|b^I| \geq 20°$, and there were none above $|b^I| = 25°$. This attests to the difficulty of detecting obscured areas at high Galactic latitudes where the stellar density decreases rapidly and fluctuations below the mean in the number of stars over a small region can mimic obscuration by dust.

Fig. 7.4 Three-color SDSS2 image (blue, red, infrared—York et al. 2000) of the dark molecular cloud Lynds 134 (also known as MBM 36). The image size is 1.5°× 1.5° centered at (RA, Dec 2000.0) = (238.40°, −4.59°). The most obscure regions of this cloud have $A_V > 6$ mag. Consequently, no stars beyond the cloud are visible through the opaque central regions; any stars projected there are in the foreground. However, note that a diffuse or translucent envelope surrounds the cloud

Contemporaneously, P.J. Melotte and K. Lundmark conducted a whole-sky search for dark nebulae which was reported by Lundmark (1926). Melotte and Lundmark used photographic plates taken during the Franklin-Adams 1911 sky survey that have an approximate limiting photographic (blue) magnitude, m_{pg}, of 15. Melotte and Lundmark attributed any five-fold or greater decrease in stellar surface density over an area of more than a few square degrees to the presence of a "dark" nebula. In this manner, they identified 1550 dark nebulae in the regions covered by the Franklin-Adams plates, with many of the regions occurring at high Galactic latitudes (see Fig. 7.5).

In the years that followed, Khavtassi (1955), Roschkovsky (1955), and Schoenberg (1964) compiled lists of dark nebulae mostly confined to the Galactic plane with few objects at $|b| \geq 20°$, but only the work of Beverly Lynds in 1962 produced a complete survey of all dark nebulae visible from the northern hemisphere. In her work, 879 fields photographed on red- and blue-sensitive photographic plates with the Palomar Observatory 48-inch Schmidt were examined for the presence of obscuration (Lynds 1962, 1968). The resulting 1801 objects (many of which are spatially related to each other) represent all the easily discernible dark clouds at $\delta > -33°$. Recent surveys of dark clouds were reviewed by Dutra and Bica (2002).

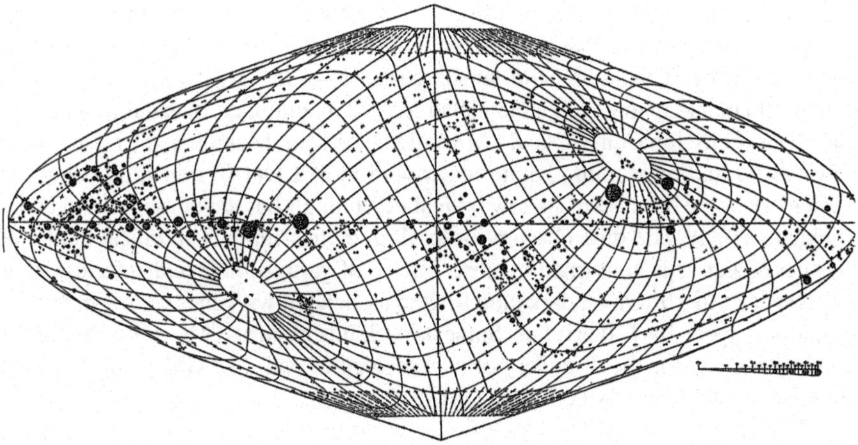

Fig. 7.5 The Melotte and Lundmark dark clouds from a paper by Lundmark (1926). The *dark clouds* denoted on map by *black circles* are based on regions of decreased stellar density identified from photographic plates taken during the Franklin-Adams 1911 sky survey. The map is centered at old Galactic coordinates $(\ell^I, b^I) = (180.0°, 0.0°)$ and the superimposed equatorial coordinate grid is epoch 1900. The wedge at lower right displays the size of the obscured region from 1° to 5°

Of particular note are the studies by Feitzinger and Stüwe (1984) and Hartley et al. (1986), both based on looking for obscuration on the ESO/SERC plates, and that of Clemens and Barvainis (1988) using the Palomar Sky Survey prints.

These studies identified dark clouds by their obscuration of background stars. The visual extinction through the most opaque regions of a dark cloud is typically more than five magnitudes and the dust responsible for the obscuration is typical of regions where hydrogen column densities are greater than 10^{21} cm^{-2}. The overwhelming fraction of their gas is in molecular form and their masses range from a few tenths to hundreds of solar masses. Like GMCs, dark clouds are distributed mainly along the Galactic plane, although they have a wider latitude distribution than the GMCs. When interstellar radio emission lines were discovered in the 1960s, dark clouds were among the first objects studied (*e.g.*, Heiles 1968; Palmer et al. 1969; Penzias et al. 1972).[6]

Initial surveys of the more important astrophysical molecular species in dark clouds included OH by Cudaback and Heiles (1969) and Crutcher (1973); H_2CO by Dieter (1973); and CO by Dickman (1975). Others, for example, Snell (1979), studied the molecular properties of a limited sample of dark clouds. In all these studies, the Lynds list provided the sources which the various observers studied. It soon became apparent that all dark clouds with extinction greater than a few

[6]The drop in HI column density as determined from the 21 cm line in the direction of dark clouds (as compared to their surroundings) implied that the gas thought to be present in the dark cloud was most likely molecular in form.

magnitudes showed molecular emission or absorption lines. In particular, Dickman (1975) reported observations of CO(1-0) emission from 63 of the 64 Lynds dark clouds he surveyed, including all four sources at $|b| \geq 25°$. By the late 1970s it was widely held that if an interstellar cloud had $A_V > 2$ magnitudes, it was almost certain that at least emission from CO and OH would be present. Moreover, the relationship between CO and extinction is quite linear with the ^{13}CO(1-0) transition being the species of choice because of its low optical depth in all but the densest cloud cores (Dickman 1978). The dust obscuration in dark clouds implies significant column densities of gas as is confirmed by observations of various molecular species. With core densities of 10^4 cm^{-3}, dark clouds were associated with sites of low mass star formation, especially T Tauri stars. Some dark clouds, such as TMC-1 are among the principal regions for the discovery of new molecules and, like GMCs, dark clouds comprise the dense component of the ISM.

7.4.2 Translucent Molecular Clouds

According to the van Dishoeck and Black hierarchy, interstellar clouds with extinctions between 1 and 5 magnitudes are called "translucent" molecular clouds. This category is not based only on extinction; from the chemical point of view it is in this regime where the bulk of the carbon changes from C^+ to CO. This drives the CO/H_2 abundance from $\sim 10^{-6}$ to 10^{-4}. Effectively, this meant that the CO column density in these types of clouds was high enough to allow detection as mm-wave receivers improved in the early 1980s. Some of the Lynds "dark" clouds actually have peak extinctions *less than* 5 magnitudes (e.g., L1642—see extinction map by Sandell et al. 1981) so that, technically, they are likely translucents. However, the first real systematic study of translucent clouds came as a by-product of the discovery of the high-latitude molecular clouds (see Chap. 9). Most of these have $A_V < 5$ mag and so can be considered translucent or diffuse molecular clouds. The high-latitude molecular surveys by Magnani et al. (1985) revealed another distinction between dark and translucent clouds. The latter are not gravitationally bound and consequently are dissipating on the sound-crossing time scale of 10^6 yr. For instance, comparing the SDSS red/blue/infrared image of L134 (coincidentally also a high-latitude molecular cloud with a Galactic latitude of +36°) shown in Fig. 7.4 with a typical translucent high-latitude cloud, MBM 54 (see Fig. 7.6), one can easily see the different dynamics at work in both types of cloud. L134 is compact and dense, while MBM 54 is less dense (you can see more stars through it) and more ragged in morphology.

One of the problems in mapping translucent high-latitude clouds is their proximity to us. Although not significantly larger than dark clouds, their average distance tends to be a few hundred pc or less so that they often subtend several square degrees in the sky. Thus, a fully-sampled map of a translucent cloud such as MBM 16 can take nearly 10^4 spectra even at the relatively low spatial resolution of $8'$ (see Fig. 5.2). Despite the large time investment, several complete maps

Fig. 7.6 Same type image as Fig. 7.4, but of the high-latitude molecular cloud MBM 54 from the SDSS2 survey. This is a translucent cloud in contrast to dark cloud in Fig. 7.4; note both the filamentarity of MBM 54 and that stars are clearly visible through the structure The image is 1.5° by 1.5° centered at (RA, Dec 2000.0) = (347.32°, 18.48°). The large spiral galaxy in the background is NGC 7497

of translucent clouds over very large regions have been made (e.g., Pound and Goodman 1997; Yamamoto et al. 2003; Shore et al. 2003, 2005). These maps show that the translucent clouds have clumpy structure like dark clouds, but with generally lower column densities ($\sim 10^{20}$ cm^{-3} vs. $\geq 10^{21}$ cm^{-3}).

Comprehensive searches for star-formation in translucents with peak $A_V \leq 3$ mag have not yielded any viable candidates (McGehee 2008). However, some researchers claim that signs of star forming activity are present. For example, Kun (1992) found Hα emitting stars projected on or near known translucent clouds, but it should be noted that not all such objects are T Tauri stars or pre-main sequences stars. Li et al. (2000) found three x-ray emitting stars near MBM clouds using the ROSAT point source catalog, but the follow-up work to determine whether these are or pre-main sequence objects has not yet been done. For the moment, no unambiguous sign of star formation has been found in any translucent cloud with A_V less than 3 magnitudes in its most obscured region. The cloud known as Lynds 1453/4/7/8 or MBM 12 does have significant star formation, actually harboring a small T-association (Luhman 2001), but extinction measurements for this region reveal a maximum value of 6.8 mag (Schlegel et al. 1998). According to the van Dishoeck and Black criterion this is a dark cloud, so its harboring a T-association is not so surprising. Figure 7.7 makes clear that there is a clear separation between

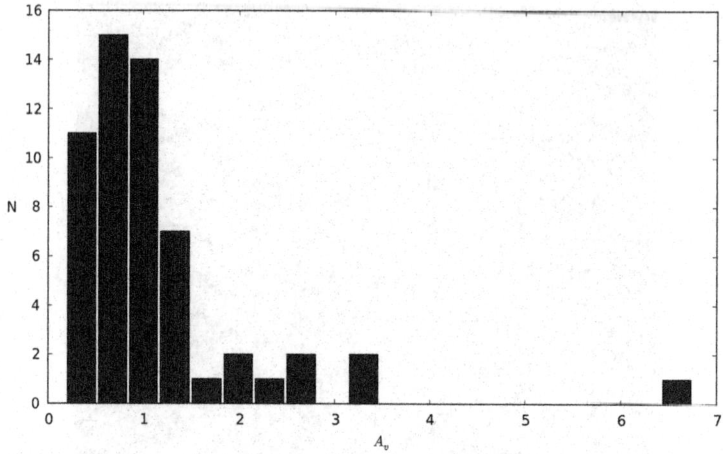

Fig. 7.7 Histogram of the high-latitude molecular clouds from the MBM catalog as a function of A_V. The peak extinction is derived from E(B-V) from the Schlegel et al. (1998) dust maps using R = 3.1. The cloud at the far right of the histogram with the greatest A_V value is MBM 12 which harbors a small T association (Luhman 2001) and is also a Lynds (1962) dark cloud complex (L1453-4, L1457-8)

MBM 12 and the other MBM clouds as far as peak extinction. L1642 or MBM 20 has two binary T Tauri stars in its core region (Sandell et al. 1987). Schlegel, Finkbeiner, and Davis determine a peak A_V of only 2.2 mag, but the actual peak extinction is likely significantly higher (see Sandell et al. 1981). A thorough review of the star-formation situation at high Galactic latitude is available in McGehee 2008.

That the translucent clouds would not be star-forming sites might have been inferred from their dynamic status. Virial analyses of the high-latitude cloud showed from the beginning that many of these clouds were not gravitationally bound (e.g., Blitz et al. 1984). Exceptions such as L1642 that are gravitationally bound (Liljestrom 1991) are also forming stars. However, aside from a few objects, the bulk of the known translucent clouds are not forming stars and have dynamics dominated by turbulence. The issue of turbulence in these clouds will be discussed further in Chap. 11.

In some respects, the distinction between translucents and diffuse clouds is academic. The classification is based on the maximum extinction in the cloud, but it ignores the filling factor for regions at that maximum extinction. A translucent core region has a diffuse molecular envelope—in all cases, and this envelope often contains a considerable amount of matter, sometimes rivaling the mass of the core region (e.g., Cotten and Magnani 2013). Dark clouds and GMCs also have diffuse molecular envelopes, something that is often ignored in discussion of these objects. We will examine this question in detail in Chap. 8.

7.4.3 Diffuse Molecular Clouds

Diffuse molecular clouds are hard to characterize because throughout the latter part of the 20th century they could only be probed in absorption along the infinitesimally small solid angle to the background continuum source, almost always a star but sometimes an extragalactic source. Based on studies of color excess over large areas (*e.g.*, Knude 1979) diffuse clouds are thought to be a few pc in size. Originally, they were considered to have extinctions less than 1 or 2 visual magnitudes and were thought to range in physical properties from what is sometimes referred to as a "standard cloud" (Spitzer 1978), to objects with significantly more extinction and column density. Spitzer characterized standard clouds statistically as having $N_H \sim 3 \times 10^{20}$ cm^{-2} and an interception rate in the disk of 6.2 clouds per kiloparsec (Spitzer 1985). Many of the sightlines typical of a standard cloud show molecular absorption lines even if the bulk of the gas is atomic. Spitzer also described a "large cloud" which has a visual extinction of nearly 1 magnitude and an interception rate of 0.8 per kiloparsec. Given current astrochemical models, these objects have a significant portion of their gas content in molecular form. With densities of order 1 cm^{-3} (Spitzer 1978) for standard clouds and perhaps an order of magnitude more for Spitzer's large clouds, these objects have masses in the 0.01–10^1 $M_\odot$ range.

In 1965, Beverly Lynds compiled a catalog of reflection or emission areas as determined from the blue and red POSS plates (Lynds 1965).[7] There are many more of these Lynds Bright Nebulae (LBN) at high Galactic latitude than there are Lynds dark nebulae. Both van den Bergh (1966) and Sandage (1976) remarked that some of the bright nebulae at high latitudes could be dust clouds reflecting the integrated star light of the Galactic plane below them. Because these clouds are rich in dust, it is likely that they contain significant atomic hydrogen and, possibly, molecular hydrogen. The area studied by Sandage comprises one of the richer regions of high-latitude molecular gas, the Ursa Major cloud complex (mapped in part by Pound and Goodman 1997). Although it is certain that some of the emission nebulosity coming from regions containing high-latitude molecular clouds is due to reflected star light, there are also indications that some of the red emission may be produced by luminescence in very small hydrogenated carbon grains (Szomoru and Guhathakurta 1998—see §3.5.4).

It is difficult to determine when exactly a diffuse cloud stops being "molecular" and is, instead, an atomic cloud with some trace amount of molecules. So where do we draw the line between diffuse molecular and atomic clouds? As discussed earlier in this Chapter, we subjectively choose a cutoff for a diffuse molecular cloud of $n(H_2)/n(H_{total}) \sim$ a few percent. Our definition differs from that of, e.g., Liszt et al. (2010) who identify diffuse clouds as having at least 25% of the hydrogen in molecular form. Our more liberal definition allows us to consider as diffuse molecular clouds those objects with $N(H_2)$ as low as a few $\times$ 10^{19} cm^{-2}. Here,

[7]Lynds dark cloud from the 1962 catalog are denoted as L followed by their numerical designation. However, the SIMBAD database refers to them as LDN followed by a number.

detection of molecular lines by radio spectroscopy is still possible and their physical characteristics can be determined over more than an infinitesimal line of sight.

Although normally studied by optical and UV absorption lines towards background stars, it became clear even in the 1980s that some diffuse lines of sight had sufficient CO column densities for detection in emission via the CO(1-0) line at 115 GHz. About half of the MBM high-latitude clouds turn out to have peak A_V less than 1 mag (see Fig. 7.7), but, for more than 20 years they were classified by nearly everyone as translucent clouds. However, in a series of brilliant papers, Liszt, Lucas and coworkers have recently studied the molecular properties of diffuse clouds and have come to some startling conclusions.

Liszt and Wilson (1993) and Liszt (1994) began by identifying about two dozen lines of sight towards background millimeter-wave sources where CO(1-0) absorption was detectable by synthesis instruments. In a series of papers, they systematically studied the molecular properties of these diffuse lines of sight using both absorption and emission lines. The formation, fractionation, and excitation of CO in diffuse clouds was treated at length by Liszt (2007). Other species studied include HCO^+ and HCN (Lucas and Liszt 1994; Liszt and Lucas 1994); H_2CO (Liszt and Lucas 1995); OH (Liszt and Lucas 1996); C_2H and C_3H_2 (Lucas and Liszt 2000); CH (Liszt and Lucas 2002).

In addition to studying the chemistry of diffuse clouds, Liszt et al. (2010) and Liszt and Pety (2012) identified a set of diffuse clouds with very strong CO(1-0) lines (antenna temperatures as high as 10 K) but with CO column densities $\lesssim 10^{16}$ cm^{-2}. Clouds with these properties do not fit the diffuse/translucent/dark categorization. Such strong CO emission and relatively modest abundances are not plausible in the traditional scheme. Liszt, Lucas, and Pety conclude that CO(1-0) emission does not trace well the molecular distribution of diffuse clouds because the molecule represents a small fraction of the carbon in the gas phase. Instead, it highlights where the CO chemistry is converting the carbon primarily into CO rather than C^+ or C^0. In addition, radiative transfer effects can produce abnormally strong CO(1-0). Line formation effects in low density regions can cause strong emission lines if the gas is subthermally excited; the low densities allow radiative processes to dominate over collisional de-excitation so the photons scatter about in the gas until they escape. This picture effectively changes the conventional wisdom that the CO(1-0) line traces almost exclusively the dense cold gas in the Galaxy (i.e., primarily the dark clouds and GMCs). Instead, strong CO(1-0) lines may arise in what is clearly diffuse molecular gas. Besides diffuse molecular clouds, the envelopes of GMCs and dark clouds are warmer, lower in pressure, and may occupy a larger fraction of the volume than the denser, colder gas. These envelopes are well traced by mid-IR emission (see Chap. 6), but not always by CO(1-0) (see Sect. 8.4). Thus, Liszt et al. (2010) conclude that a CO(1-0) map is really a map of CO abundance and chemistry, and only secondarily a map of the dense gas (and hence the mass). In their words, "The CO sky is mostly an image of the CO chemistry."

Liszt et al. (2010) find that the CO-H_2 conversion factor (see Chap. 8) is similar in diffuse and dark clouds because in diffuse clouds, a drop in the CO/H_2 abundance is compensated by an increase in the CO line intensity per CO molecule because of chemistry and radiative transfer considerations. The findings of Liszt, Lucas, and coworkers highlight the complexity of interpreting CO observations in diffuse and translucent regions. While further, in-depth, study of molecular clouds with A_V less than 1 magnitude is needed, it may be that the diffuse/translucent/dark paradigm for categorizing small clouds is just too approximate.

7.5 What Is a Molecular Cloud?

If GMCs and dark clouds were the only types, one could attempt to define a molecular cloud as a self-gravitating entities, isolated both spatially and kinematically from the column of hydrogen in their direction. The contrast in density between the molecular entity and the surrounding medium is between 10^2 and 10^3, nearly that of a baseball plowing through air. But, even in these cases, the cloud construct is too simplistic. GMCs have diffuse molecular and atomic envelopes which are sometimes filamentary, and they are not disconnected from the surrounding CNM. With the exception of Bok globules, dark clouds also have envelopes and filaments that seem to blend into the background medium. Suddenly, the analogy to a baseball moving through air becomes more complicated. And this does not even consider the role of the magnetic field in linking these structures to their environs.

The situation is even worse if one tries to include translucent and diffuse clouds within the definition. For one thing, self-gravity is often unimportant in structuring these objects. Instead, turbulence seems to dominate in their structure and dynamics. Diffuse and translucent clouds are much more filamentary than dark clouds or GMCs (compare Figs. 7.4 with 7.6). Moreover the density contrast between the molecular gas and the surrounding atomic medium is at least one order of magnitude lower than for GMCs and dark clouds. Diffuse and translucent clouds are not so much baseballs as they are the denser, colder portions of large-scale atomic flows. Given their relatively short lifetimes (10^6 yr), they are transient structures, forming and dissipating more like smoke than baseballs. To make progress in understanding the relationship between dark cloud and GMCs versus diffuse and translucent clouds, the role of turbulence in these objects must be addressed. We will revisit the question of what is a molecular cloud after we discuss turbulence in Chap. 11.

References

Barnard, E.E. 1919, ApJ, 49, 1
Barnard, E.E. 1927, *Catalogue of 349 dark objects in the sky*
Bergin, E.A. and Tafalla, M. 2007, ARAA, 45, 339
Blitz, L. 1979, PhD Thesis, Columbia University
Blitz, L. and Shu, F.H. 1980, ApJ, 238, 148

Blitz, L., Magnani, L., and Mundy, L. 1984, ApJ, 282, L9
Blitz, L. 1991*a*, in *Molecular Astrophysics*, ed. T.W. Hartquist (Cambridge: Cambridge U. Press), 35
Blitz, L. 1991*b*, in *The Physics of Star Formation and Early Stellar Evolution*, ed. C.G. Lada and N.D. Kylafis (Dordrecht: Kluwer), 3
Blitz, L. and Williams, J.P. 1999, in *The Origins of Stars and Planetary Systems*, eds. C. Lada and N.D. Kylafis (Kluwer), 3
Bok, B.J., Cordwell, C.S., and Cromwell, R.H. 1971, in *Dark nebula, Globules and Protostars*, ed. B.T. Lynds, (Tucson: U. of Arizona Press), 33
Bok, B.J. 1977, PASP, 89, 597
Burton, W.B., Gordon, M.A., Bania, T.M., and Lockman F.J. 1975, ApJ, 202, 30
Burton, W.B. and Gordon, M.A. 1978, A&A, 63, 7
Carpenter, J.M., Snell, R.L., and Schloerb, F.P. 1995, ApJ, 445, 246
Clemens, D.P., Sanders, D.B., and Scoville, N.Z. 1988, ApJ, 327, 139
Clemens, D.P. and Barvainis, R. 1988, ApJS, 68, 257
Cohen, R.S. and Thaddeus, P. 1977, ApJ, 217, L155
Combes, F. 1991, ARAA, 29, 195
Cotten, D.L. and Magnani, L. 2013, MNRAS, 436, 1152
Crutcher, R.M. 1973, ApJ, 185, 857
Cudaback, D.D. and Heiles, C. 1969, ApJ, 155, 21
Dame, T.M., Elmegreen, B.G., Cohen, R.S., and Thaddeus 1986, ApJ, 305, 892
Dame, T.M., et al. 1987, ApJ, 322, 706
Dame, T.M., Hartmann, D., and Thaddeus, P. 2001, ApJ, 547, 792
Dickey, J. and Lockman, F.J, 1990, ARAA, 28, 215
Dickman, R.L. 1975, ApJ, 202, 50
Dickman, R.L. 1978, ApJS, 37, 407
Dieter, N. H. 1973, ApJ, 183, 449
Digel, S., De Geus, E., and Thaddeus, P. 1994, ApJ,
Dutra, C.M. and Bica, E. 2002, A&A, 383, 631
Elmegreen, B.G. 1982a, 253, 634
Elmegreen, B.G. 1982b, 253, 655
Elmegreen, B.G. 1990, in *The Evolution of the Interstellar Medium*, ed. L. Blitz (San Francisco: ASP Press), 247
Elmegreen, B.G. and Palous, J. 2006, Triggered Star Formation in a Turbulent ISM, IAU Symposium 237, eds. B.G. Elmegreen and J. Palous, (Cambridge: Cambridge University Press)
Feitzinger, J.V. and Stüwe, J.A. 1984, A&AS, 58, 365
Franco, J., Shore, S.N., and Tenorio-Tagle, G. 1994, ApJ, 436, 795
Friberg, P. and Hjalmarson, Å. 1990, in *Molecular Astrophysics*, ed. T.W. Hartquist (Cambridge, Cambridge U. Press), 3
Genzel, R. 1991, in *Molecular Clouds*, ed. R.A. James and T.J. Millar, (Cambridge: Cambridge U. Press), 75
Gillmon, K. and Shull, M.J. 2006, ApJ, 636, 908
Goldsmith, P.F. and Li, D. 2005, ApJ, 622, 938
Hartley, M., Tritton, S.B., Manchester, R.N., Smith, R.M., and Goss, W.M. 1986, A&A, 63, 27
Heiles, C. 1968, ApJ, 151, 919
Hennebelle, P. and Falgarone, E. 2012, A&ApRv, 20, 55
Heyer, M.H. et al. 1998, ApJS, 115, 241
Heyer, M.H. and Dame, T.M. 2015, ARAA, 53, 583
Israel, F.P., et al. 1984, A&A, 134, 396
Khavtassi, J. Sh. 1955, Bull. Abastumani Obs., No. 18
Knude, J. 1979, A&A, 38, 407
Kun, M. 1992, A&A, 92, 875
Kutner, M.L. and Mead, K.N. 1981, ApJ, 249, 15
Kwan, J. 1979, ApJ, 229, 567

Lada, E.A. 1990, PhD Thesis, University of Texas
LaRosa, T.N., Shore, S.N., and Magnani, L. 1999, 512, 761
Li, J.Z., Hu, J.Y., and Chen, W.P. 2000, A&A, 356, 157
Liljestrom, T. 1991, A&A, 1991, 244, 483
Liszt, H.S. 1984, ComAp, 10, 137
Liszt, H.S., and Wilson, R.W. 1993, ApJ, 403, 663
Liszt, H. 1994, ApJ, 429, 638
Liszt, H.S. and Lucas, R. 1994, ApJ, 431, 131
Liszt, H. and Lucas, R. 1995, A&A, 299, 847
Liszt, H. and Lucas, R. 1996, A&A, 314, 917
Liszt, H. and Lucas, R. 2002, A&A, 391, 693
Liszt, H.S. 2007, A&A, 291, 300
Liszt, H.S., Pety, J., and Lucas, R. 2010, A&A, 518, A45
Liszt, H.S. and Pety, J. 2012, A&A 541, A58
Lucas, R. and Liszt, H. 1994, A&A, 282, 5
Lucas, R. and Liszt, H.S. 2000, A&A, 358, 1069
Lundmark, K. 1926, Upsala Medd., No. 12
Luhman, K.L. 2001, ApJ, 560, 287
Lynds, B.T. 1962, ApJ, 7, 1
Lynds, B.T. 1965, ApJ, 12, 163
Lynds, B.T. 1968, in *Nebulae and Interstellar Matter*, eds. B.M. Middlehurst and L.H. Aller, (Chicago: University of Chicago Press)
Magnani, L., Blitz, L., and Mundy, L. 1985, ApJ, 295, 402
McGehee, P.M. 2008, in *Handbook of Star Forming Regions, Volume II: The Southern Sky*, ASP Monograph Publications, Vol. 5, ed. B. Reipurt, 813.
Mead, K.N. and Kutner, M.L. 1988, ApJ, 330, 399
Mouschovias, T., Shu, F., and Woodward, P. 1974, A&A, 33, 73
Palmer, P., Zuckerman, B., Buhl, D., and Snyder, L.E. 1969, ApJ, 156, L147
Parker, E.N. 1966, ApJ, 145, 811
Penzias, A.A., Solomon, P.M., Jefferts, K.B., and Wilson, R.W. 1972, ApJ, 174, L43
Planck Collaboration, Planck 2013 Results XIII, 2014, A&A, 571, A13
Pound, M.W. and Goodman, A.A. 1997, ApJ, 482, 334
Robinson, B.J., et al. 1984, JRASC, 78, 211
Roschkovsky, D.A. 1955, Contr. Alma-Ata 1, No. 1–2, 136
Sandage, A. 1976, AJ, 81, 154
Sandell, G., Johansson, L.E.B., Rieu, N.Q., and Mattila, K. 1981, A&A, 97, 317
Sandell, G., Reipurth, B., and Gahm, G. 1987, A&A, 181, 283
Sanders, D.B., Solomon, P.M., and Scoville, N.Z. 1984, ApJ, 276, 182
Sanders, D.B., Scoville, N.Z, and Solomon, P.M. 1985, ApJ, 289, 373
Schlegel, D.J., Finkbeiner, D.P., and Davis, M. 1998, ApJ, 500, 525
Schoenberg, E. 1964, Veroffentl. Sternw. Muchen, Vol. 5, No. 21
Scoville, N.Z. and Solomon, P.M. 1975, ApJL, 199, L105
Scoville, N.Z. and Hersh, K. 1979, ApJ, 229, 578
Scoville, N.Z. 2014, in Secular Evolution of Galaxies, ed. Jesús Falcón Barroso and Johan H. Knapen, (Cambridge: Cambridge U. Press), 491
Scoville, N.Z. and Sanders, D.B. 1987, in Interstellar Processes, eds. Hollenbach, D.J. and Thronson, H.A. Jr. (Dordrecht: Reidel)
Shore, S.N., Magnani, L., LaRosa, T.N., and McCarthy, M.N. 2003, ApJ, 593, 413
Shore, S.N., LaRosa, T.N., Chastain, R.J., and Magnani, L. 2005, A&A, 197, 206
Snell, R.L. 1979, PhD Thesis, University of Texas, Austin
Snow, T.P. 2005, in *Astrochemistry: Recent Successes and Current Challenges*, ed. D.C. Lis, G.A. Blake, and E. Herbst (Cambridge: Cambridge U. Press), 175
Solomon, P.M., Stark, A.A., and Sanders, D.B. 1983, ApJ, 267, L29

Solomon, P.M. and Rivolo, A.R. 1987, in *The Galaxy*, eds. Gilmore, G. and Carswell, B. (Dordrecht: Reidel)
Spitzer, L., Jr. 1978, *Physical Processes in the Interstellar Medium*, (New York: Wiley-Interscience)
Spitzer, L., Jr. 1985, ApJ, 290, 21
Stark, A.A. 1979, PhD Thesis, Princeton University
Szomoru, A. and Guhathakurta, P. 1998, ApJ, 494, 93
Ungerechts, H. and Thaddeus, P. 1987, ApJS, 63, 645
van den Bergh, S. 1966
Van Dishoeck, E.F. and Black, J.H. 1988, ApJ, 334, 771
Wakker, B.P. 2006, ApJS, 163, 282
Williams, J.P., Blitz, L., and McKee, C.F. 1999, in *Protostars and Planets IV*, eds. V. Mannings and A. Boss (Tucson: University of Arizona Press), 97
Wilson, R.W., Jefferts, K.B., and Penzias, A.A. 1970, ApJ, 161, L43
Woodward, P.R. 1976, ApJ, 207, 466
Wouterloot, J.G.A. and Brand, J. 1989, A&AS, 80, 149
Wouterloot, J.G.A., Brand, J., Burton, W.B., and Kwee, K.K. 1990, A&A, 230, 21
Yamamoto, H., Onishi, T., Mizuno, A., and Fukui, Y. 2003, ApJ, 592, 217
York, D.G. et al. 2000, AJ, 120, 1579

Chapter 8
The Relationship Between CO and H_2

Every story of conversion is a story of blessed defeat.

C.S. Lewis

Abstract The CO(1-0) transition is the most important molecular emission line for studying molecular gas in the ISM. Unfortunately, it is nearly always optically thick, and so determining N(CO) and, more importantly, N(H_2) is complicated. Over the years several empirical techniques have been worked out for using the CO(1-0) line to determine the physically relevant quantity, N(H_2). In this chapter we discuss these techniques with particular emphasis on determining the conversion factor (called X_{CO}) for diffuse and translucent molecular gas. The chapter closes with some comments on using the CO(1-0) line to trace dark gas.

8.1 Introduction

The physically relevant quantity, N(H_2), is not obtained directly from the observed velocity-integrated CO(1-0) main-beam temperature. The transition is often optically thick in $^{12}C^{16}O$. It must be derived by applying a conversion factor X_{CO}. In addition, in diffuse and translucent regions CO is not the principal repository of carbon so chemical effects cannot be ignored. Because of the complications produced by line formation and astrochemistry, it is not fruitful to derive the conversion factor(s) between W_{CO} and N(H_2) from theoretical considerations, and we will focus primarily on the empirical techniques. Given the difficulties, the relevant question—especially for diffuse molecular gas—becomes how reliably does CO(1-0) trace the molecular gas? In the 1970s and 1980s, this was rarely addressed. The consensus was that the CO(1-0) traces virtually all the molecular mass and a map of CO is *de facto* a map of H_2. In the last decade, this idea has changed and significant amounts of molecular gas are thought not to be traced by CO(1-0) observations. This is the central issue of this chapter. The "missing" molecular gas is thought to comprise anywhere from 10% to 100% of the

L. Magnani, S.N. Shore, *A Dirty Window*, Astrophysics and Space Science Library 442, DOI 10.1007/978-3-662-54350-4_8

CO-traceable gas and has come to be known as "dark" molecular gas.[1] We will examine the evidence for this component at the end of the chapter and revisit the efficacy of the CO(1-0) transition in tracing molecular gas.

8.2 The CO-H_2 Conversion Factor

To understand the physical state of molecular clouds either N(H_2) or n(H_2) must be measured. The velocity-integrated CO(1-0) main beam antenna temperature ($\int T_{mb}$ dv $\equiv W_{CO}$) is assumed to be linearly related to the molecular hydrogen column density by a *constant* called the X-factor or X_{CO}. The definition is

$$X_{CO} = N(H_2)/W_{CO} \quad cm^{-2}[K\ km\ s^{-1}]^{-1} \quad . \tag{8.1}$$

Because the CO J=1 level is populated by collisions with H_2, there should be a relation between the number density of the two species, at least in principle.

A single value of X_{CO} for different types of clouds, or even galaxies, is a commonplace assumption despite many papers which describe in detail how the conversion factor is expected to be sensitive to cloud size (Dickman et al. 1986), physical parameters (Kutner and Leung 1985; Maloney and Black 1988; Taylor et al. 1993), and type of cloud (de Vries et al. 1987; Magnani and Onello 1995). The value of X_{CO} has been shown to vary significantly across the face of a single molecular cloud (Magnani et al. 1998; Liszt et al. 2010). Despite these problems, most researchers just use a value for X_{CO} between 1–2 $\times$ 10^{20} cm^{-2} [K km s^{-1}]$^{-1}$ (we will drop these units in the remainder of the book). Moreover, some theoretical work has been done to justify why X_{CO} should be constant in most GMCs in our Galaxy (e.g., Dickman et al. 1986; Wall 2006), and why that value shouldn't hold in other galaxies (Maloney and Black 1988).

We describe in this chapter how X_{CO} is obtained and what problems and pitfalls are expected from the various methods. For reviews of this subject see Bolatto et al. (2013) and Combes (1991). In contrast to those papers, we concentrate on X_{CO} determinations in diffuse molecular gas.

Historically, there are six principal ways to determine the value of X_{CO} for a cloud or an ensemble of clouds: (1) extinction; (2) ^{13}CO at 110 GHz; (3) "virialization"; (4) gamma-ray emission; (5) far infrared emission; (6) and CH at 3.3 GHz. Each of these methods uses a surrogate to trace N(H_2) and couples the observations of the surrogate tracer with CO(1-0) observations to yield X_{CO}. We now discuss each technique.

[1] Although the term is reminiscent of dark matter, dark gas is baryonic and has nothing to do with the former.

8.2.1 X_{CO} from Extinction

The first and earliest method for calibrating X_{CO} uses the visual extinction determined from star counts as the surrogate for N(H_2). The critical assumption here is that the gas-to-dust ratio is known and constant in the objects in question. Originally, the empirical gas-to-dust relationship obtained by Savage et al. (1977) and Bohlin et al. (1978) was used (see Sect. 6.3), although there are other proposed relations Once the empirical relationship between the total column density of hydrogen, N(H_{total}) = N(HI) + 2N(H_2) and E(B-V). Then, the ratio of total to selective extinction, R_V, assumed to be ≈3.1, is used to convert reddening to the visual extinction. Once the calibration between A_V and N(H_{total}) is established, star counts usually produce A_V for an obscured region and, if N(HI) observations of the same region are made, N(H_2) follows.

This method was used to determine X_{CO} in translucent and dark molecular clouds with average values ranging from 0.8 to 6.7 × 10^{20} (e.g., Dickman 1978; Frerking et al. 1982; Magnani et al. 1988). There are, however, several caveats. The original E(B-V) to N(H_{total}) conversion was for low-extinction clouds, although the method is most often used for clouds with $A_V > 4$ mag. Moreover, for the method to yield correct values, the extinction associated with the dust in the HI component along the line of sight must be distinguished from that associated with molecular gas, thus 21 cm or Lyman series spectroscopic observations are necessary. Although N(HI) values for the entire sky are available they tend to sample much larger angular size scales than the CO or A_V measurements.[2] A source of uncertainty in this method also arises because the original calibration was for N(H_{total}) with E(B-V). To convert the latter quantity to the visual extinction, A_V, requires and assumption about R_V, so a significant uncertainty is added to the calibration. Given the difficulty in determining extinction in the diffuse ISM, the original technique is not very useful for the types of molecular cloud we are most interested in.

A recent improvement to the extinction method was pioneered by Lada et al. (1994) and refined by Lombardi and Alves (2001). The basic idea is to use near infrared color excess measurements towards background stars to determine the dust column density produced by an intervening interstellar cloud. This technique allows for much higher resolution and accurate extinction determinations and with the plethora of stellar near infrared data (the 2MASS survey, the WISE satellite, etc.), this technique has supplanted star counts as the way to determine extinction. Once a robust estimate of A_V at high extinction is in hand, determining X_{CO} follows readily. Lombardi et al. (2006) derive $X_{CO} = 2.91 \times 10^{20}$ in the range $0.9 \le A_V \le 5.4$ mag.

[2]Even the 305 m Arecibo radiotelescope has a beam size of ∼ 4' at 1420 MHz.

8.2.2 X_{CO} Using $^{13}CO(1\text{-}0)$

Another early method for calibrating X_{CO} was to use the ^{13}CO(1-0) transition at 110.201354 GHz. This line is often optically thin away from the densest parts in GMCs and dark clouds. Thus, direct observations of W(^{13}CO) (the velocity-integrated main-beam antenna temperature) give reliable values for N(^{13}CO). An empirically calibrated relationship between W(^{13}CO) and N(H_2) (usually deriving N(H_2) from extinction measurements) is used to get N(H_2) and comparisons of this quantity to W_{CO} lead immediately to X_{CO}. (e.g., Dickman 1978). The problem with this technique is that the assumption that the ^{13}CO(1-0) line is optically thin may not be a very good one—even in diffuse molecular clouds. Although this method was used extensively in the late 1970s, it was rather quickly superseded by the following two techniques.

8.2.3 X_{CO} from Virial Equilibrium Considerations

The basic idea behind this technique is to assume that molecular clouds are in virial equilibrium. One then obtains the kinetic energy of the cloud by mapping a spectroscopic tracer that provides the velocity dispersion from the linewidths. The size of the cloud is needed in the calculation and requires a distance determination. This is the greatest uncertainty in the calculation, but if all goes well, the mass is then obtained from the standard virial relation described in Sect. 4.3.1. The average N(H_2) follows from the virial mass, M_{VT}, and area of the cloud, and comparing this column density to the *average* value of W_{CO} gives an *average* value of X_{CO} for the cloud.

The idea that GMCs are in virial equilibrium arose from a famous study by Larson (1981) which also proposed the linewidth-size relation for these objects; $\Delta v \propto r^{0.5}$. When M_{VT} is compared to the mass of the cloud determined by other means, an assessment of the virial state of the cloud can be made. Since the very first observations of GMCs, their mass is always found to be close to the virial value so by the early 1980s, they were considered to be in virial equilibrium as a matter of course. If a linewidth-size relation holds, then the mass is proportional to R^2 implying that N(H_2) is constant. This was considered a key property of GMCs that most likely had to do with how these objects form and/or how they regulate themselves.

The last comprehensive study of the virial state of GMCs was carried out by Solomon et al. (1987) who determined that a cloud's CO luminosity, L_{CO} (a directly observable parameter defined as the velocity integrated CO(1-0) brightness temperature integrated over the entire cloud) is related to M_{VT} by the relation

$$M_{VT} = 39(L_{CO})^{0.81} \quad M_{\odot} \quad . \tag{8.2}$$

Were GMCs in virial equilibrium, the exponent would be exactly unity by the following argument. The mass of the cloud can be expressed as

$$M = \mu m_H N(H_2) A \tag{8.3}$$

where μ is the mean molecular weight in the cloud, m_H is the mass of the hydrogen atom, and A is the area of the cloud. Ignoring geometrical factors and constants, we can express this relationship as:

$$M \propto N(H_2) R^2 \propto X_{CO} W_{CO} R^2 \quad . \tag{8.4}$$

where R is a characteristic size scale for the cloud. Since L_{CO} is $\propto$ W_{CO} R^2, it follows that

$$M \propto X_{CO} L_{CO} \quad . \tag{8.5}$$

For constant X_{CO} in GMCs, the mass of the cloud should be directly proportional to L_{CO}. Solomon et al. (1987) found a nearly linear relation, so the mass of the cloud is basically M_{VT} and thus clouds are virialized. But because GMCs were already considered to be virialized, they used their results to determine X_{CO} and not to verify the virial equilibrium of clouds. Since the actual exponent derived by Solomon et al. (1987) was not unity, the actual value of X_{CO} depends on the cloud mass and Solomon et al. derived 3×10^{20} for their sample of GMCs. It is worth noting that Issa et al. (1990) argued strongly for a lower value of X_{CO} in GMCs: $\sim$ 1.0–1.5 $\times\ 10^{20}$, and questioned many basic premises of the virial argument, including the relationship between linewidth and size of the cloud. For many years there was a sort of Holy War between proponents of a value of X_{CO} around 1×10^{20} and those who championed values 2 to 4 times larger.

One of the complications of the virial method is determining a representative value for σ, a measure of the gas motions and thus the kinetic energy throughout the cloud. The ideal way would be to take a composite CO spectrum over the entire cloud and determine σ from a Gaussian fit to the profile. Although this is often unfeasible given mapping timescales and the non-Gaussian nature of the line profiles, a satisfactory estimate can nevertheless be made by taking the velocity dispersion of the Gaussian-fit centroids and adding that to the average linewidths in quadrature:

$$\sigma_{tot}^2 = \sigma_c^2 + \sigma_\ell^2 \quad , \tag{8.6}$$

where σ_{tot} is the total velocity dispersion of the gas (the quantity we want to use for σ in the virial equilibrium equation), σ_c is the dispersion of the line centroids and

σ_ℓ is the average of the dispersion of the linewidths. Kleiner and Dickman (1985) called σ_{tot} the "parent" dispersion.

Of course, the virial method is most often used to obtain the mass of the cloud rather than derive X_{CO}. Since it assumes strict mechanical equilibrium, this method is not of much use in diffuse and translucent clouds which in most instances do not fulfill this condition. Even for GMCs, it is important to note the criticism by Maloney (1988, 1990) who showed convincingly that the relation derived by Solomon et al. (1987) can be derived by merely assuming the existence of a linewidth-size relation like that found empirically in GMCs by Larson (1981). Moreover, the linewidth-size relation is a natural consequence of having pressure confined clouds (see Chap. 11). Thus, a CO-luminosity-virial mass relation does *not* necessarily mean that the clouds are virialized.

8.2.4 X_{CO} *from Gamma-Rays*

A diffuse, high-energy gamma-ray background at $E_\gamma > 30$ MeV has long been known to exist (Kraushaar et al. 1972). While a portion is unresolved extragalactic sources, the rest is produced primarily by cosmic rays in our Galaxy colliding with hydrogen nucleons and inverse Compton scattering photons from the interstellar radiation field (Sect. 5.7.1). By comparing the diffuse gamma-ray emission with CO and HI data from large-scale surveys, X_{CO} can be derived. The gamma-ray emission is usually modeled as a linear combination of N(HI) and W_{CO} with the free parameters being the cosmic-ray distribution and the value of X_{CO} [Bloemen (1989) gives a thorough review of this technique]. In reality, it is not so straightforward. In addition to the instrumental uncertainties, the inverse Compton contribution must be modeled and removed, the gamma-ray emissivities per nucleon likely vary with galactocentric distance, and the HI data can also be problematic. Despite these difficulties, the typical value of X_{CO} obtained from this method for the inner Galaxy is 2.3×10^{20}, consistent with what is derived from the methods described above. Similarly, a value of $2.6 \pm 1.2 \times 10^{20}$ was obtained by Bloemen et al. (1984) for the large complex of clouds in Orion and Monoceros. Problems explaining the radial gradient of gamma-rays (too small given the decrease of cosmic-ray sources in the Outer Galaxy) have led some authors to a systematic increase in X_{CO} with galactocentric distance (Strong et al. 2004).

Although the technique is free from many of the assumptions plaguing the other methods, there are limitations to the types of individual clouds that can be detected and for which X_{CO} can be derived. Stacy et al. (1991) showed that the EGRET instrument aboard GRO was not sensitive enough to detect gamma-rays from diffuse and translucent molecular clouds in a specific high-latitude region. This situation is changing. The LAT instrument on *Fermi* (see Sect. 5.7) is more than an order of magnitude more sensitive that EGRET and is able to image the larger high-latitude clouds. Emission from regions with $\sim 10^4$ $M_\odot$ is detectable and has been used to calibrate X_{CO} in the Cepheus and Polaris Flare regions (Ackermann et al. 2012) finding X_{CO} equal to 0.6×10^{20}.

8.2.5 X_{CO} from Far-Infrared Emission

The connection between gas and dust was exploited in Sect. 8.2.1 to obtain X_{CO} from dust column densities obtained from absorption measurements of starlight using a standard gas-to-dust ratio. The large-scale infrared surveys of the Galaxy (see Chap. 5) provide dust column density determinations from its thermal emission. Various studies have shown that the IRAS 100 μm radiance [I(100)-units of MJy sr^{-1})], is linearly correlated with extinction (de Vries and Le Poole 1985; Laureijs et al. 1987, 1989; Heithausen and Mebold 1989). This emission is optically thin and, if there are no local heating sources, I(100) should be proportional to the line of sight $N(H_{total})$. The atomic hydrogen component is obtained from 21 cm measurements and the molecular component is from CO(1-0) observations.[3] This relationship is expressed as:

$$I(100) = a_1 N(HI) + 2a_2 N(H_2) + I_{bg}(100) \quad . \tag{8.7}$$

$I_{bg}(100)$ is from non-solar system foreground and background emission. The constants a_1 and a_2 are the hydrogen emissivities per hydrogen nucleon in atomic and molecular gas, respectively. N(HI) is the column associated with the cloud (as defined by the molecular velocity distribution). For the atomic portion of the diffuse ISM, Boulanger and Pérault (1988) have shown that equation (8.7) is valid (with $a_2 = 0$). Using the definition of the X-factor, we can write 8.7 as

$$X_{CO} = \frac{1}{2a_2 W_{CO}}[I(100) - I_{bg}(100) - a_1 N(HI)] \quad . \tag{8.8}$$

The value of a_1 can be derived using data from regions free of molecular gas (e.g., Boulanger et al. 1985) so all quantities with the exception of a_2 can be measured from the combination of infrared, HI, and CO(1-0) observations. de Vries et al. (1987) were the first to use this method. They assumed a_1 and a_2 are identical and derived $X_{CO} = 0.6 \pm 0.1 \times 10^{20}$ for a region rich in translucent clouds near the North Celestial Pole.

[3]The separation of the fraction of far infrared emission that is associated with the target cloud from any unrelated foreground or background gas is effected spectroscopically. The velocity ranges can be chosen from the line profile measurements. We ignore the ionized component but this, in principle, can be included from X-ray observations and diffuse Hα emission.

It is, however, neither obvious nor likely that the two coefficients are equal. There is little consensus and few studies on this issue. On the one hand, Weiland et al. (1986) find little to no evidence for change in the grain characteristics between atomic and molecular gas. On the other hand, the dust emissivity in high-latitude molecular clouds could be larger than in the atomic phase by a factor ≈ 3 (Planck Collaboration, Planck Early Results 2011, Paper XIX), and Wagle et al. (2015) find evidence of grains changing sizes across atomic-molecular transition regions.

Although used primarily for translucent clouds, this method was also used by Parkinson et al. (1987) and Bloemen et al. (1990) to determine X_{CO} for molecular clouds in the inner Galaxy. In principle, this technique is more convenient for diffuse regions than the gamma-ray method because their extended infrared emission is now relatively easy to measure from the high-resolution, infrared data sets available online. Reach et al. (1998) applied this technique to the infrared-excess clouds in their survey that had corresponding CO emission and derived a values of $X_{CO} = (1.3 \pm 0.2) \times 10^{20}$ for high-latitude molecular clouds (see Chap. 9).

8.2.6 X_{CO} Using the CH 3335 MHz Line

The CH 3335 MHz transition was discussed in Sect. 3.4.3. Its utility as a tracer of H_2 is based on the linear relationship between N(CH) and N(H_2) for molecular regions with less than 5 magnitudes of extinction; a regime that encompasses diffuse and translucent regions, and the outer regions of GMCs and dark clouds. Because this line is optically thin, it is straightforward to observationally determine N(CH):

$$N(CH) = 2.82 \times 10^{14}[T_{ex}\eta_f\eta_b(T_{ex} - T_{bg})^{-1})] \int T_A dv \quad cm^{-2} \tag{8.9}$$

where T_{ex} and T_{bg} are the excitation and background temperatures, respectively, η_f is the filling-fraction of CH in the telescope beam, η_b is the beam efficiency of the telescope at 3.3 GHz, and the integral is just W_{CH}, the velocity-integrated antenna temperature in units of K km s^{-1}. To determine N(CH) from equation 8.9, the excitation temperature of the transition must be known. Since the line was first detected, it has always been assumed that the transition is weakly masing from both observational (e.g., Rydbeck et al. 1976; Genzel et al. 1979), and theoretical considerations (Bertojo et al. 1976), so that $|T_{ex}| \gg T_{bg}$ and the factor $T_{ex}/(T_{ex}-T_{bg})$ can be set to unity. Thus, N(CH) $\propto W_{CH}$. It is important to note, however, that the negative excitation temperatures were determined from on-off radio measurements (see Sect. 4.2.3) which are notoriously difficult if the beamsizes are large. Even with the 305 meter Arecibo radiotelescope, the beam size at 3 GHz is $1.3' \times 1.6'$ so the assumption that T_{ex} on-source is the same as off-source may not hold. To further muddy the waters, the only direct determinations of T_{ex} for the 3.3 GHz lines involve measurements of the upper and lower $X^2\Pi_{1/2}$ doublet levels from optical absorption lines (Lien 1984) towards diffuse clouds. These T_{ex} values are indeed inverted, but small, ~ -1 or -2 K. This shifts $T_{ex}/(T_{ex} - T_{bg})$ away from unity significantly changing N(CH) from when $|T_{ex}| \gg T_{bg}$. Despite this concern, years of CH observations with analysis assuming $|T_{ex}| \gg T_{bg}$ have yielded results for N(H_2) for cloud masses that are consistent with those obtained from other tracers. For regions with $A_V \lesssim 5$ mag, the CH 3335 MHz line seems to yield robust estimates of N(CH).

Once N(CH) is determined, the key step is to get N(H_2). The first comprehensive analysis was by Mattila (1986) who derived a relationship between N(CH) and A_B, making the conversion to N(H_2) relatively straightforward. Subsequent work by Magnani and Onello (1995) and, in particular, Liszt and Lucas (2002) established that N(CH)/N(H_2) is $\sim 4 \times 10^{-8}$ with a variance of 45%. Thus, observations of W_{CH} and W_{CO} readily yield X_{CO} (Magnani and Onello 1995). Applying this method to translucent clouds gives $X_{CO} = 0.3 - 6.8 \times 10^{20}$. A sample of dark clouds yields a similar result, $X_{CO} = 0.8 - 8.6 \times 10^{20}$, but the linearity of N(CH) with N(H_2) fails at $A_V \gtrsim 5$ mag.

Magnani et al. (1998) applied this technique not to a sample of clouds, but over many lines of sight in a diffuse and a translucent cloud (MBM 40 and 16, respectively). Once again, they found a varying value of X_{CO} ranging from 0.7–17.3 $\times$ 10^{20}. It should have been clear from the beginning that there would be a significant variation of the CO-H_2 conversion factor for these objects since E(B-V) ranges by less than an order of magnitude over regions where W_{CO} varies by nearly a factor of 100. Either the gas-to-dust ratio varies wildly or the CO/H_2 abundance does; theoretical considerations favor the latter interpretation (e.g., van Dishoeck and Black 1988). An intesting trend in X_{CO} has been noted by Magnani et al. (1998) and Cotten and Magnani (2013). In diffuse and translucent clouds X_{CO} increases with decreasing W_{CO}. Figure 8.1 illustrates this for a set of points in the diffuse cloud MBM 40. The variation of X_{CO} in diffuse clouds forces us to consider that W_{CO} may not be a linear tracer of N(H_2). As discussed in Sect. 7.4.3, the work of Liszt and Lucas on CO-bright diffuse clouds underscores this and makes a strong case for radiative transfer effects producing the intensity of the CO(1-0) line in diffuse clouds.

Perhaps the most problematic issue with this technique is the difficulty in obtaining the CH data in diffuse molecular gas. It often takes several hours of integration to get a marginal detection (typical values of W_{CH} in diffuse molecular clouds are of order 10 mK km s^{-1}). Coupled with the historical dearth of 3 GHz receivers at radio observatories (Arecibo and Onsala being notable exceptions) this technique is not widely used, despite the possibility that it may be the best method for determining X_{CO} in diffuse molecular gas.

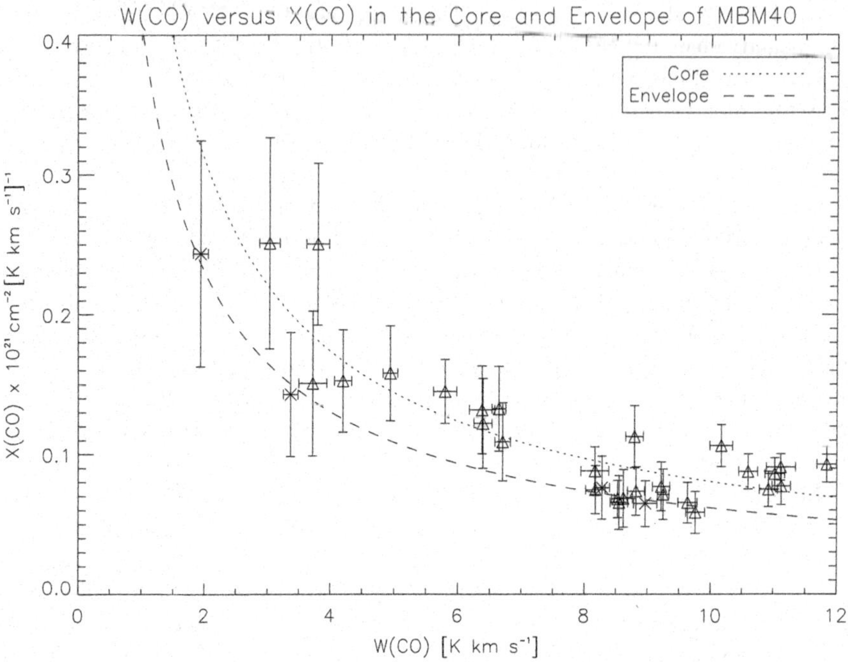

Fig. 8.1 X_{CO} versus W_{CO} for the envelope region and the core of the diffuse cloud MBM 40. The envelope consists of four points and are labeled with X. The curve fit for the core is of the form $X_{CO} = 4.5 \times 10^{20}\ (W_{CO})^{0.78}$ (with a coefficient of determination of 0.54) and for the envelope it is of the form $X_{CO} = 2.6 \times 10^{20}\ (W_{CO})^{0.71}$ (with a coefficient of determination of 0.97). The figure is reproduced from Cotten and Magnani (2013)

8.3 Do the Galactic CO Surveys Trace Most of the Molecular Gas?

In Chap. 7 we discussed the different types of molecular clouds. It is safe to say that most of what we know about these objects has come from observations of the lowest CO rotational transitions. Statements to the effect that CO(1-0) surveys are *de facto* surveys of the H_2 distribution are often heard at meetings or during talks. Diffuse molecular gas, if mentioned at all, has been termed "chaff" and estimates of its overall contribution to the molecular gas inventory are usually at or below the 10% level. But does the CO(1-0) line really trace all, or even most, of the molecular gas in the Galaxy?

There is considerable evidence that the current CO surveys have missed some portion of the molecular gas in the *diffuse* ISM. Grenier et al. (2005) claimed that as much as half of the total molecular inventory of the Galaxy is not accounted for by the CO surveys and is *spectroscopically undetectable*. So, they detected this gas using gamma-ray and infrared emission and named it "dark gas". But long before there were indications of the incompleteness of the large-scale CO spectroscopic surveys as far as diffuse molecular gas was concerned.

In the late 1970s, Lars E.B. Johansson conducted a CH 3335 MHz survey of the Galactic plane using the Onsala Observatory 26 m radiotelescope and found that the molecular distribution determined by CH observations is more extended than that from the large-scale CO(1-0) surveys mentioned in the previous chapter. He found that the CH distribution is "in between" that of the CO and HI 21 cm. But, because he published his study in an Onsala publication (Johansson 1979), it is not well-known. We reproduce in Fig. 8.2 the "smoking gun" plot from his paper.

As is clear from the figure, the distribution of CH from its number density peaks at the Galaxy's molecular ring, but does not decrease as sharply as CO (what is plotted for the CO and HI is the mass surface density). There has been no follow-up study most likely because of the difficulty in observing the weak CH 3.3 GHz line. On much smaller scales, the CH 3.3 GHz line appears to trace molecular gas where CO emission is undetected (Federman and Willson 1982; Sandell et al. 1987; Magnani and Onello 1993). Similar results have been noted for the HCO^+(1-0) transition [Lucas and Liszt (1994); Hogerheijde et al. (1995); and Liszt and Lucas (1998)], and for the 18 cm OH main lines [Wannier et al. (1993); Barriault et al. (2010*b*); Cotten et al. (2012); Allen et al. (2012); Allen et al. (2015)].

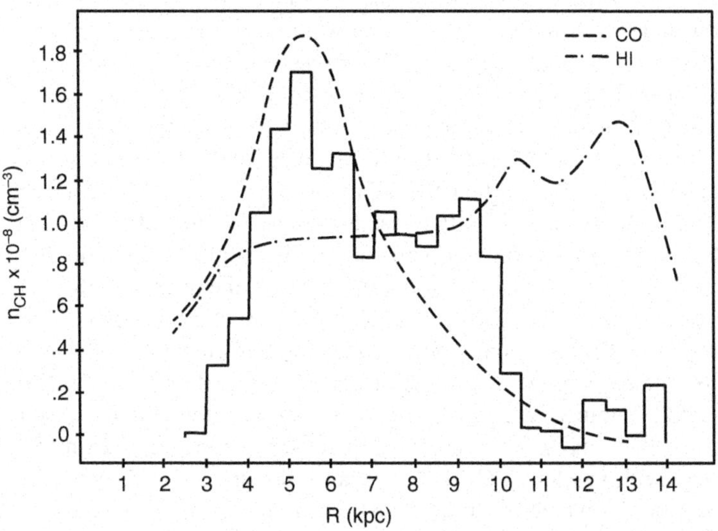

Fig. 8.2 The galactocentric distribution of molecular gas along the Galactic plane as determined from CH observations at 3335 MHz made by L.E.B. Johansson (1979) using the Onsala Observatory 26 m radiotelescope. The *solid line* is the CH distribution with the y-axis representing the volume density of CH in units of 1×10^{-8} cm^{-3}. The *dashed line* and the *dashed-dotted line* are the CO and HI mass surface density distributions, respectively, normalized to the peak of the CH emission in the Molecular Ring (4–6 kpc in this figure; the Solar Circle here is at 10 kpc)

Bloemen et al. (1986) used gamma-ray data from COS-B to show that there is more gas mass in the Galaxy than indicated by CO and HI surveys. This result was corroborated by Strong and Mattox (1996) using EGRET, and Abdo et al. (2010) using the LAT instrument on *Fermi*. A similar approach was taken by Grenier et al. (2005) but we will discuss those results specifically in the next section.

Ironically, CO observations can be used indirectly to address the incompleteness of the large scale CO surveys. Polk et al. (1988) and Chiar et al. (1994) looked along the Galactic plane and found ratios of $^{12}CO/^{13}CO$ greater than those found in GMCs (6.7 vs. 3.0) and CO(2-1)/CO(1-0) ratios of ~ 0.7, respectively. Both ratios are typical of non-star-forming, diffuse molecular gas. Interpreting the observed ratios as arising from a mixture of lower opacity translucent regions and GMCs led to the surprising conclusion in both studies that *Approximately half the molecular gas could be contained in clouds with relatively low opacity, or, possibly, in the extended envelopes of GMCs*. These regions tended to be missed by the large-scale CO surveys as the regions in question have relatively weak CO emission. Along completely different lines, Dame and Thaddeus (1994) established the existence of a new molecular gas component in the Inner Galaxy with a significantly greater scale height than that of GMCs and with a $^{12}CO/^{13}CO$ ratio greater than that of GMCs and more in keeping with that of translucent clouds.

The evidence for significant low-density, diffuse molecular gas was not exclusively confined to spectroscopic means. Infrared maps of the sky, first from the IRAS and then from the COBE satellite, were used in conjunction with existing CO and HI surveys to identify regions of excess infrared emission. These regions showed higher infrared intensity than expected from a purely atomic medium and so the presence of molecular gas not traced by the CO(1-0) line was raised by several studies: Reach, Koo, and Heiles (1994); Boulanger et al. (1998); and Reach et al. (1998). This last study, in particular, used a color-corrected and re-calibrated version of the IRAS 60 and 100 μm data to establish that the amount of molecular gas at high Galactic latitudes in the northern Galactic hemisphere was virtually the same as in the south. This conclusion is very different from the results of two high-latitude CO surveys conducted by Hartmann et al. (1998) and Magnani et al. (2000) which are discussed in the next chapter. Briefly, these authors found an order of magnitude more molecular gas in the south than in the north based on CO(1-0) observations of over 15,000 pointings at $|b| \geq 30°$. Both the CO and infrared results can be consistent if there exists a considerable quantity of molecular gas in the northern hemisphere with too low a CO/H_2 abundance for detection via the CO(1-0) line when the latter is mapped with the usual sensitivity (i.e., ~ 0.1 K rms in antenna temperature).

A similar approach by Barriault et al. (2010*a*) indicated that CO might not be tracing all the molecular gas in a high-latitude region and follow-up observations of the OH 18 cm main lines by Barriault et al. (2010*b*) seemed to confirm this assertion (more on this in Sect. 8.4.2). A similar conclusion was reached by Cotten et al. (2012) as they partially mapped the diffuse cloud MBM 40 in OH 1667 MHz and found a molecular mass twice that obtained from the CO maps. About half of the emission resides in an extended envelope around the cloud. However—and this is a point that has not been adequately noted when discussing this issue—there is nothing sacrosanct about choosing an rms level of 0.1 K in antenna temperature for CO(1-0) observations of molecular clouds. Cotten and Magnani (2013) made systematically more sensitive CO observations of MBM 40 and derived virtually the same molecular mass as that from the OH observation. In other words, CO(1-0) observations may match the results from the OH 1667 MHz line—if more sensitive than usual CO(1-0) observations are made. We will examine this issue in more detail is Sect. 8.4.1.

Finally, studies of self-absorbed HI profiles (sometimes known as HISA—HI Self-Absorption) have shown that they are often associated with molecular gas (e.g., Jackson et al. 2002; Li and Goldsmith 2003). Regions with HISA appear to be the gravitationally unbound portions of cold molecular regions; a description that sounds very much like a typical high-latitude cloud. However, HISA regions are often associated with dark cloud complexes (such as the Taurus/Aurigae/Perseus clouds) and a study of HISA regions in Perseus implies that significant amounts of cold molecular gas, not always detectable in CO, exist there (Klaassen et al. 2005).

In summary, there is considerable observational evidence that the large scale CO(1-0) surveys may have missed more molecular gas than is conventionally believed. In the cases described above, the ignored molecular gas is diffuse but detectable—if not by more sensitive CO(1-0) observations, then by other tracers. It is likely that not much molecular fraction is tied up in this component because of the diffuse nature of the gas. Estimates of the overall molecular mass in diffuse gas are difficult to come by, but it is hard to imagine that it would amount to more than 10%–20% of the overall molecular mass inventory of the Galaxy. In contrast, Grenier et al. (2005) claim that about half of the molecular gas in the Galaxy can *not* be detected spectroscopically. We turn to this issue in the next section.

8.4 On the Question of Dark Gas

The existence of "dark" molecular gas (at first, molecular gas that could not be detected spectroscopically; later, molecular gas not traced by the CO(1-0) line) was proposed based on evidence that this line was not tracing all of the molecular gas in various regions. This has nothing to do with dark matter but is meant to represent a component of the ISM that was spectroscopically undetectable (e.g., Wolfire et al. 2010). The dark gas surrounded all nearby CO clouds and their paper produced a flurry of studies several years later that detected the presence of this dark gas.

It is clear from the basic idea of PDRs that a region with substantial H_2 but undetectable levels of N(CO) should exist surrounding the molecular region traditionally defined by CO emission. The carbon in that region is atomic in either neutral or ionized form. The real question concerns the extent of this region and how much molecular mass it contains compared to that traced by CO. Spectroscopic observations of such a region are necessary along with gamma-ray and infrared imaging because spectral lines provide critical velocity information that can reveal a cloud's distance and the kinematics of its outer regions.

In light of these considerations, the CII $^2P_{3/2} -^2 P_{1/2}$ transition is an obvious choice for detecting the dark gas. Langer et al. (2010) used the *Herschel* HIFI instrument to observe the [CII] fine structure line at 158 μm in sixteen lines of sight traversing the inner Galaxy. These intersected 146 interstellar clouds and 29 diffuse clouds ($A_V < 1.3$ mag). CO(1-0) observations were made and HI was obtained from public databases. [CII] emission was detected at velocities not seen in CO. Because the [CII] fine structure line has an excitation temperature of 30 K, Langer et al. merely claimed to have detected "warm" dark gas and did not weigh in on the overall fraction of dark molecular gas. They did say explicitly, *In conclusion, the warm"dark gas" is no longer "dark", indeed it shines quite brightly in* C^+. At this point, the definition of "dark gas" began shifting to "relatively cold gas that is not detectable by CO(1-0) emission".

Abdo et al. (2010) used the LAT on the *Fermi* satellite to survey gamma-ray emission from the Gould Belt clouds in Cassiopeia, Cepheus, and the Polaris Flare and, using an analysis similar to that by Grenier et al., determined that dark molecular gas was present in amounts up to ~ 50% of the molecular mass traced by CO. However, it is important to note that many of the Gould Belt clouds are of the translucent variety (e.g., the Polaris Flare) and that in these clouds, the size of the region with H_2 and undetectable CO is likely to be much larger than for the GMCs along the Galactic plane.

Finally, observations by the *Planck* satellite show a dust component traced by its mm and submm continuum emission does not have corresponding CO or HI emission (Planck Collaboration, Planck Early Results 2011, Paper XIX). The amount of gas hidden in this component in the solar neighborhood was reported to have a mass equivalent to 28% of the atomic mass and 118% of the CO emitting gas in the same region. However, this analysis excludes regions with $|b| \leq 10°$. In other words, the result applies to local gas and avoids dealing with GMCs which dominate the Galactic plane.

Taken together, these observations all indicate that nondetected gas is a significant component of the Galactic molecular mass, at least as far as the diffuse and translucent clouds are concerned. However, all these studies compare their H_2 tracers to CO from conventional cloud mapping. Historically, given the size of nearby molecular clouds, thousands or tens of thousands of spectra are required to map a typical molecular cloud [see, e.g., Fig. 5.2, which required over 9000 CO(1-0) spectra or the IRAM key project - Falgarone et al. (1998)]. Consequently, there is a trade-off between the area mapped and the sensitivity of the observations. What happens when significantly more sensitive CO(1-0) observation are made?

8.4.1 Sensitive CO(1-0) Observations

One of the regions studied by Grenier et al. (2005) is at high latitudes ($50° \leq \ell \leq 140°$ and $-60 \leq b \leq -26°$) and shows a dark to non-dark ratio of molecular gas of 1.3. However, to derive the CO-detected molecular content, the authors included only some of the known high-latitude molecular clouds in the region. They ignored the Southern Hemisphere Galactic survey (Magnani et al. 2000) which indicated that there are clouds in the region such as the MBM 3–5 complex. Including the masses of these objects lowers the ratio of dark-to-conventional molecular gas, although it still allows a sizeable fraction of the molecular gas to be dark. In addition the technique used by Grenier et al., tracing molecular gas using gamma-ray and infrared emission, is subject to some concerns. The gamma-ray emission requires removing point sources and backgrounds such as unresolved extragalactic sources and the inverse Compton scattering of the CMB. The infrared data may not give a reliable measure of the amount of gas present in a region because of three principal reasons: (1) variations in the gas-to-dust ratio; (2) miscalculating N(HI) because of the optically thin approximation for the 21 cm line; and (3) non-homogeneous grain size distributions which can change the emissivity and optical depth from region to region. Even if all these issues are favorably resolved for the dark gas hypothesis, its original definition—spectroscopically undetectable molecular gas—is wrong even independent of [CII] observations. FUSE observations (see Chap. 5) spectroscopically detect *all* of this gas in absorption. What makes *mapping* it difficult is the lack of sufficient background stars. In arriving at their conclusions they also had to assume that more sensitive observations of CO(1-0) would not reveal additional molecular gas. Referring to the Grenier et al. study, Chastain et al. (2006) summarized the main problem with using CO data from the literature to trace low-density, diffuse molecular gas:

The authors (i.e., Grenier et al.) *used several interstellar gas tracers to identify large regions of cold dust and gas. Although they state that these regions contain molecular gas not detectable in CO, the shell described in this paper is visible in their "dark" molecular gas map and does contain low-level CO(1-0) emission. Moreover, such a correspondence between CO detections and structures identified by the gamma-ray method described by Grenier et al. is not unique. A comparison of their map with the CO maps of Hartmann et al. (1998) and Magnani et al. (2000) indicates that many of their "dark" molecular interstellar structures have CO counterparts. The CO(1-0) line is a very robust tracer of molecular gas, although it may be necessary to integrate to lower rms values than is normally done for large-scale Galactic plane surveys.*

Chastain et al. (2006) detected CO(1-0) and CO(2-1) emission in a large, shell-like, molecular structure in the Pegasus-Pisces region. We reproduce two figures from that paper (Figs. 8.3 and 8.4) to underscore how faint some of the CO(1-0) lines are in diffuse regions and, perhaps more importantly, to show that CO is sporadic at E(B-V) values as low as 0.1–0.2 mag. Either the gas-to-dust ratio and/or the CO excitation conditions are changing over scales of a few hundredths of a parsec.

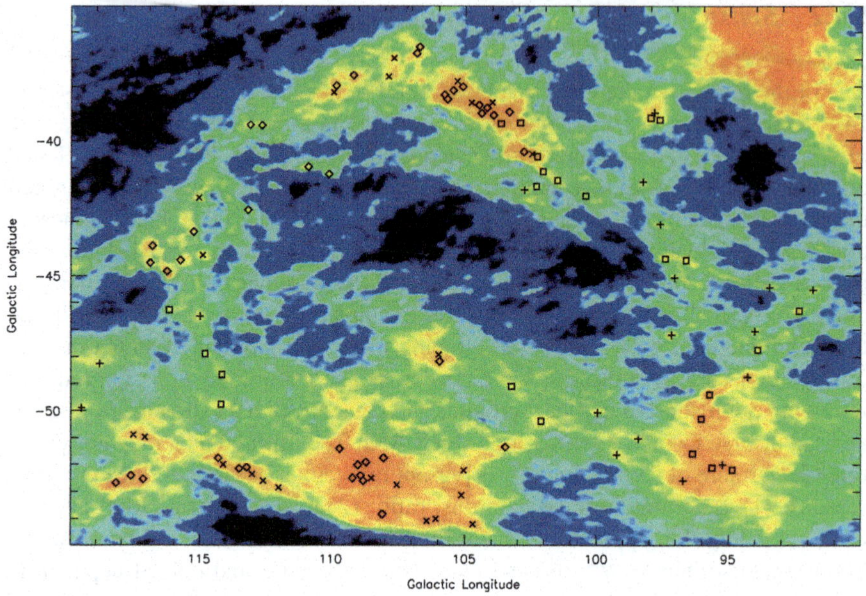

Fig. 8.3 The Pegasus-Pisces shell described by Chastain et al. (2006). This is Figure 3 from that paper. The color image is from the SFD E(B-V) maps with the color excess ranging from 0.0 to 0.5 mag. The plus signs and exes represent detections of the CO(1-0) and CO(2-1) lines, respectively, and the *squares and diamonds* represent non-detections. The scale is histogram equalized to enhance contrast. Note that detections and non-detections occur haphazardly in region with identical color excesses and thus dust column densities. This implies that either the dust-to-gas ratio, the excitation conditions, or the astrochemistry is varying over scales of a few hundredths of a parsec

Thus, sensitive observations can reveal the presence of molecular gas that would otherwise be missed. An indication of how much gas might reside in these low-emission regions was made by Cotten and Magnani (2013) who observed 103 lines of sight in MBM 40. Conventional CO(1-0) mapping of the cloud by Chastain (2008) gave a molecular mass of 11 $M_{\odot}$. Although over 25,000 spectra from the FCRAO 14 m radio telescope were used to map the cloud (Fig. 5.5), the 1-σ rms noise was relatively high; 0.7 K. The Cotten and Magnani (2013) mass, 32 $M_{\odot}$, was based on 1-σ rms values for CO(1-0) often as low as 0.08 K. The outermost regions of the cloud where the color excess was less than 0.12 mag contains nearly a third of the overall cloud mass. But there may be even more CO(1-0) emission than they found in their survey of that cloud. Figure 8.5 shows how a CO(1-0) observation with an rms value of 0.04 K reveals emission at the dust-defined edge of MBM 40. There is likely little mass tied up in this region given detections with $T_R^* \sim 0.15$ K, and it may not be worthwhile to map a cloud to this level to pick up just 5%–10% of the molecular emission. Nevertheless, the point is that sensitive observations can reveal molecular gas in regions with much lower extinction than previously thought. Claims of regions with spectroscopically undetectable dark gas should be taken with

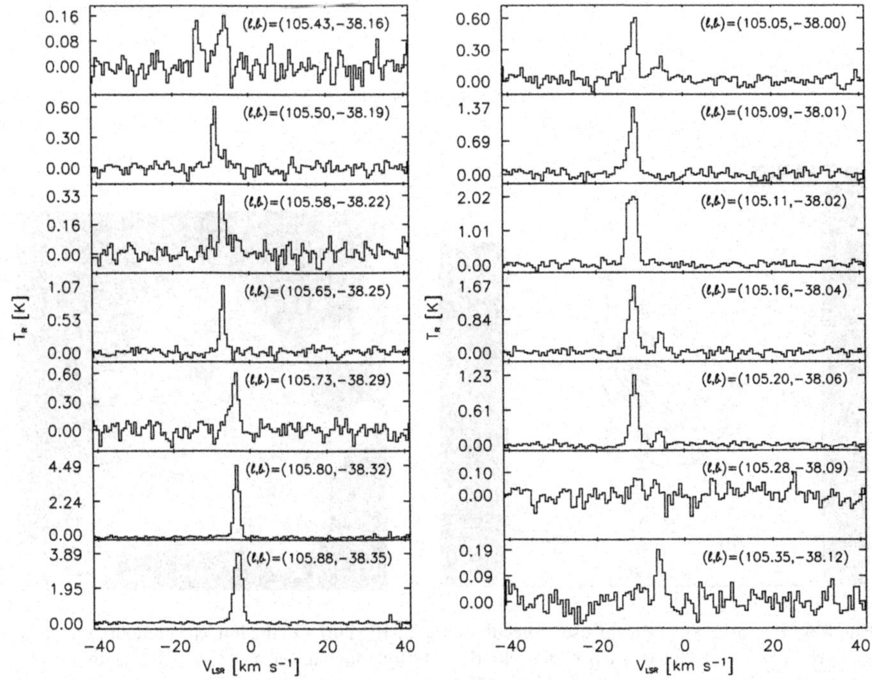

Fig. 8.4 This is figure 5 from Chastain et al. (2006) and shows CO(1-0) observations across a strip in the top right portion of the shell shown in Fig. 8.3. In this region, the color excess varies from 0.04 to 0.29 mag, a factor of 7, while the peak CO line antenna temperature, T_R^* varies from 0.15 to 3.9 K, a factor of 26. It is important to note that several of the detections in this figure would be difficult or impossible to identify if the observations had a 1-σ rms noise levels of 0.1 K

a grain of salt unless the CO observations have 1-σ rms noise values ~20 mK, or better. Of course, this result was for a diffuse molecular cloud. It will not be the same for GMCs or even dark molecular clouds. The bulk of their mass resides in the denser regions which are *completely* traced by the CO(1-0) line. Highly sensitive observations of GMCs will almost certainly not change significantly the mass of these objects because the envelope and interclump regions are likely to not contain much molecular mass.

8.4.2 OH as the Best Tracer of Dark Gas?

Indications that the OH ground-state, hyperfine, main lines at 1665 and 1667 MHz might be exceptional tracers of molecular gas in low-density environments come from the large scale mapping described in Jan Wouterloot's thesis (1981). The idea that perhaps these emission lines trace gas otherwise undetectable in CO(1-0) was first broached by Wannier et al. (1993) when they studied the warm neutral

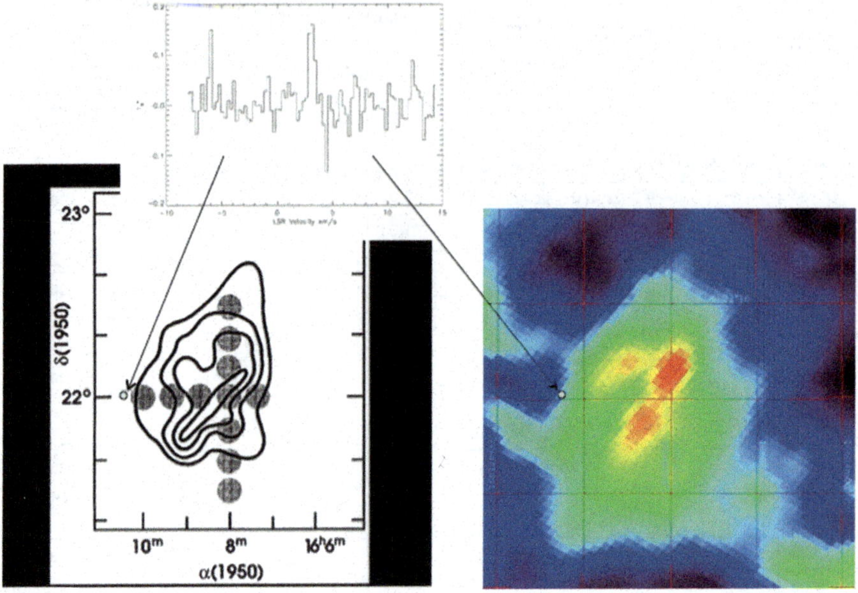

Fig. 8.5 Two aspects of the diffuse molecular cloud, MBM 40. On the left, the contours represent a poorly-sampled CO(1-0) map of the cloud made by Magnani et al. (1985), but with 1σ rms noise values that ranged from 0.03–0.1 K in T_A^*. The *gray circles* are pointings made with the old NRAO 140 m radiotelescope at 3.3 GHz to search for CH emission (Magnani et al. 1998). All the observed positions (*gray circles on the bottom left image*) showed clear CH lines—including the two southernmost ones beyond the lowest CO(1-0) contour (T_A^* = 0.5 K). On the right, at the same scale is a color E(B-V) map from the Schlegel, Finkbeiner, and Davis (1998) data. The *green color* has an E(B-V) ranging from 0.12–0.18 mag and the thin, outermost, aquamarine layer has a value of $\approx$ 0.11 mag. The *small dot* in both images is the beam size of the Arizona Radio Observatory 12 m radiotelescope at 115 GHz ($\approx$ 1′). The spectrum is a sensitive CO(1-0) observation with 1σ rms of 0.04 K in T_R^*. The weak signal at $\sim$ 3 km s^{-1} is real and is at the LSR velocity of the molecular gas in the denser regions of the cloud

haloes around small molecular clouds. Until the appearance of the dark gas issue this remained no more than a curiosity. It makes sense that OH would be a great choice along with [CII] for probing the PDR-like region outside of the CO-defined molecular cloud. OH is one of the first molecules formed and the excitation of the hyperfine states is very favorable under CNM conditions. Column densities of OH at 10^{13} cm^{-2} can be detected using the 1665 and 1667 MHz lines, whereas CO column densities at that level would be undetectable in emission for the lowest rotational transitions.

Recently, four papers describe how the OH main lines in different environments might be tracing invisible or difficult to detect molecular gas. Barriault et al. (2010*a*) observed two diffuse, high-latitude molecular clouds in the 1665 and 1667 MHz transitions and found no correlation between the column densities of CO and OH. They conclude that OH might be a more robust tracer of H_2 in these regions than

CO(1-0). Allen et al. (2012) sparsely mapped a 4° × 4° region centered on $\ell \approx$ 108° and $b \approx$ 5°. They found weak ($T_A \approx$ 20–40 mK) widespread OH 1667 MHz emission similar in extent to the local HI. Most importantly, less than 10% of those features were associated with CO(1-0) emission. A follow-up paper by Allen et al. (2015) came to the same conclusions. Finally, another study of a high-latitude diffuse molecular cloud by Cotten et al. (2012) found weak OH 1667 MHz emission in regions with E(B-V) as low as 0.12 mag.

In summary, sensitive spectroscopic observations can pick up a significant fraction of the dark gas. When coupled with [CII] 158 μm observations, the dark gas is not really dark at all and, most importantly, the spectroscopic observations allow velocity information to be determined.

References

Abdo, A.A. et al. 2010, ApJ, 710, 133
Ackermann, M. et al. 2012, ApJ, 755, 22
Allen, R.J., Ivette Rodríguez, M., Black, J.H., and Booth, R.S.. 2012, AJ, 143, 97
Allen, R.J., Hogg, D.E., and Engelke, P.D. 2015, AJ, 149, 123
Barriault, L., Joncas, G., Lockman, F.J., and Martin, P.G. 2010*a*, MNRAS, 407, 2645
Barriault, L. et al. 2010*b*, MNRAS, 406, 2713
Bertojo, M., Cheung, A.C., and Townes, C.H. 1976, ApJ, 208, 914
Bloemen, J.B.G.M. et al. 1984, A&A, 139, 37
Bloemen, J.B.G.M. et al. 1986, A&A, 154, 25
Bloemen, H.[4] 1989, ARAA, 27, 469
Bloemen, J.B.G.M., Deul, E.R. and Thaddeus, P. 1990, A&A, 233, 437
Bohlin, R.C., Savage, B.D., and Drake, J.F. 1978, 224, 132
Bolatto, A.D., Wolfire, M., and Leroy, A.K. 2013, ARAA, 51, 207
Boulanger, F., Baud, B., and van Albada, G.D. 1985, A&A, 144, 9
Boulanger, F. and Pérault, M. 1988, ApJ, 330, 964
Boulanger, F., Bronfman, L., Dame, T.M., and Thaddeus, P. 1998, A&A, 333, 709
Chastain, R.J., Shelton, R.L., Raley, E.A., and Magnani, L. 2006, AJ, 132, 1964
Chastain, R.J. 2008, PhD thesis, University of Georgia
Chiar, J.E., Kutner, M.L., Verter, F., Leous, J. 1994, ApJ, 431, 658
Combes, F. 1991, ARAA, 29, 195
Cotten, D.L., Magnani, L., Wennerstrom, E.A., Douglas, K.A., and Onello, J.S. 2012, AJ, 144,163
Cotten, D.L. and Magnani, L. 2013, MNRAS, 436, 1152
Dame, T.M. and Thaddeus, P. 1994, ApJ, 436, 173
Dame, T.M., Hartmann, D., and Thaddeus, P. 2001, ApJ, 547, 792
de Vries, C.P. and Le Poole, R.S. 1985, A&A, 145, 7
de Vries, H.W., Heithausen, A., and Thaddeus, P. 1987, ApJ 319, 723
Dickman, R.L. 1978, ApJS, 37,407
Dickman, R.L., Snell, R.L., and Schloerb, F. P. 1986, ApJ, 309, 326
Falgarone, E., et al. 1998, A&A, 331, 669
Federman, S.R. and Willson, R.W. 1982, ApJ, 260, 124
Frerking, M.A., Langer, W.D., and Wilson, R.W. 1982, ApJ, 262, 590

[4]H. Bloemen and J.B.G.M. Bloemen are the same person.

Genzel, R., Downes, D., Pauls, T., Wilson, T.L., and Bieging, J. 1979, A&A, 73, 253
Grenier, I.A., Casandjian, J.-M., and Terrier, R. 2005, Science, 307, 1292
Hartmann, D., Magnani, L., and Thaddeus, P. 1998, ApJ, 492, 205
Heithausen, A. and Mebold, U. 1989, A&A, 214, 347
Hogerheijde, M.R., et al. 1995, ApJ, 441, L93
Issa, M.R., MacLaren, I., and Wolfendale, A.W. 1990, ApJ, 352, 132
Jackson, J.M., et al. 2002, ApJ, 566, 81
Johansson, L.E.B. 1979, Research Lab. Electronics and Onsala Space Obs. Research Rep. 136 (Onsala: Onsala Space Obs.)
Klaassen, P.D., Plume, R., Gibson, S.J., Taylor, A.R., and Brunt, C.M. 2005, ApJ, 631, 1001
Kleiner, S.C. and Dickman, R.L. 1985, ApJ, 295, 466
Kraushaar, W.L. et al. 1972, ApJ, 177, 341
Kutner, M.L. and Leung, C.M. 1985, ApJ, 291, 188
Lada, C., Lada, E.A., Clemens, D.P., and Bally, J. 1994, ApJ, 429, 694.
Langer, W.D. et al. 2010, A&A, 521, L17
Larson, R.B. 1981, MNRAS, 194, 809
Laureijs, R.J., Mattila, K., and Schnur, G. 1987, A&A, 184, 269
Laureijs, R.J., Chlewicki, G., Wesselius, P.R., and Clark, F.O. 1989, A&A, 220, 226
Li, D. and Goldsmith, P.F. 2003, ApJ, 585, 823
Lien, D.J. 1984, ApJ, 284, 578
Liszt, H.S. and Lucas, R. 1998, A&A, 339, 561
Liszt, H. and Lucas, R. 2002, A&A, 391, 693
Liszt, H.S., Pety, J., and Lucas, R. 2010, A&A, 518, 45
Lombardi, M. and Alves, J. 2001, A&A, 377, 1023
Lombardi, M., Alves, J., and Lada, C.J. 2006, A&A, 454, 781
Lucas, R. and Liszt, H.S. 1994, A&A, 282, L5
Magnani, L., Blitz, L., and Mundy, L. 1985, ApJ, 295, 402
Magnani, L., Blitz, L., and Wouterloot, J.G.A. 1988, ApJ, 326, 909
Magnani, L. and Onello, J.S. 1993, ApJ, 408, 559
Magnani, L. and Onello, J.S. 1995, ApJ, 443, 169
Magnani, L. et al. 1998, ApJ, 504, 290
Magnani, L., Hartmann, D., Holcomb, S.L., Smith, L.E., and Thaddeus, P. 2000, ApJ, 535, 167
Maloney, P. and Black, J.H. 1988, ApJ, 325, 389
Maloney, P. 1988, ApJ, 334, 761
Maloney, P. 1990, ApJ, 348, L9
Mattila, K. 1986, A&A, 160, 157
Parkinson, M., Richardson, K.M., and Wolfendale, A.W. 1987, QJRAS, 28, 277
Planck Collaboration, Planck Early Results XIX, 2011, A&A, 536, A19
Polk, K.S., Knapp, G.R., Stark, A.A., and Wilson, R.W. 1988, ApJ, 332, 432
Reach, W.T., Koo, B.-C., and Heiles, C. 1994, ApJ, 429, 672
Reach, W.T., Wall, W.F., and Odegard, N. 1998, ApJ 507, 507
Rydbeck, O.E.H. et al. 1976, ApJS, 31, 333
Sandell, G., Stevens, M.A., and Heiles, C. 1987, A&A, 179, 25
Savage, B.D., Bohlin, R.C., Drake, J.F., and Budich, W.F. 1977, ApJ, 216, 291
Schlegel, D.J., Finkbeiner, D.P., and Davis, M. 1998, ApJ, 500, 525
Solomon, P.M., Rivolo, A.R., Barrett, J., and Yahil, A. 1987, ApJ, 319, 730
Stacy, J.G. et al. 1991, Proceedings of the 22nd International Cosmic Ray Conference, 11–23 August, 1991. Dublin, Ireland. Volume 1, 141
Strong, A.W. and Mattox, J.R. 1996, A&A, 308, L21
Strong, A.W., Moskalenko, I.V., Reimer, O., Digel, S., and Diehl, R. 2004, A&A, 422, 47
Taylor, S.D. Hartquist, T.W., and Williams, D.A 1993, MNRAS, 264, 929
van Dishoeck, E.F. and Black, J.H. 1988, ApJ, 334, 771
Wagle, G.A., Troland, T.H., Ferland, G.J., and Abel, N.P. 2015, ApJ, 809, 17

Wall, W.F. 2006, RMxAAm 42m 117
Wannier, P.G. et al. 1993, ApJ, 402 585
Weiland, J.P. et al. 1986, ApJ, 306, 101
Wolfire, M.G., Hollenbach, D., and McKee, C.F. 2010, ApJ, 716, 1191
Wouterloot, J.G.A. 1981, PhD Thesis, Leiden University

Chapter 9
Surveys for High-Latitude Molecular Clouds

Surveys aren't the most important thing in astronomy - they're the only thing. Jerry Ostriker, with apologies to Vince Lombardi and Red Sanders.

Abstract Searches for high latitude molecular gas have followed different strategies and been subjects of different systematics and biases. To show how these affect the emergent physical picture, we describe in some detail a few surveys and overview the development of the searches in the millimeter wave tracers.

9.1 Introduction

The vast majority of molecular clouds are located below $|b| = 15°$. Given the thinness of the disk, when one sees higher clouds, they are likely nearby (e.g., the Taurus/Aurigae dark clouds ($\sim$140 pc), the ρ Ophiuchi clouds ($\sim$130 pc), and even the Orion GMC at $\sim$0.5 kpc is fairly close). Throughout the 1970s and early 1980s, searches for molecular gas at high latitudes were deemed a waste of time because little molecular gas was expected there, but by the mid 1980s, a population of small molecular clouds numbering less than 200 was identified at $|b| \geq 20°$. Of these *high-latitude molecular clouds*, 57 were cataloged by Magnani et al. (1985) and became known as the MBM clouds (numbered MBM 01–MBM 57).[1] It became clear after a few years that these objects are for the most part small, diffuse and translucent, molecular clouds. There are a few dark clouds in the mix that had been previously identified (e.g., Lynds 1642 = MBM 20), but most of the objects in the MBM catalog were new and constituted the largest population of diffuse and translucent clouds identified to that date by CO(1-0). Of course, there are plenty of diffuse and translucent molecular clouds at $|b| < 20°$, but the long sightlines through the Galactic plane at low latitudes usually intersect more distant GMCs and dark cloud complexes so that any smaller foreground or background diffuse and translucent clouds are hard to pick out. The initial interest in these objects wasn't

[1] MBM 10 is not a molecular cloud. It was included in the original catalog but the signal turned out to be the telluric line—see Sect. 4.2.2.

L. Magnani, S.N. Shore, *A Dirty Window*, Astrophysics and Space Science Library 442, DOI 10.1007/978-3-662-54350-4_9

so much their peculiar latitudes but, instead, they represented a different population than the usual dark clouds and GMCs.

Because the high-latitude clouds are close, anywhere from ~100 pc to a few hundred pc away, they can be studied at the highest spatial resolution. The few high-latitude molecular clouds that are forming stars (see McGehee 2008 for a review) are among the nearest birthing stars to the Sun. However, their dynamical state is more important than their proximity: the vast majority of high-latitude clouds are not in virial equilibrium. Their kinetic energies are anywhere from a few to hundreds of times greater than their gravitational potential energies, thus, unless an exterior medium is confining them, they are breaking up on a sound crossing time. They seem to be condensing out of the atomic CNM and then rapidly dissipating. It was clear from the beginning that, unlike their denser brethren, these objects are structured primarily by turbulence rather than gravity and, as such, they are likely the best laboratory for studying the effects of turbulence in the diffuse ISM.

There have been many searches for high-latitude molecular gas, but not all of them are familiar to researchers in other fields. Many claims about the distribution and amount of molecular gas in the last 10 years have ignored previous work. So, in this chapter, we will describe the major searches and surveys and their biases and limitations. We will use chronological order only to set in context the evolution of the sample selection criteria and of the methods employed for the searches.

9.2 The Initial Searches

The large scale CO surveys of the Galaxy Sect. 7.2.1 all concerned the midplane distribution of GMCs. Despite the marked concentration of molecular gas towards the Galactic midplane, the presence of CO emission in Fig. 7.1 away from the midplane raises the question of the existence of gas outside the latitude boundaries of the map (approximately ± 35°), the so-called "high" Galactic latitudes. The demarcation line where "high" begins depends very much on the particular observer; values ranging from $|b| = 10°$ to $30°$ have been used at various times. In this book we will define high latitude to begin at $|b| = 25°$. The reason for this, admittedly subjective, choice is to avoid including in the high-latitude molecular gas inventory the southern extension of the Taurus/Aurigae dark clouds below the Galactic plane and the northern extension of the ρ Ophiuchi dark clouds above the plane. With this definition, virtually all molecular radio astronomers in the 1970s were of the opinion that regions with $|b| \geq 25°$ were so deficient in molecular gas as to make surveys not worthwhile, despite the fact that from the very first observations of the CO(1-0) line it was clear that *some* molecular gas was present there.

Table 9.1 lists the 19 Lynds dark clouds known before 1983 to be located at high Galactic latitudes. Dickman (1975) had detected CO emission from four of them: L169, L183, L1642, and L1778. The first detection of a high-latitude molecular cloud was reported by Penzias et al. (1972) who described CO observations of 12 dark clouds including L134 at $b = 35.8°$. But in addition to the CO line detections in

Table 9.1 Lynds dark clouds at $|b| \geq 25°$

Number	ℓ	b	Area (sq. deg.)	Opacity[a]	Complex[b]
L134	3.98	35.94	0.220	5	85
L169	5.09	36.97	0.860	5	85
L183	5.50	36.81	0.240	5	85
L184	5.50	36.81	0.240	5	85
L238	7.49	25.12	0.054	3	87
L1289	121.55	26.09	0.012	2	346
L1292	121.74	26.12	0.466	1	346
L1311	125.25	32.00	1.660	1	348
L1321	126.71	31.85	1.760	1	347
L1453	158.42	−33.93	0.066	4	159
L1454	158.66	−33.89	0.860	3	159
L1457	158.87	−34.58	0.262	4	159
L1458	158.90	−33.76	0.056	4	159
L1569	189.35	−36.87	0.631	2	103
L1615	203.28	−25.27	0.084	2	74
L1616	203.42	−25.05	0.016	5	74
L1642	210.77	−36.72	0.372	3	73
L1778	358.85	36.92	0.086	5	86
L1780	359.01	36.78	0.197	4	86

[a]The opacity is a subjective measure which ranges from a value of 1 for those nebulae barely detectable on the POSS plates to 6 for the most obscured regions
[b]A complex groups together dark clouds that appear to be related

a few Lynds dark clouds at $|b| \geq 25°$, there were at least five other early indications that appreciable quantities of high-latitude molecular gas might exist: (1) the bright nebulae noted by Lynds (1965); (2) the absorption observations of the 18 cm OH lines and the 21 cm HI line in the direction of extragalactic continuum sources (Dickey et al. 1981; Kazès and Crovisier 1981); (3) the CO(1-0) emission detected by Knapp and Jura (1976) along lines of sight previously studied only via optical absorption lines; (4) the Blitz et al. (1982) catalog of CO emission from Sharpless HII regions; and (5) the two populations of quasars present in a study of color excess of quasars compared to the atomic hydrogen column density in their directions (Teerikorpi 1981). We will examine each of these indicators in turn.

We discussed the Lynds Bright Nebulae (LBN) in Sect. 6.4. There are more of these at high Galactic latitude than there are Lynds dark nebulae. Because these clouds are rich in dust, they likely contain some molecular gas, but this was just a suspicion.

Dickey et al. (1981) searched for OH *absorption* lines at 18 cm in the direction of 58 extragalactic sources and detected OH absorption and/or emission in 16 directions, including two at $b = -38.2°$ and $-33.6°$. Similarly, Kazès and Crovisier (1981) searched for CO(1-0) emission in 76 lines of sight showing strong HI 21 cm absorption. CO(1-0) was detected from 21 directions including four at $|b| \geq$

25° (two of the four detections were the same as those in the Dickey, Crovisier, and Kazès sample). The molecular gas along these four high-latitude lines of sight not coinciding with any Lynds dark clouds was thought to arise from diffuse, low column density clouds. To our knowledge, these clouds still have not been mapped although two of the lines of sight are almost surely part of the MBM 53–55 high-latitude cloud complex. It is important to note that, at that time, the astronomy community which studied optical and ultraviolet interstellar absorption lines included few radio astronomers, so it is not surprising that the detections by Crovisier, Dickey, and Kazès were not followed up.

A few years earlier, Knapp and Jura (1976) detected CO(1-0) emission from interstellar clouds along lines of sight to stars which showed CO absorption in the ultraviolet. Although some of the detections were questioned by Lada and Blitz (1988), none had $|b| \geq 24°$. Similarly, a study by Dickman et al. (1983) included lines of sight to other stars known to exhibit CO absorption lines, but none of those were above $|b| = 21°$. However, the fact that several intermediate-latitude sight lines *outside of known Lynds dark clouds* had sufficiently high CO column densities to be detected by radio techniques should have provoked more interest in this type of low-extinction, molecule-rich, interstellar clouds. It did not.

Three clear detections of CO(1-0) emission at high-latitudes from objects other than Lynds dark clouds were made by Blitz et al. (1982) during the compilation of their catalog of CO radial velocities toward Galactic HII regions, chosen from the Sharpless (1959) catalog. Of the more than 300 regions observed, six objects at $|b| \geq$ 25° showed CO emission. Despite the fact that these objects had been cataloged by Sharpless because they are associated with red nebulosity on the POSS plates, none of the six detections is actually an HII region. The red emission is produced either by reflected starlight or photoluminescence of carbon grains (see Sect. 3.5.4). Fich et al. (1990) detect Hα emission from most of these objects, but the profiles are wide in velocity and are detections of the WIM, not of high-latitude HII regions.

Finally, using a completely different line of reasoning, Teerikorpi (1981) showed that a substantial number of molecular clouds has to exist at high latitudes. His novel method consisted of examining residuals in the (B-V) magnitude of quasars compared to their *unreddened* color derived from theoretical considerations (in other words a type of quasar color excess: [(B-V) $-$ (B-V)$_{Zo}$] where (B-V)$_{Zo}$ is the unreddened quasar color). A plot of the residuals of B-V versus N(HI) showed a two-branch structure in the quasar population, with the quasars in the directions of the local spiral arms showing a greater residual than quasars in the direction away from spiral arms. Teerikorpi concluded that the more reddened branch of the quasar population is caused by greater absorption in the local Galactic spiral arm. If the gas to dust ratio is constant, this extra absorption can not be produced by HI associated dust and must come from dust associated with molecular hydrogen. He obtained for these $2N(H_2)/N(H_{total}) \sim 1, A_V \geq 0.5$ magnitudes, linear size less than 2 pc if $n >$ 50 cm^{-3}, $R_V \approx 3.3$ for the dust component, and a surface filling factor of 40% of the sight lines at moderate latitudes. This intriguing argument was completely ignored by the molecular cloud observing community.

These detections and lines of reasoning pointed to some molecular gas residing at high latitudes, but the only actual published search or survey for CO emission in that territory came up empty (Dewey et al. 1983). They searched for CO(1-0) emission along 68 lines of sight around $b \sim 59°$ and $b \sim 84°$. This area had been chosen based on its high reddening over several square degrees determined from the Shane and WIrtanen (1967) galaxy counts. It is likely, however, that many of the early CO observers conducted high-latitude CO searches while they were at the telescope surveying the Galactic plane. Yet, with the exception of the Dewey et al. survey, the results were not published. Evidently, the repeated failure to detect high-latitude CO emission spread by word-of-mouth and added to the lore that such gas was not to be found.

A successful search for high-latitude CO emission was conducted by Michel Fich and Leo Blitz in January of 1982 yielding detections toward the North Galactic Pole (M. Fich, private communication), but the exact coordinates were forgotten and the results were never published. This success, however, inspired the Blitz, Magnani, and Mundy high-latitude survey conducted in 1983 and published in 1984.

Thus, by the summer of 1983, the inventory of molecular gas with sufficient column density to provide a CO(1-0) signal at $|b| \geq 25°$ included 7 separate Lynds (1962) dark cloud complexes (comprising the 19 Lynds dark clouds in Table 9.1) with visual extinctions in their core regions in all cases less than one magnitude, four lines of sight to diffuse, unmapped interstellar clouds, and three lines of sight toward regions of red nebulosity on the POSS plates. Despite a few additional, indirect, indications to the contrary, there appeared to be very little dust and extinction at high Galactic latitudes. The IRAS mission (Low et al. 1984) changed this perception with the discovery of the infrared cirrus primarily at 60 and 100 μm (see Sect. 5.3.1).

9.3 The MBM Survey

The 1970s and 1980s were the watershed period in Galactic molecular astronomy and surveys were an essential part of the development. The probing of the high latitude sky was seen as a risky undertaking, one that would likely come up empty, and for that reason was not a high priority for most researchers. In this section, where some of the description will be necessarily based on personal experiences, we are neither nostalgically reminiscing nor indulging in vapid autobiography. By describing the methods used to discover high-latitude clouds we can illustrate how current data sets, especially in the optical and near infrared, can be used to find new objects, and we can dispel the common misconception that the clouds were found by surveying the IRAS cirrus. More important, the methods carry implicit biases that can skew the results if not examined more closely. Thus, concurrently with, but independent of the IRAS mission, Leo Blitz at the University of Maryland initiated a project to systematically search for high-latitude molecular gas. As outlined above, the Blitz et al. (1982) survey and unpublished observations by Blitz and Fich had hinted at the presence of undiscovered molecular clouds at high Galactic latitudes.

Blitz assigned Loris Magnani, then a graduate student at the University of Maryland, the task of poring over prints of the POSS plates and the Whiteoak extension to the POSS to find all regions at $|b| \geq 20°$ showing any sign of obscuration by dust. This was the traditional method for finding high dust column density regions which the earliest molecular surveys had shown to often contain molecular gas. The choice $|b| \geq 20°$ was originally made because at these latitudes there did not appear to be significant molecular gas associated with lower latitude clouds. The northern and southern extension of the Taurus-Aurigae and Ophiuchus dark cloud complex later forced the "boundary" to be shifted to be $|b| \geq 25°$, where only the Polaris Flare shows a possible connection to lower latitudes.

Each red and blue print of the POSS plates ($6° \times 6°$) was examined for regions of size $5'$ or or larger showing: (1) a noticeable drop in stellar density, (2) any emission nebulosity with a sharp boundary, possibly indicating the edge of a molecular cloud (the emission is produced primarily by reflecting the integrated starlight of the Galactic plane (however, some of the emission could be due to red luminescence Sect. 3.5.4), or (3) any change in the stellar density from the blue to the red print possibly indicating reddening produced by dust. All obvious regions of obscuration (*i.e.*, those with A_V greater than a few magnitudes) had been found by Lynds (1962); thus, Magnani had to effectively identify those regions which had obscuration equivalent to $\sim$ 1 magnitude of visual extinction or less. At $|b| \geq 50°$, candidate regions become especially difficult to identify because of the significant drop in stellar density and the ensuing relatively larger stellar surface density fluctuations. The lack of blue prints in the Whiteoak extension to the POSS also complicated the search in the range $-44° < \delta < -37°$. In many instances, candidate regions extended over a large area of a print, or several regions were in close proximity so that several search positions for a given candidate region had to be included. A total of 458 candidate regions were selected and 439 were searched for CO(1-0) emission during November and December 1983 using the University of Texas Millimeter Wave Observatory's (MWO) 5-m radiotelescope (see Vanden Bout et al. 2012). The initial observations were made in frequency-switched mode and the presence of the telluric CO(1-0) line in every spectrum caused considerable confusion at the beginning of the run. However, by the end of the run, it was clear that the number of detections was much greater than had been anticipated. Emission had been detected from 105 of the 439 observed candidate positions—a detection rate of 24%.

The initial results of the first survey were published in a letter in *The Astrophysical Journal* (Blitz et al. 1984). The detections were divided into objects in the $20° \leq |b| < 30°$ range and those with $|b| \geq 30°$. Of the clouds at $|b| \geq 30°$, CO emission was detected from 37 of 287 observed positions (13%), with the detections grouped into 29 distinct clouds. Eleven clouds were mapped in a very undersampled pattern (typically $10'$–$20'$ grids with a $2.3'$ beam). Some of the Lynds clouds in the sample had been partially mapped by others (e.g., L1642 by Sandell et al. 1981) and five clouds in the vicinity of and including L134 had been mapped in CO by Joe Montani and Mark Morris with the Columbia 1.2 meter mm-wave radiotelescope (now at Harvard), although the data were never published. With the Montani and Morris results, a total of 16 clouds at $|b| \geq 30°$ had sufficient mapping data to permit

Table 9.2 Properties of high-latitude molecular clouds based on the initial CO survey by Blitz et al. (1984)

Quantity	Range or value	Mean
CO antenna temperature in K	0.6–6.8	2.6
CO linewidth (FWHM) in km s^{-1}	1.0–5.7	1.9
Velocity dispersion of individual clouds in km s^{-1}	0.2–2.2	0.7
Cloud-to-cloud velocity dispersion in km s^{-1}	5.7 ± 1.2	...
Projected cloud area in square degrees	0.1–5.4	1.3
Mean distance of ensemble in pc	...	100
Mean size in pc	...	2.0
Mass in solar masses	2–260	66
H_2 density in cm^{-3}	46–250	170
$N(H_2)/N(HI)$	0.2–30	5
Age in years	$\leq 10^6$	...

Blitz et al. (1984) to characterize some of the basic CO and physical properties of these "high-latitude molecular clouds" (see Table 9.2). Blitz et al. (1984) also opined that these objects had to be associated with the IRAS cirrus clouds which had been identified concurrently by Low et al. (1984).

The distances to the clouds were obtained statistically using the velocity dispersion of the cloud ensemble and the local mass surface density (see Chap. 10). With a mean distance of 100 pc, the initial survey established that at least a portion of the infrared cirrus was local; a point made explicitly by Gautier (1986) and Weiland et al. (1986). With a distance estimate, many of the observed properties of the clouds could be converted to physical properties. It was thus clear that the high-latitude clouds are small, low-density, low-extinction clouds quite unlike GMCs or dark clouds.

Besides the discovery of substantial number of new molecular clouds, the Blitz et al. (1984) paper made a strong assertion: most of the high-latitude clouds are not gravitationally bound. Many high-latitude clouds have centroid velocity dispersions which imply virial masses larger by 1–2 orders of magnitude than the masses derived from the CO emissivity. Before Blitz et al. (1984) no one had explicitly stated in the literature that at least some molecular clouds are unbound. A few researchers (Mahoney et al. 1976; Clark et al. 1977) had noticed that the kinetic energy of some clouds appears to be substantially greater than their gravitational potential term, but this discrepancy was attributed to rotation or other kinematic effects (*i.e.*, orbiting fragments). The large velocity dispersions in the handful of mapped clouds was the first clear indication that the majority of the high-latitude clouds could represent a population different from GMCs and dark clouds. At this point in time, the diffuse/translucent nature of the high-latitude clouds was not yet appreciated. They were assumed have molecular properties like dark clouds with somehow lower extinctions (attributed to different grain properties). Some of the more savvy astronomers referred to them as "CO-rich diffuse clouds", but their nature was still uncertain.

Many of the detections are associated with regions of marked emission, especially on the POSS red prints. Magnani then went through the prints a second time and subsequently identified 35 additional regions of apparent obscuration, focussing on the presence of low-level blue and red emission. Many of the new candidates had previously been noted by Sharpless (1959) and Lynds (1965). The final search catalog constituted 493 candidate regions, with over 400 completely new lines of sight.

The second set of observing runs consisted of a few runs at the MWO during the first half of 1984. The observing setup was similar to that of the first run and is further described by Magnani et al. (1985). Of the 493 positions from the search catalog, 488 had been observed by June, 1984. Some of the five unobserved positions were eventually detected (they were not observed because the Texas 5 meter telescope could not point to the North Celestial Pole). CO(1-0) emission was found from a total of 133 positions. Mapping showed that some of these were from the same contiguous structure so 124 clouds were identified. At $20° \leq |b| < 25°$, 67 positions were detected (68% detection rate), and 57 (14%) at $|b| \geq 25°$. The relatively high detection rate in the $20° \leq |b| < 25°$ region was largely due to the many candidate positions near or in the Taurus-Auriga and Ophiuchus cloud complexes and, finally, only those objects at $|b| \geq 25°$ were called *high-latitude* molecular clouds. Tables 1 and 2 in the MBM paper list all the detections from the initial observing runs.

The distribution of the original detections in Galactic coordinates at $|b| \geq 20°$ is shown in Fig. 9.1. The clustering of detections at $b \sim -20°$ and $\ell \sim 155°$-$180°$ and $b \sim +20°$ and $\ell \sim 350°$ to $5°$ indicated that the sample was contaminated by portions of the Taurus/Perseus and Scorpius/Ophiuchus dark cloud systems. A north-south asymmetry (29%–71% at $|b| \geq 25°$) was noticeable. The implications of this asymmetry for the location of the Sun with respect to the Galactic midplane will be discussed in Chap. 10.

Seventy percent of the objects detected at $|b| \geq 25°$ are located within one degree of a Lynds bright nebula (LBN; Lynds 1965). The LBN at high latitudes are prominent on both the blue and red POSS prints and are most likely dusty nebulae reflecting the starlight of the Galactic plane. Similarly, some of the high-latitude clouds are at the same locations as Sharpless HII regions (Sharpless 1959). The molecular clouds are also associated with HI features as determined from the Heiles and Habing (1974) HI survey.

The MBM survey as described by Magnani et al. (1985) established that a significant population of high-latitude molecular clouds existed, that the clouds as an ensemble were relatively nearby, and that many of them were not gravitationally bound. The last significant step in understanding their nature was to determine their connection with the IRAS infrared cirrus.

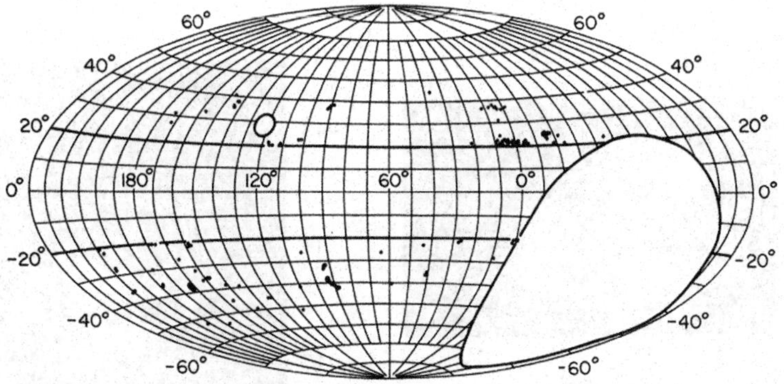

Fig. 9.1 Distribution of CO(1-0) detections in Galactic coordinates and Aitoff projection obtained during the MBM high-latitude survey in 1983–4. The data were taken at the MWO of the University of Texas with the 5-m radiotelescope at McDonald Observatory. The beamsize at 115 GHz was 2.3′. The region in white at $\ell \sim 120°$ and $b \sim +30°$ is centered on the North Celestial Pole and could not be reached with the Texas instrument. The large white region mostly in the southern Galactic hemisphere either did not rise above the horizon or culminated at too low an altitude to be observed from McDonald Observatory. Figure from Magnani et al. (1985)

9.4 The IRAS Cirrus: High Latitude Cloud Connection

As soon as the IRAS data were released, a comparison between the infrared images at 100 μm and the dozen mapped high-latitude clouds immediately confirmed the speculation by Blitz et al. (1984) that there was a correspondence between some of the infrared cirrus and the molecular clouds. Figure 9.2, reproduced from Weiland et al. (1986) shows the striking similarity between the dust and gas tracers. The high-latitude clouds have cirrus counterparts, but many regions with similarly high 100 μm emission do not show CO(1-0). In the subsequent two decades, determining how to use the infrared data to predict the presence of molecular gas in a given region became a small cottage industry. Some of the various efforts in this direction are discussed in below.

The most important contribution of the groundbreaking Weiland et al. (1986) paper was establishing that at least some of the cirrus emission is interstellar. While many had suspected this, the lack of velocity information in the infrared images precluded any definitive assignment of a location. Thanks to the CO data and the distance estimates to the high-latitude clouds, the cirrus became firmly established as a feature of the local ISM.

Weiland et al. (1986) also established that some regions displaying CO emission had the same infrared properties as others that do not. The dust temperature in the CO-rich clouds is ∼20 K and the relatively low infrared luminosity-to-mass ratio indicates that there are no significant embedded heating sources. A linear relationship exists between W_{CO} and the 100 μm intensity [also noted at about the same time by Cor de Vries (1986)] but with a significantly higher slope than the value obtained by Hauser et al. (1984) in a survey of the central part of the

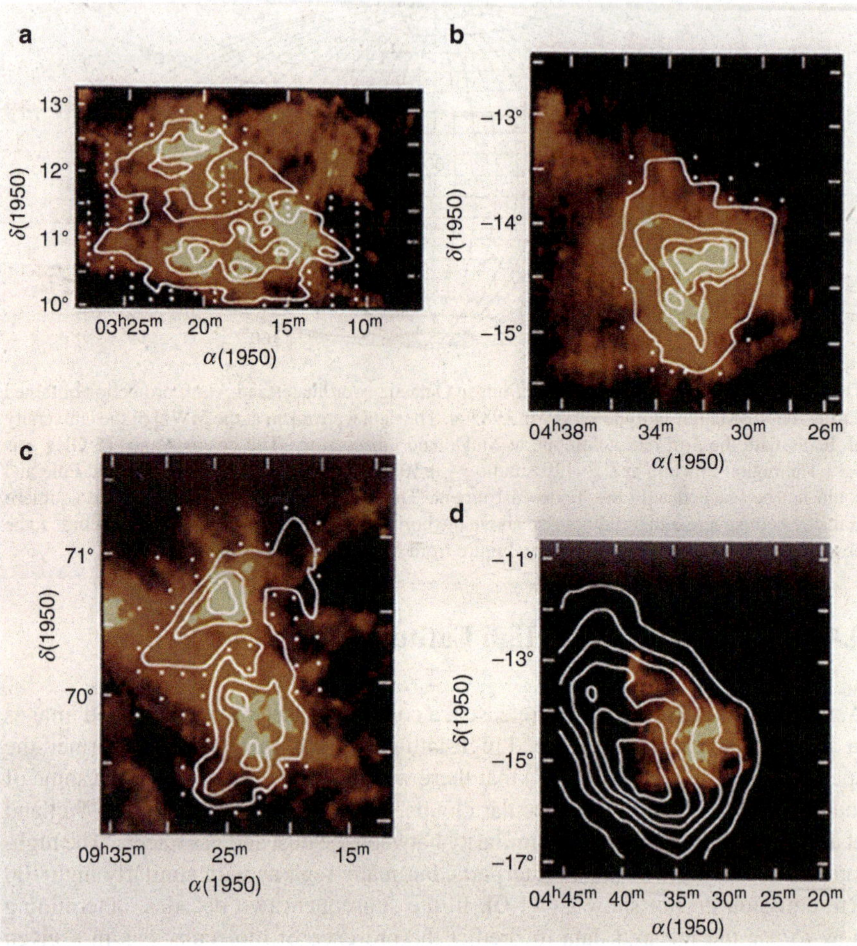

Fig. 9.2 Relationship between IRAS 100μm emission and CO(1-0) mapping for three high-latitude molecular clouds from the MBM survey (MBM 30, 16, 20; clockwise from the bottom left). The bottom right panel shows the relationship between the IRAS emission and atomic hydrogen in MBM 20. Unlike the first three panels, the peak of the atomic hydrogen is offset from the peak of the infrared emission. This figure first appeared in Weiland et al. (1986)

Galaxy. The latter property was discussed as indicating that high-latitude clouds have anomalously large CO emission relative to the far-infrared cirrus clouds. Finally, they found excess infrared emission at 12 and 25 μm for the CO-rich infrared cirrus clouds, a property that had been noted for the cirrus in general (e.g., Boulanger et al. 1985). The nature of this emission was attributed either to the presence of very small dust particles (e.g., Draine and Anderson 1985) or to polycyclic aromatic hydrocarbons (PAHs; Puget et al. 1985; Allamandola et al. 1985).

The Weiland et al. (1986) paper, and similar work by the Dutch infrared/molecular group led by Harm Habing were the first salvos of many subsequent infrared-CO comparisons for the high-latitude clouds. It is worth noting, however, that some of the questions raised by the Weiland et al. (1986) paper such as: (1) what induces the transition between predominantly atomic vs. predominantly molecular regions? (2) why is the CO/IR ratio much larger in the high-latitude clouds than is generally the case for the Galaxy or external galaxies? (3) what is the precise nature and evolutionary state of the "very small grain" component? have not been completely answered even three decades later. To explore further the infrared-molecular connection, a more complete census of the molecular content of the high-latitude sky was needed.

9.5 Post-MBM CO Surveys at High Latitudes

Shortly after the MBM survey, Magnani et al. (1986) attempted a blind survey over random regions to determine the fractional completeness of their survey, ϵ, sampling $10^\circ \times 10^\circ$ grids in one degree increments in ℓ and b for a total of 100 points per grid. Seventeen grids covering 1750 points (three of the grids were not completed, and one was sampled in RA and Dec and included more than 100 points) in the northern Galactic hemisphere resulted in only four detections, while 8 grids covering 750 points in the southern Galactic hemisphere produced five detections.

The results of the survey confirmed that the southern Galactic hemisphere is richer in high-latitude clouds than the north and yielded a value of ϵ of 0.55—implying that nearly twice as many clouds of the MBM type were still to be found. From this blind survey, estimates of the basic parameters of high latitude molecular gas could be determined: the total number of high latitude clouds is 120^{+40}_{-30}, the total local molecular mass $\sim 5 \times 10^3$ $M_\odot$, and the mass surface density ~ 0.2 $M_\odot$ pc^{-2}. High-latitude molecular gas was estimated to contribute about 5–10% of the overall local molecular mass.

Other groups began to search for high latitude clouds. Keto and Myers (1986) searched for high-latitude clouds from the southern hemisphere focussing on the portion of the Galaxy that was not visible from the north. Using the Columbia University 1.2 m antenna on Cerro Tololo, Chile, they detected CO(1-0) emission from 15 high-latitude clouds ($|b| \geq 20^\circ$) and from 3 small clouds associated with the Chamaeleon dark cloud complex. The objects were found in a similar manner to MBM objects: looking for regions of obscuration or reflection using the ESO sky survey plates, so the Keto and Myers survey is subject to the same selection biases as the MBM survey. Nevertheless the Keto and Myers survey showed that regions not surveyed by MBM contain high-latitude clouds, and that these objects are not gravitationally bound. However, Keto and Myers (1986) did raise the possibility that the clouds might be confined by the external pressure of the ISM.

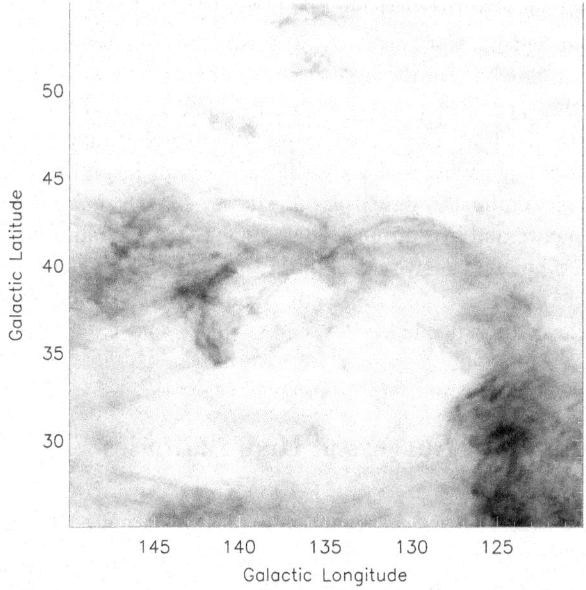

Fig. 9.3 Dust map from Schlegel et al. (1998) of the infrared cirrus clouds near the North Celestial Pole. Initial detections of several molecular clouds in the region were made by Magnani et al. (1985), Heithausen (1987), and de Vries et al. (1987). These clouds are sometimes known as the North Celestial Pole Loop or the Ursa Major clouds. The prominent structure in the southwest portion of the image is the top of the Polaris Flare. Image made with the Skyview Virtual Observatory

de Vries et al. (1987) mapped a large region ($\sim 10° \times 8°$) in Ursa Major and showed that six of the MBM clouds are almost certainly related. MBM 27–32 are located in this region, embedded in a large infrared loop (sometimes called the North Celestial Pole Loop—see Figs. 9.3 and 9.4, and Heithausen (1987)). Using the Columbia 1.2 m mm-wave radiotelescope,[2] they convincingly showed that individual high-latitude clouds from the MBM survey are parts of large scale filamentary structures, something that was not noticed by MBM because of their severely undersampled maps. The high-latitude clouds in this region have come to be known as the Ursa Major clouds and half of them were the subject, a decade later, of a more comprehensive, higher resolution study by Pound and Goodman (1997). One of the most important results by de Vries et al. (1987) was an attempt to calibrate the CO-H_2 conversion factor by analyzing the relationship between the infrared emission from the dust in the clouds, the atomic gas contribution from the 21 cm HI line, and the molecular contribution from the CO data.

[2]Now at the Harvard-Smithsonian Center for Astrophysics.

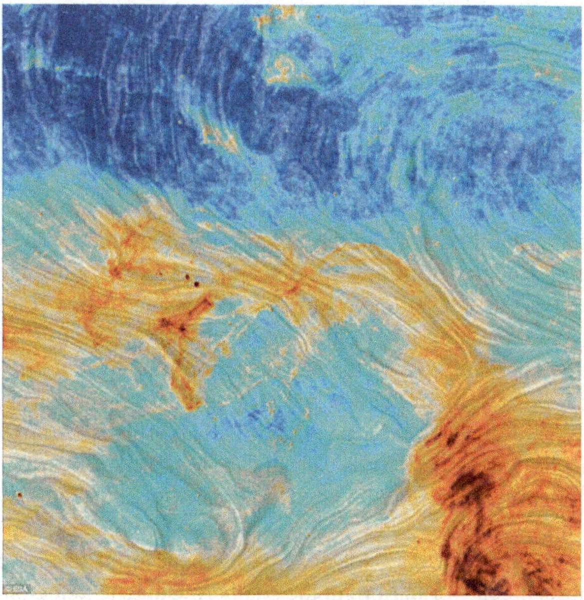

Fig. 9.4 Image from the *Planck* satellite of approximately the same region as in Fig. 9.4. This image is centered on $(\ell, b = 135^\circ, +40^\circ)$ and is $30^\circ \times 30^\circ$. The *colors* represent dust emission with the warmer colors representing greater emission. The "waves" show the orientation of the magnetic field in the region from the 353 GHz Planck polarization data using the Linear Integral Convolution technique devised by Diego Falceta-Goncalves. Image: ESA

A different approach was taken by Heiles et al. (1988). Instead of selecting regions based on obscuration, they identified 26 isolated, degree-sized IRAS cirrus clouds and then observed their HI emission to define S_{100}/N_{HI}, the ratio of the infrared surface brightness at 100 μm to the atomic hydrogen column density. Heiles, Reach, and Koo found that regions with high values of S_{100}/N_{HI} contained sufficient molecular gas for the detection of the CO(1-0) line. They found that CO emission can be detected even in clouds with hydrogen column density as low as $\sim 2.4 \times 10^{20}$ cm^{-2}, and identified 11 high-latitude clouds, later increased to a total of 14 by Reach et al. (1994). The CO-emitting regions in these objects tend to be compact and smaller than the clouds in the MBM survey. Moreover, their latitude distribution differs markedly from that of the MBM clouds. This is partly due to the search criteria employed by Heiles, et al., but may also highlight a real difference in cloud populations. In a somewhat similar vein, Clemens and Barvainis (1988) optically selected a sample of round or elliptical, small, obscured regions. These objects are mostly at lower latitudes and are similar to Bok globules. Five of the objects are at $|b| \geq 25^\circ$ and exhibit CO emission so they can be considered high-latitude clouds.

Heithausen et al. (1993) surveyed 620 square degrees in the region $117° \leq \ell \leq 160°$ and $16° \leq b \leq 44°$. They found that 13% of the region showed CO(1-0) emission. Their sampling was quite good: mostly 15′ with an 8.7′ beam. with 245 square degrees mapped at 0.5° resolution. Although their survey dipped significantly below the Blitz et al. surveys in latitude, a 25° minimum latitude still produced a surface filling fraction of 9.2%. However, the large feature known as the Polaris Flare dominated the CO detections: Of the nearly 80 square degrees with CO emission, 40.21 of them were from the Polaris Flare. Excluding the Polaris Flare, the surface filling fraction decreases to $\sim$ 4.1%, still significantly greater than the Magnani et al. (1986) blind survey. Along with the surface filling fraction, the mass surface density of molecular gas in this region also increases to 0.86 $M_\odot$ pc^{-2} if one assumes that even at $b > 44°$ the CO detection rate does not decrease, and to 0.47 $M_\odot$ pc^{-2} if there is no CO at $b > 44°$. It was clear by comparing these results to those of Magnani et al. (1986), or even MBM, that the region in the northern Galactic hemisphere near the North Celestial Pole is relatively rich in molecular gas, but whether similar high-latitude regions existed in either hemisphere was not known at the time.

A search for molecular clouds toward intermediate-to-high latitude IRAS point sources (signposts of star formation) in the southern Galactic hemisphere was made by Yonekura et al. (1999). Using the NANTEN 4-m millimeter-wave telescope to survey 29 IRAS-selected targets, they detected CO(1-0) emission from five of them, with three sources at $|b| \geq 24.7°$. However, all 3 objects had been previously detected by Knapp et al. (1977), MBM, and Bally et al. (1991). A more extensive survey was made, using the same telescope, by Onishi et al. (2001), who searched the far-infrared excess clouds identified by Reach et al. (1998) using the DIRBE instrument on the COBE satellite. Onishi et al. observed those regions which rose at least 30 degrees above the horizon at Las Campanas, Chile, the site of the telescope, 68 out of 153 candidates. They observed all 68 detecting CO emission from 32 and often mapping them systematically. The immediate consequence of this work was to increase the number of known high-latitude molecular clouds significantly. Figure 9.5 shows all the CO detections or clouds at $|b| \geq 30°$ that were known at the end of 2011 before the *Planck* satellite and the systematic high-latitude CO survey by the Thaddeus/Dame group.

The studies described above all used CO(1-0) *emission* line as a surrogate for tracing molecular hydrogen, but CO(1-0) *absorption* can also be used. At 115 GHz, CO(1-0) absorption lines can be detected against the continuum of background quasars or active galaxies that have sufficient fluxes at 2.6 mm to allow detection of molecular line absorption. A search for continuum sources suitable for millimeter absorption studies was made by Liszt and Wilson (1993) and was discussed in the context of diffuse molecular clouds in Chap. 7. From the survey point of view, over 100 sources were selected for CO(1-0) emission observation, 64 of which were at $|b| \geq 20°$. Since Liszt and Wilson were searching for lines of sight with CO emission, this result was an almost blind high-latitude CO survey in the spirit of Magnani et al. (1996). Of the 64 high-latitudes sources, only one (B2251+158) showed clear CO absorption. This line of sight to the quasar 3C454.3 had been

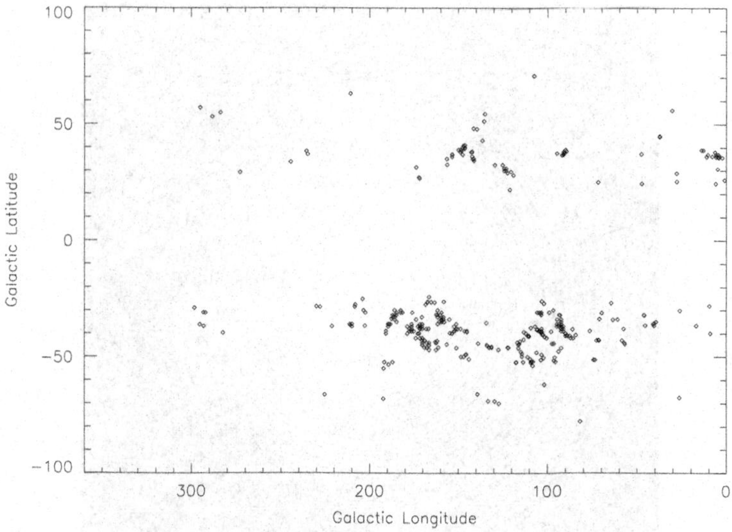

Fig. 9.5 All known spectroscopic CO(1-0) detections at Galactic latitudes $|b| \geq 25°$ as of 2011. See text for details

known as a source of molecular emission as early as the OH survey of Kazès and Crovisier (1981) and is likely associated with the envelope or environs of the nearby MBM 53–55 complex. The low detection rate at $|b| \geq 20°$ corroborates the results of Magnani et al. (1986) and the original Magnani et al. (1985) results. A smaller survey for CO(1-0) absorption at somewhat higher Galactic latitudes was conducted by Liszt and Lucas (1998). The lines of sight included had been previously studied in other molecular tracers. Although they detected CO absorption in 8 of the 9 sources they observed, only one, B2251+158, was at what we would consider high-latitude. An earlier survey for HCO^+ absorption towards extragalactic sources by Lucas and Liszt (1996) had better high-latitude coverage (9 lines of sight), and produced only two detections (B2145+067 and B2251+158). Although, in principle, absorption line observations can detect molecular gas, a more thorough survey at high-latitudes would have to be conducted to yield results comparable to the CO emission surveys.

9.6 The Georgia-Harvard CfA High Latitude Survey

In the late 1990s a truly unbiased CO survey of the high-latitude sky was made using the Harvard Center for Astrophysics 1.2-m millimeter wave telescope. This is the instrument used for the wide-scale Galactic plane survey by Thaddeus and Dame (see Fig. 7.1). The northern and southern sky at $|b| \geq 30°$ which culminated at least 30° above the horizon from Cambridge, Massachusetts was surveyed every degree in Galactic latitude and longitude. The survey pattern for the northern and southern skies is shown in Figs. 9.6 and 9.7.

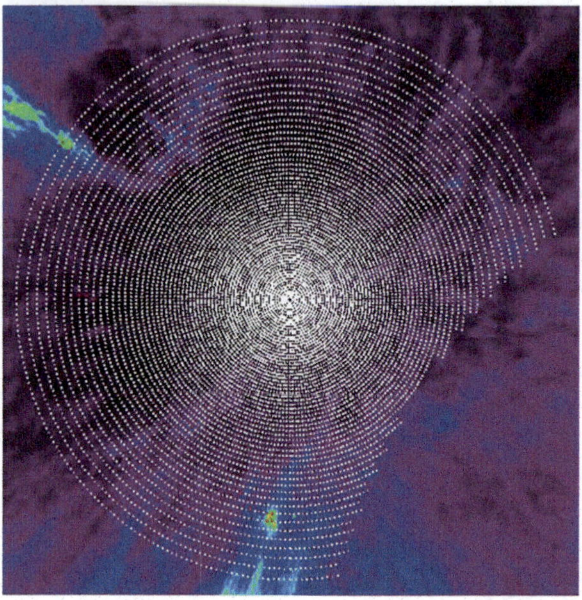

Fig. 9.6 Polar projection of the mapping grid for the northern Galactic hemisphere for the Georgia-Harvard CfA High-Latitude CO survey. The *color image* represents IRAS 100 μm emission while the white dots are the grid pattern observed with the 1.2 m Harvard-Smithsonian center for Astrophysics millimeter-wave telescope. The results of this survey were reported by Hartmann et al. (1998)

The results from the surveys are presented in Hartmann et al. (1998) and Magnani et al. (2000). In the Northern Galactic Hemisphere (NGH) 10,562 out of a maximum of 11,478 points at $|b| \geq 30°$ were observed. The missing points did not rise above 30° altitude from Cambridge, Massachusetts, the site of the 1.2 m radiotelescope. Only 26 lines of sight showed evidence of CO(1-0) emission at $T_{mb} > 0.3$ K in antenna temperature (3σ rms). This was not surprising as the NGH does not have many high-latitude molecular clouds.

In the Southern Galactic Hemisphere (SGH) 4982 points were observed with the same sensitivity. There were 133 detections, 75 of which coincide with previously known molecular clouds. The average centroid velocity of the detections is -1.67 ± 0.58 km s^{-1} attesting again to the local nature of the clouds. The average T_A^*, line width (FWHM), and W_{CO} are 0.83 K, 2.08 km s^{-1}, and 1.81 K km s^{-1}. Both the average antenna temperature and W_{CO} are lower than in the MBM survey which was biased towards larger, more opaque clouds. The number of CO detections in both hemispheres is significantly lower at $|b| \geq 50°$ than would be expected from extrapolating the molecular in-plane results from the Dame et al. (2001) survey. This may be an indication that supernovae from the Scorpius/Ophiuchus associations have "blown through" the disk and cleared out most of the denser neutral material at the highest latitudes. This is not a new idea. The Galactic Fountain model (see

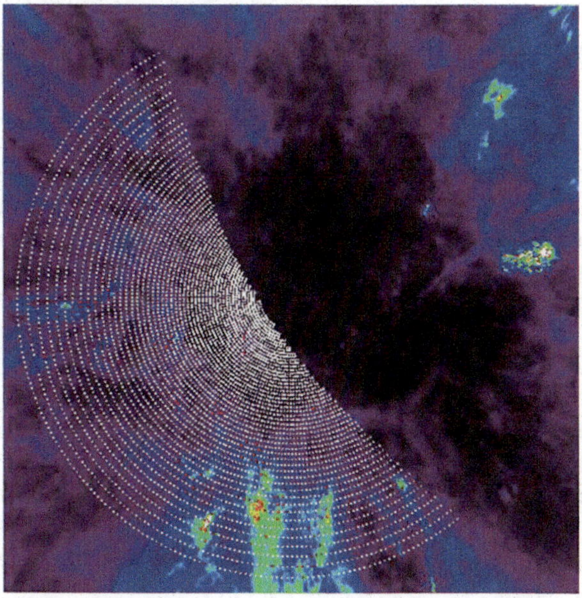

Fig. 9.7 Same as Fig. 9.7 but for the southern Galactic hemisphere. More information on this survey can be found in the paper by Magnani et al. (2000)

Chap. 1) can produce blowouts of hot gas from the plane to the halo, and Heiles (1993) wrote an interesting paper titled "The Worm-Ionized ISM" where supernovae produced chimneys or "worms" (so-called because of their morphology) of hot gas that vent to the halo. Whether this explains the marked drop in molecular clouds at $|b| \geq 50°$ or that something else is at work requires further investigation.

Both surveys calculated a surface filling fraction for molecular gas, although this quantity is uncertain without a model for the distribution. However, taken at face value, ϵ quantifies how easy it may be to detect CO emission in random directions for $|b| \geq 30°$. For the NGH, $\epsilon = 0.0025$ and 0.03 for the SGH. This significant discrepancy between NGH and SGH was explained in the original MBM survey as a displacement of the Sun above the Galactic midplane (see Magnani et al. 1996, for details). However, it is important to note that this result, i.e., that there is an order of magnitude more molecular gas in the SGH than the NGH at high latitudes, was called into question by a couple of surveys for molecular gas which used infrared emission from dust as the tracer for molecular gas. We discuss these in the next section. Finally, the mass surface density of molecular gas in the NGH is 0.025 $M_{\odot}$ pc^{-2} and 0.09 $M_{\odot}$ pc^{-2} in the SGH. Comparing these values to the local value of molecular gas from all sources (1-2 $M_{\odot}$ pc^{-2}) indicates that the results of the Georgia-Harvard CfA high-latitude CO surveys are consistent with previous work: The high-latitude molecular clouds contribute about 10% to the local inventory of molecular gas.

There is a certain artificiality about ϵ and mass surface density estimates given for a 30° cutoff in latitude. If the cutoff was lowered to 25°, a good portion of the Polaris Flare (Heithausen and Thaddeus 1990) would be included in the survey and both numbers would increase. If the survey includes all points above 20°, the entire Polaris Flare is covered along with much of the Ophiuchus dark cloud complex (De Geus 1988), increasing both ϵ and mass surface density. Nevertheless, comparing the northern vs. southern Galactic hemispheres for a given cutoff, the north has significantly less molecular gas than the south, for CO(1-0) emission surveys with 0.1 K, 1σ sensitivity.

The bulk of the detections are clustered around the MBM 53–55 region ($\ell \sim 90°$, $b \sim -35°$) and south of the Taurus/Auriga dark clouds ($\ell \sim 170°$, $b \sim -40°$) There may also be an enormous partial loop structure extending from southwest of the Taurus dark clouds, extending to $\ell \sim 130°$, $b \sim$ -45°), and curling back up east of the MBM 53–55 region. This feature was noted by Bhatt (2000) who connected it to the Ursa Major clouds in the north and attributed their formation to a single (dynamical) event. Even if the whole loop postulated by Bhatt is not a real feature, the lower portion could still be a connected feature. A velocity-coded map of the southern detections is shown in Fig. 9.8 and a gradient may be present in this feature.

The mass surface density value of 0.09 $M_\odot$ pc^{-2} was derived by Magnani et al. (2000) using $X_{CO} = 1 \times 10^{20}$. A factor of two increase in X_{CO} brings the mass surface density estimate more in line with the result from Magnani et al. (1996) derived from an inventory of the known high-latitude clouds (at that time), encompassing a sample of objects identified using very different selection criteria. This encouraging concordance implies that the two NGH and SGH surveys are an

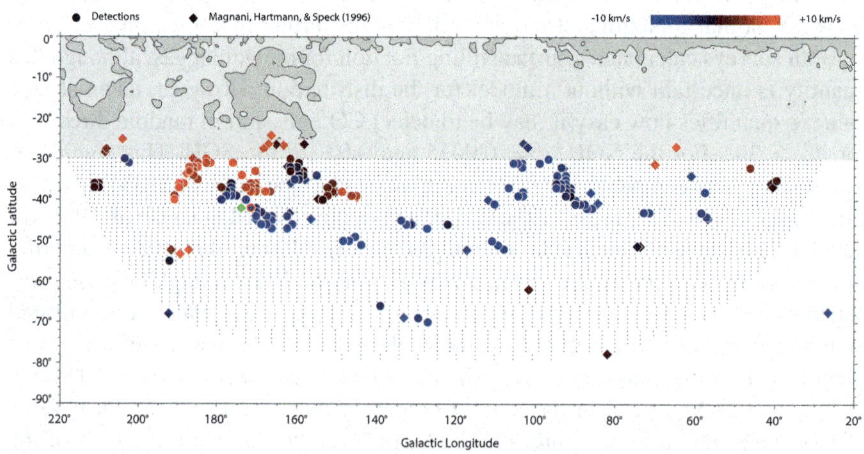

Fig. 9.8 Detections of CO(1-0) emission from the SGH. Color coded for LSR velocity. The *color bar* is at the top right of the figure. The *green diamond* represents a cloud at a velocity of $\sim$ 12 km s^{-1}

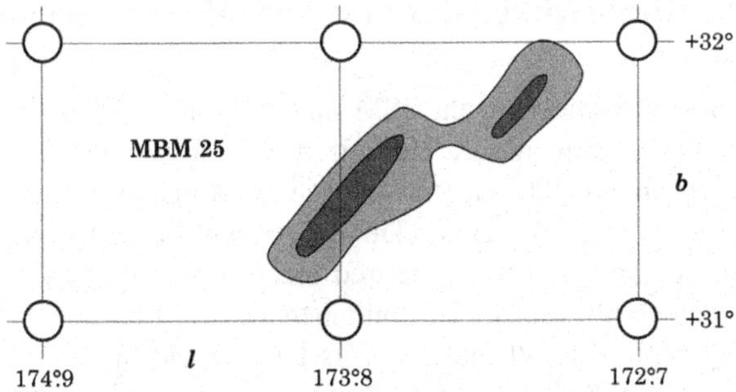

Fig. 9.9 High-latitude cloud (MBM 25) that was missed by the sampling grid pattern from the Georgia-Harvard-CfA survey shown in Fig. 9.7. This figure is from Hartmann et al. (1998)

excellent snapshot of the general properties of the high-latitude molecular clouds detected by the previous non-systematic searches.

However, there are three worrisome aspects of the results from the Georgia-Harvard CfA high-latitude CO surveys: (1) The search grid was chosen on the basis of the typical size cloud from Magnani et al. (1985). While all the "big fish" were certain to be caught in the sampling net, it is clear that some of the smaller objects would be missed. A particularly telling example is shown in Fig. 9.9. The question of the size distribution of high-latitude clouds has never been satisfactorily addressed because of the lack of complete mapping data for the majority of the objects. Even among the mapped clouds, many researchers have focused on the core regions and ignored the more extended, low CO intensity regions. Indications of that a population of smaller clouds likely exists is clear from the surveys of Keto and Myers (1986), Heiles et al. (1988), and Heithausen (2002). A systematic search for this population of objects at high-latitude (or even at low latitude) has never been carried out. (2) Similarly, a population of intermediate-velocity molecular clouds is known to exist at high-latitudes (Magnani and Smith 2010; Heiles et al. 1988). The Draco cloud is a famous, but atypical example of this population (Mebold et al. 1985; Magnani et al. 1985), but the other half-dozen or so known clouds are smaller and are likely much more distant than the typical high-latitude clouds. How many of these clouds exist, and where they are located with respect to the Galactic plane is still unknown. (3) The surveys targeting regions with excess infrared emission have always determined that there was more molecular gas than the CO(1-0) surveys detect. In particular Reach et al. (1998) claim that roughly equal amount of molecular gas are present at intermediate and high-latitude in the NGH and SGH, in disagreement with the Georgia-Harvard CfA surveys results. We now examine this last point in greater detail.

9.7 High-Latitude Molecular Gas via Infrared Techniques

Gas and dust are well-mixed in the ISM and HI column density maps of large areas are relatively easy to make. Thus, if one removes the dust emission in a band anywhere from 60–300 μm that is due to grains associated with the HI, the remainder should trace H_2. This relatively straightforward approach was made possible by the opening up of the infrared electromagnetic window in the early 1980s. Désert et al. (1988) used this technique to identify over 500 "infrared excess" regions at $|b| \geq 5°$. Could all these objects be detected in CO(1-0)? Blitz et al. (1990) answered "no", detecting only 13% in a survey of 201 DBB clouds, The undetected DBB objects seem to be clouds with enough H_2 for significant column densities, but not enough molecular hydrogen to sufficiently self-shield to yield CO detections (i.e., regions composed primarily of dark gas as discussed in Sect. 8.4). Others arrived at this conclusion from similar types of observations (e.g., Reach et al. 1994; Meyerdierks and Heithausen 1996).

The most complete work of this type was done by Reach et al. (1998) who improved on the method of Desert, Bazell, and Boulanger by using COBE data (in particular, the 240 μm DIRBE data) to pick out the colder infrared clouds. In addition, the HI data that they used were of higher quality and better calibrated than what was used in the Désert, Bazell, and Boulanger catalog. The RWO catalog consists of 60 infrared excess regions associated with previously identified molecular clouds either from the MBM catalog or the Magnani et al. (1996) compilation, and 81 objects not associated with known molecular clouds at that time. All of these objects are at $|b| \geq 20°$. The surface mass density of the infrared excess clouds is similar to that of the original MBM CO survey: 0.3 $M_{\odot}$ pc^{-2}, but their results differ significantly from those of the Georgia-Harvard CfA high-latitude CO surveys in one key aspect: Their SGH mass surface density is similar to Magnani et al. (2000), but an order of magnitude greater in the NGH than what was obtained by Hartmann et al. (1998). In other words, the North/South asymmetry in the CO distribution originally noted in Blitz et al. (1984) and confirmed by the Georgia-Harvard CfA blind surveys is not evident in the RWO infrared excess clouds.

As described above, a search for CO emission from a sample of RWO objects was made in 2001 by Onishi et al.. They observed 68 infrared-excess objects (all the RWO sources that were visible from the site of the telescope used for the survey, the NANTEN 4 m mm-wave radiotelescope) from the RWO catalog and detected CO emission from 32, a significantly higher percentage than Blitz et al. found from the DBB catalog. This underscores how the temperature correction in the RWO catalog made a significant difference in identifying infrared regions that might contain H_2. Similarly, the temperature corrected dust maps from Schlegel et al. (1998) (Sect. 6.7.1) are an excellent tool for identifying regions in the diffuse ISM where molecular gas might be present.

References

Allamandola, L.J., Tielens, A.G.G.M., and Barker, J.R. 1985, ApJ, 290, L25
Bally, J., Langer, W.D., WIlson, R.W., Stark, A.A., and Pound, M.W. 1991, in *Fragmentation of Molecular Clouds and Star Formation*, eds. E. Falgarone, F. Boulanger, and G. Duvert, (Dordrecht: Kluwer), 11
Bhatt, H.C. 2000, A&A, 362, 715
Blitz, L., Fich, M., and Stark, A.A. 1982, ApJS, 49, 183
Blitz, L., Magnani, L., and Mundy, L. 1984, ApJ, 282, L9
Blitz, L., Bazell, D., and Désert, F.X. 1990, ApJ, 352, L13
Boulanger, F., Baud, B., and van Albada, G.D. 1985, A&A, 144, L9
Clark, F.O., Giguere, P.T., and Crutcher, R.M. 1977, ApJ, 215, 511
Clemens, D.P. and Barvainis, R. 1988, ApJS, 68, 257
Dame, T.M., Hartmann, D., and Thaddeus, P. 2001, ApJ, 547, 792
Désert, F.X., Bazell, D., and Boulanger, F. 1988, ApJ, 334, 815
De Geus, E.J. 1988, PhD Thesis, Leiden University
de Vries, C.P. 1986, Ph.D. Thesis, Leiden University
de Vries, H.W., Heithausen, A., Thaddeus, P. 1987, 319, 723
Dewey, R.J., Uson, J.M., Weisberg, J.M., Wilkinson, D.T., and Stark, A.A. 1983, AJ, 88, 1832
Dickey, J.M., Crovisier, J., and Kazès, I. 1981, A&A, 98, 271
Dickman, R.L. 1975, ApJ, 202, 50
Dickman, R.L., Somerville, W.B., Whittet, D.C.B., McNally, D., and Blades, J.C. 1983, ApJS, 53, 55
Draine, B.T. and Anderson, N. 1985, ApJ, 292, 494
Fich, M., Dahl, G.P., and Treffers, R.R. 1990, AJ, 99, 622
Gautier, T.N. 1986, in *Light on dark matter*, ed. F.P. Israel, (Dordrecht, D. Reidel Publishing Co.) p. 49
Hartmann, D., Magnani, L., and Thaddeus, P. 1998, ApJ, 492, 205
Hauser, M.G. et al. 1984, ApJ, 285, 74
Heiles, C. and Habing, H.J. 1974, A&AS, 193, 31
Heiles, C., Reach, W.T., and Koo, B.-C. 1988, 332, 313
Heiles, C. 1993, in *Star Formation, Galaxies, and the Interstellar Medium*, eds. J. Franco, F. Ferrini, and G. Tenorio-Tagle (Cambridge: Cambridge U. Press), 245
Heithausen, A. 1987, Ph.D. Thesis, University of Bonn
Heithausen, A. and Thaddeus, P. 1990, ApJ, 353, 49
Heithausen, A., Stacy, J.G., de Vries, H.W., Mebold, U., and Thaddeus, P. 1993, A&A, 268, 265
Heithausen, A. 2002, A&A, 2002, 393
Kazès, I. and Crovisier, J. 1981, A&A, 101, 401
Keto, E.R. and Myers, P.C. 1986, ApJ, 304, 466
Knapp, G.R. and Jura 1976, ApJ, 209, 782
Knapp, G.R., Kuiper, T.B.H., Knapp, S.L., and Brown, R.L. 1977, ApJ, 214, 78
Lada, E.A. and Blitz, L, 1988, ApJ, 326, L69
Liszt, H.S. and Wilson, R.W. 1993, ApJ, 403, 663
Liszt, H.S and Lucas, R. 1998, A&A, 339, 561
Low, F.J. et al. 1984, ApJ, 278, L19
Lucas, R. and Liszt, H, 1996, A&A, 314, 917
Lynds, B.T. 1962, ApJS, 7, 1
Lynds, B.T. 1965, ApJS, 12, 163
Mahoney, M.J., McCutcheon, W.H., and Shuter, W.L.H. 1976, AJ, 81, 508
Magnani, L., Blitz, L., and Mundy L. 1985, ApJ, 295, 402
Magnani, L., Lada, E.A., and Blitz, L. 1986, ApJ, 301, 395
Magnani, L., Hartmann, D., and Speck, B.G. 1996, ApJS, 106, 447
Magnani, L., Hartmann, D., Holcomb, S.L., Smith, L.E., and Thaddeus, P. 2000, ApJ, 535, 167

Magnani, L. and Smith, A.J. 2010, ApJ, 722, 1685
McGehee, P.M. 2008, in *Handbook of Star Forming Regions, Volume II: The Southern Sky*, ASP Monograph Publications, Vol. 5, ed. B. Reipurt, 813.
Mebold, U. et al. 1985, A&A, 151, 427
Meyerdierks, H. and Heithausen, A. 1996, A&A, 313, 929
Onishi, T. et al. 2001, PASJ, 53, 1017
Penzias, A.A., Solomon, P.M, Jefferts, K.B., and Wilson, R.W. 1972, ApJ, 174, L43
Pound, M.W. and Goodman, A.A. 1997, ApJ, 482, 334
Puget, J.L., Leger, A., and Boulanger, F. 1985, A&A, 142, L19
Reach, W.T., Koo, B.-C., and Heiles, C. 1994, ApJ, 429, 672
Reach, W.T., Wall, W.F., and Odegard, N. 1998, ApJ, 507, 507
Sandell, G., Johansson, L.E.B., Rieu, N.Q., and Mattila, K. 1981, A&A, 97, 317
Schlegel, D.J., Finkbeiner, D.P., and Davis, M. 1998, ApJ, 500, 525
Shane, W.W. and WIrtanen, C.A. 1967, Publ. Lick Obs., 22, 1
Sharpless, S. 1959, ApJS, 4, 257
Teerikorpi, P. 1981, A&A, 98, 300
Vanden Bout, P.A., Davis, J.H., and Loren, R.B. 2012, Journal of Astronomical History and Heritage, 15(3), 232
Weiland, J.P. et al. 1986, ApJ, 306,, L101
Yonekura, Y. et al. 1999, PASJ, 51, 837

Chapter 10
Distances

I felt a tremendous distance between myself and everything real
—Hunter S. Thompson

Abstract Historically, finding the distances to objects outside the Solar System has been one of astronomy's most difficult tasks. This is especially true of things like interstellar clouds where luminosity distances play no role. In this chapter we describe some of the techniques that have been applied over the years to obtaining the distances to molecular clouds. In particular we focus on high-latitude molecular clouds to illustrate the efficacy of Wolf diagrams, spectroscopic, and photometric techniques. We also discuss the asymmetry in the high-latitude molecular cloud population with respect to the northern and southern Galactic hemispheres and the possibility that some high-latitude molecular clouds may be at the disk/halo interface.

10.1 Introduction

With the exception of stellar parallaxes, getting accurate distances to objects outside our Solar System is arguably the most difficult thing in astronomy. For the ISM, where discrete objects are limited to atomic and molecular clouds this is even more the case than for stars and galaxies. The checkered history of the distance of the high-latitude cloud MBM 12 illustrates the difficulties and wrong turns that can be expected. Originally, the cloud was thought to be about 65 pc away, making it the nearest molecular cloud of all (Hobbs et al. 1986) and, possibly, located within the Local Bubble. In the last fifteen years, the distance determinations to the cloud have been progressively increasing so that, today, the latest estimates put MBM 12 at about 275 pc (Luhman 2001). It is no longer considered to be in the Local Bubble, greatly changing both its environment and ideas about its origin and evolution. In this chapter we focus on those methods that have been used most often to tease out the distances to Galactic atomic and molecular clouds with a special emphasis on the high-latitude clouds as a *local* sample whose distances have been obtained by a variety of techniques over the last three decades.

L. Magnani, S.N. Shore, *A Dirty Window*, Astrophysics and Space Science Library 442, DOI 10.1007/978-3-662-54350-4_10

10.2 Distances to the Clouds: Galactic Rotation Curve

The usual method to obtain an interstellar cloud's distance is to spectroscopically measure the cloud's radial velocity and then to use a model for the Galactic rotation curve to link that velocity and direction to its distance from the Sun. The relationship between ℓ, r, and v_r can be obtained from

$$R_{GC} = R_o \sin(\ell) \frac{v(R_{GC})}{v_r + v_0 \sin(\ell)} \tag{10.1}$$

where R_{GC} is the galactocentric radial distance of the cloud, R_0 is the Galactocentric radius of the Sun, v_0 is the orbital velocity of the Sun around the Galactic Center (the Local Standard of Rest, LSR), $v(R_{GC})$ is the speed of objects at R_{GC} from the rotation curve of the Galaxy, and v_r is the radial velocity of the object in question (see, e.g., Binney and Merrifield 1998). For the Outer Galaxy ($R_{GC} > R_0$), there is no Galactic quadrant-related ambiguity in determining the distance; v_r decreases monotonically with the distance to negative values. However, for the Inner Galaxy, R_{GC} corresponds to two distances along the line of sight, the so-called "near" and "far" kinematic distances located on either side of the tangent point at R_{GC}. The two solutions for the distance are given by

$$d = R_0 \cos(\ell) \pm \sqrt{R_{GC}^2 - R_0^2 \sin^2(\ell)} \tag{10.2}$$

This method was widely used in the 1980s to determine the distance to GMCs in the Galactic plane. The distance ambiguity was resolved (somewhat haphazardly) by assuming that a "smaller" than usual object is at the far distance. An improvement of this technique was made by Bania and Lockman (1984) who used the presence or absence of HI self-absorption to identify whether a molecular cloud was at the near or far distance.

10.3 Distances to the Clouds: Statistical

In the solar neighborhood (within a few hundred parsecs of the Sun) at a distance $R_0 = 8$ kpc from the Galactic center the vertical mass distribution approximates a thin self-gravitating layer rather well. In the co-orbiting reference frame (LSR), stellar orbits have individual eccentricities, inclinations and total energies. Therefore, relative to the LSR, the vertical motion is like an ensemble of harmonic oscillators bound to the disk. Each star reaches its maximum height depending on its orbital properties but if drawn from a single population (both in mass and age) the vertical density distribution is a result of the stellar kinetic energy distribution. Those with the highest vertical velocities in the midplane reach the greatest distance. Since these are also be the rarest in the ensemble, their number density decreases from the plane. They are, however, still bound, so there is a turning point in their orbit and the distribution is not the same as for an infinite slab with a superimposed atmosphere.

A single population can be modeled as a sort of "gas of stars", a collisionless ensemble with a single, constant velocity dispersion. It is more correct, though, to use a triaxial velocity ellipsoid in which the angular, radial, and vertical dispersions are all different. In the Schwarzschild distribution, a set of three Gaussian functions is used to describe the statistics of the three velocity components (see, e.g., Binney and Tremaine 2008). An isotropic distribution of motions, in the corotating frame, would have $\sigma_z^2 = \frac{1}{3}\sigma^2$ and this is sometimes assumed in the literature.[1] But the *Hipparcos* (and rather soon, *Gaia*) parallaxes, proper motions, and radial velocities give a different picture. The vertical velocity dispersion of young, thin disk stars is about 10 km s^{-1}, varying by ± 2 km s^{-1} depending on spectral type (hence mass and evolutionary stage). This is about half of the radial and angular dispersions, which are also unequal (Binney and Merrifield 1998). From the virial theorem, an isothermal distribution of stars with mass density $\rho_\star(r, z)$ and one-dimensional velocity dispersion satisfies the vertical structure equation

$$\rho_\star \frac{d\varphi}{dz} = \frac{d\rho_\star \sigma_z^2}{dz} \tag{10.3}$$

where φ is the gravitational potential and we have assumed that the gas density is negligible relative to the stars. We will return to this point presently. The similarity to the equation of hydrostatic equilibrium is not an accident, for a collisionless isothermal ensemble the virial equilibrium (i.e., virialization in an extended structure) is so described. To find the gravitational potential requires the Poisson equation

$$\nabla^2 \varphi = 4\pi G \rho_\star \tag{10.4}$$

where we assume (with later justifications) that the total mass density is dominated by the stars. A slab geometry permits a reduction of the dimensionality (the orbital motion decouples from the vertical) as long as we concentrate on the solar neighborhood so equations 10.3 and 10.4 can be combined

$$\frac{d^2}{dz^2} \ln\left(\frac{\rho_\star}{\rho_{\star,0}}\right) = \frac{4\pi G \rho_{\star,0}}{\sigma_z^2} \left(\frac{\rho_\star}{\rho_{\star,0}}\right) \tag{10.5}$$

where $\rho_{\star,0}$ is the *stellar* density in the midplane and the vertical distance is scaled to the inverse length, z_0 (that also quantifies the stability limit of the layer against

[1] The z-direction is perpendicular to the Galactic plane.

gravitational collapse, see Chap. 11),

$$z_0^{-2} = \frac{4\pi G\rho_{\star,0}}{\sigma_z^2} \quad . \tag{10.6}$$

Calling $u = (\ln \rho_\star/\rho_{\star,0})$ equation 10.5 becomes

$$\frac{d^2u}{dx^2} = e^u$$

that, with $\rho_\star(0) = \rho_{\star,0}$ and $\rho(z \to \infty) \to 0$, integrates to

$$\rho_\star = \rho_{\star,0}\,\mathrm{sech}^2\left(\frac{z}{z_0}\right) \quad . \tag{10.7}$$

The vertical gravitational acceleration, K_z, follows immediately from the virial-hydrostatic equilibrium equation

$$K_z = -\sigma_z^2\frac{du}{dx} = 2\frac{\sigma_z^2}{z_0}\tanh\left(\frac{z}{z_0}\right) \quad . \tag{10.8}$$

Near the plane the density falls off like a Gaussian, at large distance (but still on scales much smaller than R_0) it decreases exponentially with height.

The gas, having a much lower volume and surface density, can be treated as a separate hydrostatic component whose "pressure gradient" is governed by K_z (that approaches a constant for $z/z_0 \gg 1$). Approximating the gas (i.e. clouds) with an isothermal pressure, $P_c = \sigma_c^2\rho_c$ and substituting this in equation 10.8 gives,

$$\ln\frac{\rho_c}{\rho_{c,0}} = -2\frac{\sigma_z^2}{\sigma_c^2}\ln\cosh\left(\frac{z}{z_0}\right) \quad , \tag{10.9}$$

so calling $p = (\sigma_z/\sigma_c)^2$ we find that the gas density varies as

$$\rho_c \sim \mathrm{sech}^{2p}\left(\frac{z}{z_0}\right) \quad . \tag{10.10}$$

Choosing the appropriate value for the gas velocity poses a conceptual problem. The stars are collisionless point masses interacting collectively to produce K_z and each population is well approximated as a dynamically distinct "isothermal" component with a characteristic velocity dispersion (the analogy with a rarified gas such as a planetary exosphere, comes to mind). In strong contrast, the gas phase is not one and unique, nor are the different phases distinct. Only GMCs behave like hypermassive "stars"; their self-gravitation makes them a dynamically distinct population moving in the Galactic gravitational field. The small clouds are neither self-gravitating nor permanent, they appear out of the diffuse gas and return to it when they dissipate. They have different densities than their surroundings, but may not differ in internal dynamics from that of the larger scale flows. While the diffuse gas, like the GMCs, orbits the Galactic center and has a mean circulation at any galactocentric distance, the random motions and turbulence have many different origins that are time and location dependent. Supernova remnants, stellar winds and their associated bubbles, outflows from stellar sources, and many other distributed drivers combine to generate a broad range of velocities. Even for turbulence, there is no single σ_c independent of length scale (see Chap. 11). It is conventional to chose $\sigma_{cz}^2 = (1/3)\sigma_c^2$, assuming isotropy over the scale of a few kiloparsecs, and then set p = constant. If special conditions produce the small molecular clouds, such as compressions at interfaces between large scale flows or condensations out of denser regions, it is not implausible that they can also have different dynamics relative to the diffuse gas. This can only be addressed by very large scale surveys in multiple tracers. A stab at this was taken for the Small Magellanic Cloud by Stanimirović et al. (1999), who produced a dramatic picture of a fractal-like medium with velocity variations following a power spectrum on many scales. This is highlighted by multiwavelength views of the different phases of the diffuse gas (see, e.g. Chu 2009). Chepurnov et al. (2010) surveyed HI at high Galactic latitude and found a power law spectrum for the velocities, a distribution that has no dispersion. Whatever the origin of the motions, using an ideal gas-like equation to characterize the stresses should be nuanced.

10.3.1 A Practical Example: High-Latitude Molecular Clouds

As an example, we discuss the average distance to the high-latitude molecular clouds discussed in Chap. 9. Equation 10.10, through

$$z_{c\frac{1}{2}} = \frac{1.7\,\langle\sigma_{cz}^2\rangle^{\frac{1}{2}}}{(2\pi G\rho_o)^{\frac{1}{2}}} \tag{10.11}$$

[see van der Kruit and Shostak (1984) for details] relates the vertical distribution of the high-latitude cloud ensemble (in terms of the half-height of the distribution) to their velocity dispersion for a given stellar gravitational potential. The greatest uncertainty in the calculation is determining the local gravitation potential to sufficient accuracy. Magnani et al. (1985) used equation 10.11 with ρ_0 expressed in terms of the more readily obtainable quantities of stellar mass surface density, $\Sigma_\star$, and the stellar scale height, $h_\star$, to determine the half-height of the population of high-latitude molecular clouds.

If the stars are distributed exponentially in z, *i.e.*, $\rho_0\ \exp(-z/h_*)$, then as a function of galactocentric distance, R, the relationship between the three quantities at $z = 0$ is given by

$$\Sigma_* = 2\rho_0 h_* \quad . \tag{10.12}$$

Magnani et al. (1985) used $\Sigma_* = 110 \pm 20$ M$_\odot$ pc^{-2} and $h_* = 600 \pm 50$ pc (from van der Kruit and Shostak 1984), equivalent to a stellar volume density of 0.09 M$_\odot$ pc^{-3}. More recent values for Σ_* and h_* are 50 ± 10 M$_\odot$ pc^{-2} and $300 \pm$ 50 pc (Bahcall 1984, 1986; Flynn and Fuchs 1994; Gould 1996), equivalent to $\rho_0 =$ 0.08 M$_\odot$ pc^{-3}. Thus, we can write equation 10.11 as

$$z_{c\frac{1}{2}} = 36.5 \, \langle \sigma_{cz}^2 \rangle^{\frac{1}{2}} \quad pc, \tag{10.13}$$

where σ_{cz} is measured in km s^{-1} and $\rho_0 = 0.08$ M$_\odot$ pc^{-3}. Following normal convention, we will assume the clouds have a Gaussian vertical distribution so that $z_{c\frac{1}{2}}$ is related to the Gaussian cloud scale height, h_c, by $z_{c\frac{1}{2}} = 2.35h_c$. Thus,

$$h_c = (15.5 \pm 2.1)\sigma_{rc} \quad pc, \tag{10.14}$$

where σ_{rc} is the radial velocity dispersion of the cloud ensemble (we are assuming $\sigma_{rc} = \sigma_{cz}$).

Magnani et al. (1985) found $\sigma_{rc} = 5.6 \pm 1.2$ km s^{-1} from their original sample for a scale height of 87 pc (the original paper quotes 81 pc because of a different ρ_0). A decade later, using a value of $\rho_0 = 0.08 M_\odot$ pc^{-3} and a compendium of all high-latitude ($|b| \geq 25°$) clouds then identified in the literature, Magnani et al. (1996) obtained $\sigma_{rc} = 9.9$ km s^{-1}. This significant increase in velocity dispersion was driven by the inclusion of seven intermediate-velocity clouds ($|v_{LSR}| \geq 20$ km s^{-1}) out of a total sample of 120 objects. It is likely that these clouds have not reached equilibrium with the gravitational potential field and should be excluded from the scale height calculation. Excluding them drops the the velocity dispersion to 5.8 km s^{-1}, virtually the same as in Magnani et al. (1985). The problem with both determinations is the subjective criteria used in compiling the original sample (see Sect. 9.3). The Georgia-Harvard/CfA high-latitude surveys are the least subjective large-scale high-latitude surveys currently available, so, a more reliable scale height determination should use σ_{rc} derived from the combined northern and southern Galactic hemisphere data. This value, 7.2 ± 1.0 km s^{-1}, for 159 detection or 7.1 $\pm$ 1.0 km s^{-1} for 56 "cloud complexes" (see Sect. 9.6) leads to a scale height of 110 pc.

A statistical distance to each cloud is obtained by determining the average value of $z(\csc(b))$ for each cloud where z is the expectation value of the vertical distance from the midplane ($z = 0.798h_c$ for a Gaussian distribution). The resulting distance for the detections in the Georgia-Harvard/CfA high-latitude surveys is 140 ± 20 pc. Corrections for differential Galactic rotation do not significantly change this value. In principle, this technique can be used to get the scale height and statistical distance to any local *distribution* of interstellar clouds.

10.4 Distances to the Clouds: Direct Measurements

The average distance to the high-latitude clouds derived in the previous section is valid only in a statistical sense. It cannot be used to determine the correct distance to any particular cloud. However, a number of high-latitude clouds have distances determined from direct methods which can be used as a check of the statistical results. Three direct methods have been employed to obtain direct distances to high-latitude clouds and will be described below: Wolf diagrams, optical spectroscopy to detect interstellar absorption lines, and photometry to identify abrupt increases in the color excess produced by absorption presumably from the cloud in question. These techniques can be and have been applied to other interstellar clouds, though here we will concentrate only on high-latitude molecular clouds.

10.4.1 The Method of Wolf Diagrams

Wolf (1923) used star counts to determine the distances to interstellar dust clouds. Basically, the method involves comparing the number of stars of a particular apparent magnitude in a field obscured by a dust cloud to the number of stars of the same magnitude in a comparison, obscuration-free field. The presence of the obscuring dust in the cloud will produce a decrease in the total number of stars of a given apparent magnitude in that direction. This decrease can be quantified in terms of the distance and extinction through the cloud if the spatial distribution of stars in that direction is known.

Following the treatment in Bok's classic 1937 book, *The Distribution of Stars in Space* (see also Trumpler and Weaver 1953), the number of stars within the apparent magnitude interval $(m, m + dm)$ in a solid angle ω in the direction of Galactic longitude and latitude (ℓ, b) is denoted by $A(\ell, b, m)dm$. This quantity is known as the apparent stellar brightness distribution. If $N(\ell, b, m)$ is the total number of stars brighter than magnitude m within the solid angle ω, then

$$N(\ell, b, m) = \int_{-\infty}^{m'} A(\ell, b, m')dm' \quad . \tag{10.15}$$

The value of $A(\ell, b, m)$ is determined by two other functions: the space density of stars of various luminosity, and the distribution of the interstellar extinction. The space density of stars of various luminosity is described by the luminosity function $\Phi_r(\ell, b, M)dM$ which is the total number of stars with absolute magnitude, M, in the interval $(M, M + dM)$ in a unit volume of space at a distance r. We will denote the interstellar extinction as a function of ℓ, b, and r as $a(\ell, b, r)$. The relation between

A, Φ_r, and a is then

$$A(\ell, b, m)dm = \omega dM \int_0^\infty r^2 \Phi_r(\ell, b, m + 5 - 5 \log r - a(\ell, b, r))dr \ . \quad (10.16)$$

If $a(\ell, b, r) = 0$ and Φ_r is distance-independent, then equation 10.16 can be solved analytically, $\log A(m) \propto 0.6m$. Because a and Φ_r are neither zero nor constant, they must be modeled. One way is described by de Vries (1986). The interstellar extinction function a is separated into a cloud contribution a_c and the extinction along the line of sight from the intervening ISM, a_g. For lines of sight free of clouds, a_g is assumed to be constant with distance (*e.g.*, 1.9 mag kpc^{-1}; Münch (1952). To obtain the distance to nearby clouds ($d \leq 500$ pc), de Vries (1986) proposed modelling Φ_r using the local stellar luminosity function in the Galactic plane, Φ_0, tabulated in many places (e.g., Allen 1973; Cox 2000). If the cloud lies away from the Galactic plane, Φ_0 can be modified by assigning a scale height $h_*(M)$ to stars of varying absolute magnitude:

$$\Phi_r(M) = \Phi_o(M) \exp\left[\frac{r \sin b}{h_*(M)}\right] \ . \quad (10.17)$$

This parameterization will also decrease the slope of $\log A(m)$ versus m plots and, additionally, it will cause the curve to flatten further as m increases.

The stellar luminosity function can be independently calibrated for the Milky Way by using tabulated values of $N(m)$ as a function of position. One such tabulation was compiled by van Rhijn in the 1920s. The "van Rhijn luminosity tables" (van Rhijn 1929) were based on star counts from photographic plates somewhat similar to the blue POSS plates in use today.

Because extinction is defined logarithmically, it is linearly proportional to the dust optical depth, τ_d. In turn, τ_d is the sum of all individual contributions along the line of sight because, unlike absorption, τ_d does not saturate. If the gas to dust ratio is constant, the individual contributions indentified in velocity can be separated in their contribution to the total line of sight optical depth. In the method elaborated by Wolf (1923), an interstellar dust cloud is represented by an infinitely thin layer producing extinction a_c at the cloud distance d. Thus, the total extinction along the line of sight is modified as follows

$$a(r) = a_g \qquad r < d \quad (10.18)$$

$$a(r) = a_g + a_c \qquad r \geq d \quad (10.19)$$

The introduction of this functional form in equation 10.15 decreases the slope of $A(m)$ over a particular magnitude interval depending on the values of a_c and d. Because $\log A(m) \propto \log N(m)$, the resulting plots of $\log N(m)$ versus m (which are called Wolf diagrams) for an obscuring cloud will differ from those for a cloud-free region.

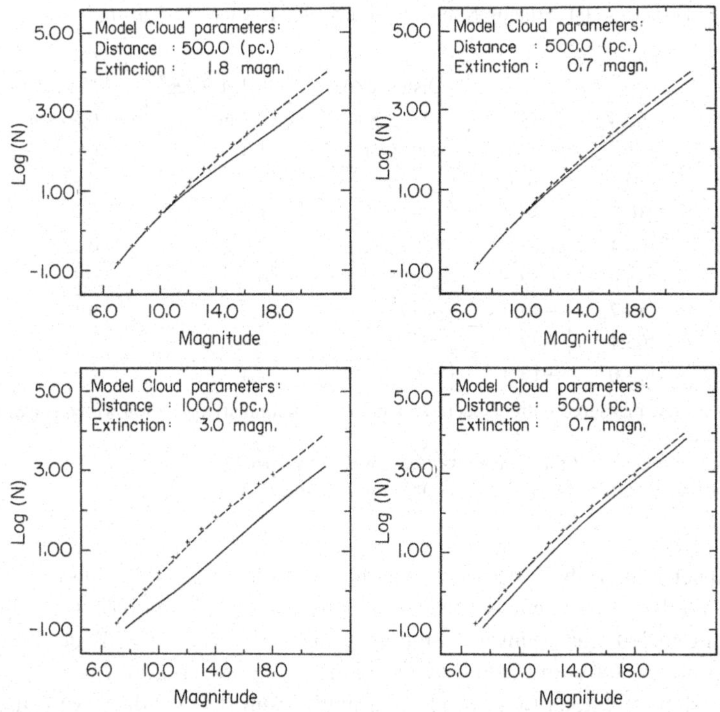

Fig. 10.1 Examples of Wolf diagrams (see text). The *dashed line* shows the calibrated stellar density function (equation 10.36) for a given line of sight and the crosses show the van Rhijn (1929) data for the line of sight. Log (N) is the total number of stars brighter than a given magnitude *m*. The *solid line* shows the effects of an obscuring cloud with the given distances and extinctions. Unless the cloud has sizeable extinctions and is also distant (like the example with d = 500 pc and A_V = 1.8 mag), the turnoff occurs at magnitudes brighter than 10 where the star counts are too sparse to give meaningful data points. Figure from Magnani and de Vries (1986)

These effects are shown in Fig. 10.1 for four obscuring clouds with varying extinctions at varying distances. Unless the cloud has considerable extinction and is also relatively distant (*e.g.*, upper left-hand corner of Fig. 10.1) the log $N(m)$ turn-off occurs at magnitudes brighter than 10 where the star counts are few in number and thus subject to significant fluctuations. A modern analogy of the Wolf method has been pioneered by Lombardi (2009) on the Pipe Nebula, using 2MASS data in the infrared.

Although Wolf diagrams work well for clouds with $A_V > 2$ magnitudes and distances greater than a few hundred parsecs, compiling the necessary star count data is a laborious task. Wolf diagrams are not often used to determine cloud distances because of the difficulty in separating faint galaxies from faint stars in the database. Recent attempts to use Wolf diagrams include Magnani and de Vries (1986), Kun (1998) and Kim et al. (2000). Magnani and de Vries focused on a sample of high-latitude clouds and since these objects are our exemplars we will

Table 10.1 Distances and extinctions to high-latitude clouds using Wolf diagrams

Cloud	ℓ	b	Distance (pc)	A_V^a 0.4–1.0 deg^2	A_V Entire cloud
MBM 07	150.4	−38.1	125 ± 50	0.7 mag	0.5 mag
MBM 16	171.7	−37.7	100 ± 50	1.4	0.9
MBM 18	189.1	−36.0	≤ 175	0.8	0.8
MBM 20	210.9	−36.5	≤ 125	1.0	0.8
MBM 26	156.4	+32.6	175 ± 50	0.6	0.7
MBM 32	146.4	+39.6	≤ 275	0.5	0.3
MBM 55	89.2	−40.9	≤ 175	0.5	0.5
–	105.1	−39.9	≤ 275	0.5	[b]
Taurus[c]	175.0	−19.0	∼100	2.4	–

[a]The extinction is measured with respect to an unobscured comparison region which is assumed to have $A_V = 0$

[b]Cloud was not mapped at the time so its extent was not known

[c]Representative region in Taurus-Auriga dark cloud complex

take a closer look at their technique. Eight POSS blue plates (7 centered on high-latitude clouds and one on a Taurus dark clouds as a control sample) were digitized and an automated star counting computer program written by C.P. de Vries was used to produce star count data down to blue magnitude 19–20 for use in creating Wolf diagrams. Regions ranging in solid angle from 0.4–1.0 deg^2 were used for the obscured cloud and for unobscured comparison areas. The results are shown in Table 10.1.

The Wolf diagram method was tested on a dark cloud region in Taurus and yielded a distance and extinction for the cloud which corresponded to accepted values for the region (Elias 1978). The seven high-latitude cloud regions all yielded distances within a few hundred parsecs of the Sun. The Wolf diagram method confirms that the high-latitude clouds examined by Magnani and de Vries (1986) are indeed nearby objects. The overall cloud extinctions are very low but are measured with respect to the lowest-extinction region on the digitized plate. This latter region is assumed to have $A_v = 0$ mag. If the lowest-extinction region on the plate has a non-zero extinction then the overall cloud extinction would increase by that amount. Some of the objects whose distances were obtained using Wolf diagrams were also examined using the interstellar absorption line method. This method and its results are discussed in the next subsection.

10.4.2 Distances from Spectroscopy

Optical and ultraviolet spectroscopy of background and foreground stars provides upper and lower cloud distances. Although the method is simple, the observations can be difficult because the column density of the gas producing the interstellar lines is quite low and high signal-to-noise and high velocity resolution data are required. It should be noted that the velocity resolution *for absolute radial velocities* needed outside of the radio range is limited to ~ 1 km s^{-1} at best.[2] Moreover, dark clouds with greater column densities are too opaque to permit the viewing of background stars. We will again use examples from studies of high-latitude molecular clouds to illustrate this method in detail.

The first high-latitude cloud distance obtained in this fashion was for MBM12. Hobbs et al. (1986) estimated a distance of 65 pc based on the lack of interstellar lines in the spectrum of the star, 47 Ari, at a distance of $\sim$ 60 pc and unequivocal absorption from the NaI D lines in spectra of ϵ Ari AB at a then-tabulated distance of $\sim$ 70 pc. A subsequent paper by Hearty et al. (2000) revised the distance to 58 ± 5 pc $< d < 90 \pm 12$ pc because of more accurate stellar distances based on *Hipparcos* data. MBM12 was thought to be among the nearest molecular clouds to the Sun and several groups speculated that the cloud complex (comprising MBM11–13) was actually immersed within the hot gas inside the Local Bubble (Hobbs et al. 1988; Pound et al. 1990). Unfortunately, all that speculation was incorrect. Subsequent observations revealed that the cloud was actually at $\sim$ 275 pc from the Sun (Luhman 2001), greatly changing its previous status and invalidating all ideas about the cloud being in the Local Bubble. There *is* an absorbing layer between 60 and 90 pc from the Sun, but that layer is atomic (likely a part of the CNM). The absorbing layer that contains the molecular gas and, thus, the cloud is at $\sim$ 275 pc along the same direction.

Many high-latitude clouds have had their distance determined by spectroscopic techniques. The most extensive work of this kind was made by Penprase (1993), who combined spectroscopic and photometric techniques (see next section) to obtain the distances to a sample of high-latitude clouds. Table 10.2 compiles a list of some high-latitude clouds with distances obtained by spectroscopic observations.

10.4.3 The Method of Photometry

It is not always necessary to resort to spectroscopic techniques to obtain accurate cloud distances. Photometric methods which identify increases in stellar reddening as a function of distance can give results which are completely consistent with spectroscopic techniques and are easier to obtain. As for spectroscopy case, observing stars in front of and behind the obscuring cloud yields reddening values that increase directly with the amount of dust in the cloud. The greater the cloud extinction, the easier to notice the jump in reddening values of the more distant stars. Franco (1988) used $ubvy\beta$ photometry to estimate the distance to L1569 (MBM18) and, in a subsequent paper, Franco (1989), he extended this analysis to an additional 5 high-latitude cloud regions. Penprase (1992) used similar

[2] With the exception of Fabry-Perot measurements of individual lines.

Table 10.2 Spectroscopic distances to high-latitude clouds

Cloud	ℓ	b	Distance (pc)	Reference(s)
MBM03	131	−46	≤ 150	Penprase (1993)
MBM10	177	−54	120–150	Penprase (1993)
MBM12	159	−34	275	Luhman (2001)
MBM16	172	−38	60–96	Hobbs et al. (1988)
MBM18	189	−36	≤ 160	Penprase et al. (1990)
MBM20	211	−37	70–220	Andreani et al. (1988)
			≤ 110	Penprase (1993)
			112–161	Hearty et al. (2000)
MBM21	208	−28	≤ 210	Penprase (1993)
MBM23–24	172	+27	139 ± 33	Grant and Burrows (1999)
MBM29–31	142	+38	100–120	Penprase (1993)
MBM40	38	+45	≤ 140	Welty et al. (1989)
			90–150	Penprase (1993)
MBM41–44	90	+38	≥ 300	Lilienthal et al. (1991)
			463–618	Gladders et al. (1998)
MBM53	93	−32	110–155	Welty et al. (1989)
MBM54	93	−37	≤ 260	Welty et al. (1989)
MBM55	89	−38	≤ 265	Welty et al. (1989)
MBM57	6	+31	50–240	Penprase (1993)
G192–67	192	−67	109 ± 14	Grant and Burrows (1999)
K272.9+29.3	273	+30	≤ 100	Penprase (1993)

photometric techniques supplemented by spectroscopy to determine the distance to about a dozen high-latitude clouds. Table 10.3 lists the clouds or cloud regions whose distances have been determined primarily by photometry or star counts. A comparison of Tables 10.1 and 10.2 shows broad general agreement and supports the statistical results obtained in Sect. 10.2: high-latitude molecular clouds are relatively nearby objects with typical distances of a few hundred parsecs. These early results established that the high-latitude molecular clouds, as a population, tend to be located within the Galactic plane though there may be a few exceptions (see Sect. 10.6).

The most comprehensive work of this kind is by Schlafly et al. (2014) who used the Pan-STARRS-1 optical photometry survey of stars to estimate distances to almost all of the clouds in the Magnani et al. (1985) catalog. They also derive distances to other well-studied molecular clouds at lower Galactic latitudes. The resulting catalog of cloud distances is the most accurate compilation with uncertainties of 5%–10% to the diffuse and translucent clouds. In Table 10.4 we compare the results using the three techniques, Wolf diagrams, spectroscopy, and photometry for a selected sample of clouds. It is expected that the picture of the Galaxy produced by *Gaia* will substantially improve and expand our knowledge of the distances.

Table 10.3 Photometric distances to high-latitude molecular clouds

Cloud	ℓ	b	Distance (pc)	Reference(s)
MBM 03–04	132	−45	90–180	Penprase (1992)
MBM 15	192	−52	110–160	Penprase (1992)
MBM 18	189	−36	140–220	Penprase (1992)
			130 ± 20	Franco (1988)
MBM 20	211	−37	100–130	Penprase (1992)
			114–230	Franco (1989)
MBM 21–22	208	−28	≥ 400	Penprase (1992)
			250–300	Franco (1989)
MBM 27–31	142	+36	130–220	Penprase (1992)
MBM 36	4	+36	100 ± 50	Mattila (1979)[a]
			110 ± 10	Franco (1989)
MBM 41	90	+38	800–1300	Penprase et al. (2000)[a]
MBM 45	10	−28	160 ± 15	Franco (1989)
MBM 46–47	41	−36	250–300	Franco (1989)
			70–130	Penprase (1992)
MBM 57	6	+31	150–210	Penprase (1992)
K272.9+29.3	273	+29	100–230	Penprase (1992)

[a] Distance estimate based on star count data

Table 10.4 Comparison of distance estimates (in pc) to high-latitude molecular clouds

	Wolf diagram	Spectroscopy	Photometry	Photometry
Cloud	Table 10.1	Table 10.2	Table 10.3	Schlafly et al. (2014)
MBM 03–04	—	≤ 150	90–180	277^{+22}_{-26}
MBM 07	125 ± 50	—	—	148^{+13}_{-12}
MBM 16	100 ± 50	60–96	—	147^{+10}_{-9}
MBM 18	≤ 175	≤ 160	130 ± 20	166^{+18}_{-17}
MBM20	≤ 125	112–161	100–130	124^{+11}_{-14}
MBM 32	≤ 275	—	—	259^{+14}_{-15}
MBM 40[a]	—	90–150	—	64^{+21}_{-25}
MBM 55	≤ 175	≤ 265	—	206^{+8}_{-6}

[a]According to Schlafly et al. (2014), this would be the nearest high-latitude molecular cloud

10.5 Asymmetries in the Distribution of High-Latitude Molecular Clouds

Even a casual glance at Fig. 7.2 indicates that the southern Galactic hemisphere is richer in molecular gas than the northern hemisphere. This was clear even from the earliest systematic searches for high-latitude molecular clouds. Magnani et al. (1986) made a first attempt at an unbiased survey at $|b| \geq 30°$ surveying in the CO(1-0) line 1750 points in the NGH and 750 points in the SGH at one degree sampling with a $2.3'$ beam (see Sect. 9.5). Despite the more than a factor of two increase in sampled points in the NGH, the detection rate was greater in the south: 5 versus 4. In the early papers by Magnani et al., the asymmetry in the north-south distribution of high-latitude clouds was interpreted in terms of a displacement of the Sun above the Galactic midplane (e.g., Magnani, Hartmann, and Speck determined a displacement of 14 pc). Such a displacement had been previously noted for many different Galactic populations: Cepheids (Blaauw 1960), OB stars (Reed 1997); disk population stars (Humphreys and Larsen 1995), and open clusters (Joshi 2005). The Georgia-Harvard/CfA high-latitude surveys (Sect. 9.6) demonstrated instead that this is statistical artifact. The real culprit is the relatively small number of clouds at $|b| \geq 30°$.

The surface filling factor, ϵ, can be expressed as the number of lines of sight with at least one CO detection divided by the number of grid points in the survey. The Georgia-Harvard/CfA surveys find $\epsilon = 0.027$ for the SGH and 0.0025 for the NGH. If the fraction of unobserved SGH has the same ϵ as the 43% of the hemisphere actually observed, the number of detections at $|b| \geq 30°$ in the south is a factor of ten larger than that in the north. However, as noted, the north-south asymmetry is extremely sensitive to the low-latitude cutoff. If a substantial number of clouds exists in the $10° \leq |b| \leq 30°$ region in both hemispheres, the overall difference between the north and south Galactic hemispheres may be fairly modest. Were the observing grids used in the surveys of both hemispheres to be extended to $|b| = 10°$ and a similar number of detections achieved in both hemispheres in those regions, then the overall values of ϵ and the north-south asymmetry factor would both change significantly.

Specifically, in the NGH there were 4924 observable gridpoints (with 1° spacing) in the region $+10° \leq b < +30°$, while in the SGH ($-10° \geq b > -30°$) this number is 4000. Thus, it is evident that extending the grid used in the Georgia-Harvard/CfA survey to intermediate latitudes would result in many detections towards the Lupus clouds, Aquila rift, and the Cepheus clouds in the NGH, and towards Orion and Taurus in the SGH. Moreover, the Dame et al. (1987) survey sampled only ~ 10% of the sky between $10° \leq |b| < 30°$. Overlaying the extension of the 1°× 1° observing grid on the Dame et al. survey, and taking all points in the range $10° \leq |b| < 30°$ with $W_{CO} \geq 0.5$ K km s^{-1} as detections, we find that there are 523 such detections in the NGH and 569 in the SGH. This amounts to 11% and 14% of the gridpoints between $10° \leq |b| < 30°$ in the NGH and SGH, respectively. As this intermediate-latitude region is very much undersampled in the Dame et al. survey, a conservative estimate for the detection rate in this region would be ~ 15% of the number of gridpoints. If we assume the values for illustrative purposes, i.e., that the detection rate in both the NGH and SGH is exactly 15% in the intermediate-latitude region, then the total number of detections in the NGH (at $b \geq +10°$) is $739 + 26 = 765$, while for the SGH (at $b \leq -10°$) this number becomes $600 + 133 = 733$. Hence, $\epsilon = 0.05$ for the NGH at $b \geq +10°$ and 0.08 for the SGH at $b \leq -10°$— i.e., the surface filling factors for the NGH and SGH do not differ greatly and the displacement of the Sun from the Galactic midplane decreases significantly. Without data at intermediate latitudes, the distribution of translucents is skewed towards the SGH.

10.6 Molecular Clouds in the Halo?

The large-scale CO surveys established that molecular gas is closely confined to the Galactic plane with a relatively small scale height of about 100 pc (e.g., Dame et al. 2001). However, at least one large GMC in Lupus has a knematic distance placing it at about 200 pc above the plane. Dame and Thaddeus in (1994) established that a "thick" molecular disk is present in the inner Galaxy with at least three times the vertical extent of the nominal Galactic CO layer. Studies of nearby spirals lend support to this conclusion (Garcia-Burillo et al. 1992; Sofue and Nakai 1993) There are a few objects at anomalous radial velocities which might be indicative of greater distance from us, perhaps even in the Galactic halo. The first of these objects is the so-called Draco Cloud complex (Goerigk et al. 1983), also known as MBM41–44. Distance estimates for this object have varied considerably over the years, from a short distance of 200 pc (Penprase 1992) using photometry, to a long 800–2500 pc (Goerigk and Mebold 1986) using photometry and star counts. Using the NaI D lines, Gladders et al. (1998) found the cloud distance at $463^{+192}_{-136} \leq d \leq 618^{+243}_{-174}$. However, Penprase et al. (2000) find a range of 800–1300 pc using B and V photometry and Wolf diagrams. Schlafly et al. (2014) did not include the Draco cloud complex, it is one of the few that were skipped by these authors.

With a latitude of +38° and an estimated distance of 400–1300 pc, the Draco Cloud is anywhere from 250–800 pc above the Galactic midplane. This could place it near the plane/halo interface, wherever that is (e.g., 1–2 kpc; Kalberla and Kerp 2009). In addition to its anomalous distance, the Draco complex has radial velocities in the −20 to −24 km s^{-1} interval, a 4σ deviation from the typical high-latitude cloud LSR velocity. The combination of larger than usual distance and LSR velocity for this complex points to a different origin than for typical high-latitude clouds. Odenwald and Rickard (1987) suggested that the infrared morphology of the complex can be understood if it is an ensemble of dense clouds are moving through an especially low density ambient interstellar medium of low Reynolds number, which might be coronal gas. Other authors (*e.g.*, Kerp et al. 1994) have discussed this object in the context of the impact of a high-velocity cloud with the Galactic disk. Whatever its true origin, the Draco Clouds need to be studied separately from the bulk of high-latitude clouds.

Other intermediate velocity clouds may be at the plane/halo interface or even within the Galactic halo itself if their anomalous velocities imply a large vertical distribution. There are five other Intermediate Velocity Molecular Clouds (IVMCs) besides the Draco complex. The first object (G135.3+54.5) was discovered by Heiles et al. (1988) and the second (G211+63) was detected by Désert et al. (1990)

although it had also been independently observed by Heiles, Reach, and Koo (though not included in their 1988 compilation). G135.3+54.5 has the most discrepant radial velocity, −45 km s^{-1} of any high-latitude cloud. They also identified a nearby object, G135.5+51.3, with $v_{LSR} \sim -48$ km s^{-1}, but this latter detection is tentative and needs confirmation. G211+63 has a v_{LSR} of −39 km s^{-1} and exhibits a double-peaked structure in the CO(2-1) transition. Moreover, Désert et al. examined the infrared morphology of G211+63 and concluded that, like the cometary globules studied by Odenwald (1988), this cloud exhibits "plumes" extending behind the two main clumps. This may be the signature of an interaction of the molecular gas with a lower density ambient gas such as that found in the halo. However, unlike the Draco Clouds, neither object has a distance estimate and locating these objects in the halo is currently based solely on their anomalous LSR velocities. Future studies will be required to resolve the question of their distance from the Sun and location in the local ISM.

The situation did not change for more than a decade until Magnani and Smith (2010) nearly doubled the inventory of intermediate-velocity molecular clouds by finding three new objects. These are all relatively near each other and may be the molecular peaks of an underlying atomic intermediate velocity cloud. How many more IVMCs are still to be found (note that all those known are in the NGH) awaits more complete dedicated searches.

10.7 The Southern Extension of the Taurus Dark Clouds

The Taurus-Auriga-Perseus system of dark clouds extends from the Galactic midplane to a latitude of −20°. The Taurus-Auriga portion of the complex has a distance of ~ 140 pc (Elias 1978) and is considered the southernmost extension of the Gould Belt. The Perseus portion of the complex is much farther away, 300–380 pc (Eklof 1959). Because the average distance to the high-latitude clouds as determined in Sect. 10.1 above is 140 pc, we will focus only on the Taurus-Auriga clouds in the rest of this section.

Along with their northern Gould Belt counterpart, the ρ Ophiuchi dark cloud complex, the Taurus-Auriga clouds are the nearest star forming regions. The complex was completely mapped in the CO(1-0) line by Ungerechts and Thaddeus (1987). It extends to a Galactic latitude of ~ −25° to −30° and includes the high-latitude cloud MBM 12 (also known as L1453, 1454, 1457, and 1458). A comparison of their mapping with the Georgia-Harvard SGH high-latitude survey (Sect. 9.6) indicates that there are many detections directly south (in latitude) of the Taurus-Auriga clouds. The main clouds are MBM 16, 18, and 19—although other small molecular clouds are present in the region. The IRAS 100 μm map of the region shows widespread infrared emission encompassing the Taurus-Auriga clouds and the high-latitude clouds to the south. Its morphology is like a "peninsula" extending southward from the plane. The infrared maps hint that all these objects are related (Magnani 1988). The distances to MBM 16, 18, and 19 given by Schlafly et al. (2014) are 147, 166, and 156 pc, respectively and are all consistent with the Taurus-Auriga complex distance. Their radial velocities are also similar to those of the Taurus-Auriga clouds, though there are objects in the region at very discrepant velocities (see Magnani et al. 2003).

References

Allen, C.W. 1973, *Astrophysical Quantities*, The Athlone Press
Andreani, P., Ferlet, R., Lallement, R., and Vidal-Madjar, A. 1988, ESO Messenger, 52, 47
Bahcall, J.N. 1984, ApJ, 276, 169
Bahcall, J.N. 1986, ARAA, 24, 577
Bania, T.M. and Lockman, F.J. 1984, ApJS, 54, 513
Binney, J. and Merrifield, M. 1998, *Galactic Astronomy*, Princeton University Press
Binney, J. and Tremaine, S. 2008, *Galactic Dynamics*, 2nd ed., Princeton University Press
Blaauw, A. 1960, MNRAS, 121, 164
Chepurnov, A., Lazarian, A., Stanimirović, S., Heiles, C., and Peek, J.E.G. 2010, ApJ, 714, 1398
Chu, Y.H. 2009, *The Magellanic System: Stars, Gas, and Galaxies*, Proceedings of the IAU, IAU Symposium, 256, 166
Cox, A.N. 2000, *Allen's Astrophysical Quantities*, AIP Press, Springer
Dame, T.M et al. 1987, ApJ, 322, 706
Dame, T.M., Hartmann, D., and Thaddeus, P. 2001, ApJ, 547, 792
Désert, F.X., Bazell, D., and Blitz, L. 1990, ApJ, 355, 51
de Vries, C.P. 1986, Ph.D. Thesis, University of Leiden
Eklof, O. 1959, Ark.Mat.Astron.Fys., 2, 213
Elias, J.H. 1978, ApJ, 224, 857
Franco, G.A.P. 1988, A&A, 202, 173
Franco, G.A.P. 1989, A&A, 223, 313
Flynn, C. and Fuchs, B. 1994, MNRAS, 270, 471
Garcia-Burillo, S., Guelin, M., Cernicharo, J., and Dahlem, M. 1992, A&A, 266, 21
Gladders, M.D. et al. 1998, ApJ, 507, L61
Goerigk, W., Mebold, U., Reif, K., Kalberla, P.M.W., and Velden, L. 1983, A&A, 120, 63
Goerigk, W. and Mebold, U. 1986, A&A, 162, 279
Gould, A. 1996, PASP, 108, 465
Grant, C.E. and Burrows, D.N. 1999, ApJ, 516, 243
Hearty, T. et al, 2000, A&A, 357, 681
Heiles, C., Reach, W.T., and Koo, B.-C. 1988, ApJ, 332, 313
Hobbs, L.M., Blitz, L., and Magnani, L. 1986, ApJ, 306, L109
Hobbs, L.M., Penprase, B.E., Welty, D.E., Blitz, L., and Magnani, L. 1988, ApJ, 327, 356
Humphreys, R.M. and Larsen, J.A. 1995, AJ, 110, 2183
Joshi, Y.C. 2005, MNRAS, 362, 1259
Kalberla, P.M.W. and Kerp, J. 2009, ARAA, 47, 27
Kerp, J., Lesch, H., and Mack, K.-H. 1994, A&A, 286, 13
Kim, H.-G., Lee, Y., Park, B.-G., and Kim, B.-G. 2000, JKAS, 33, 151
Kun, M. 1998, ApJS, 115, 59
Lilienthal, D., Wennmacher, A., Herbstmeier, U., and Mebold, U. 1991, A&A, 250, 150
Lombardi, M. 2009, A&A, 493, 735
Luhman, K.L. 2001, ApJ, 560, 287
Magnani, L., Blitz, L., and Mundy, L. 1985, ApJ, 295, 402
Magnani, L., Lada, E.A., and Blitz, L. 1986, 301, 395

Magnani, L. and de Vries, C.P. 1986, A&A, 168, 271
Magnani, L. 1988, in *The Outer Galaxy*, eds. L. Blitz and F.J. Lockman, Lecture Notes in Physics, Vol. 306, (Berlin: Springer-Verlag), 168
Magnani, L., Hartmann, D., and Speck, B.G. 1996, ApJS, 106, 447
Magnani, L. et al. 2003, ApJ, 586, 1111
Magnani, L. and Smith, A.J. 2010, ApJ, 722, 1685
Mattila, K. 1979, A&A, 78, 253
Münch, I.G. 1952, ApJ, 116, 575
Odenwald, S. and Rickard, L.-J 1987, ApJ, 318, 702
Odenwald, S. 1988, ApJ, 325. 320
Penprase, B.E., Blades, J.C., Danks, A.C., and Crane, P. 1990, ApJ, 365, 241
Penprase, B.E. 1992, ApJS, 83, 273
Penprase, B.E. 1993, ApJS, 88, 433
Penprase, B.E., Rhodes, J.D., and Harris, E.L. 2000, A&A, 364, 712
Pound, M. W., Wilson, R.W., and Bania, T.M. 1990, ApJ, 351, 165
Reed, B.C. 1997, PASP, 109, 1145
Schlafly, E.F. et al. 2014, ApJ, 786, 29
Sofue, Y. and Nakai, N. 1993, PASJ, 45, 139
Stanimirović, S., Staveley-Smith, L., Dickey, J.M., Sault, R.J., and Snowden, S.L. 1999, MNRAS, 302, 417
Trumpler, R.J. and Weaver, H.F. 1953, *Statistical Astronomy*, Dover Publications
Ungerechts, H. and Thaddeus, P. 1987, ApJS, 63, 645
van der Kruit, P.C. and Shostak, G.S. 1984, A&A, 134, 258
van Rhijn, P.J. 1929, Groningen Pub., 43, 1
Welty, D.E., Hobbs, L.M., Penprase, B.E., and Blitz, L. 1989, ApJ, 346, 232
Wolf, M. 1923, Astron. Nachr., 219, 109

Chapter 11
Dynamical Considerations: Instabilities and Turbulence

I am an old man now, and when I die and go to Heaven there are two matters on which I hope enlightenment. One is quantum electrodynamics and the other is turbulence of fluids. About the former, I am really rather optimistic. Horace Lamb (author of Hydrodynamics and a referee of Reynold's papers; cited by Goldstein, S. 1969, ARFM, 1, 23)

Abstract Models are the roadmaps through the observational thicket. Despite its enormous range properties, a relatively few basic physical processes govern much of the structure and evolution of the ISM. In this chapter, we review some of the most important. We begin with a reprise of the issue of virialization. Then we examine some instabilities to which the ISM is subject. The core of this chapter is the structural, energetic, and dynamical importance of large scale motions and turbulence in the diffuse medium and some of the issues we see as fruitful lines for further investigations.

11.1 Introduction

The space between the stars is an arena in which the different processes play out in a dynamic, multiphase medium. The combined effects of motions induced on a wide range of length and time scales are difficult to disentangle and often combine from very separate locales along lines of sight. Passing between the extremes requires considering hydrodynamic and thermodynamic instabilities, gravitational effects, and most importantly, turbulence. We need models of the underlying physics that governs this complexity to understand the structures and states of these regions. In this chapter we discuss some of the dominant processes. To comprehend the origin of the spectroscopic environmental tracers requires models. Astrochemical modeling must be placed in context. Instabilities driven by the huge variations in optical depth and densities evident in even relatively compact structures require treating the teleconnections across the ISM. In short, this chapter is presented in the spirit of the Green movement, "act locally, think globally".

L. Magnani, S.N. Shore, *A Dirty Window*, Astrophysics and Space Science Library 442, DOI 10.1007/978-3-662-54350-4_11

11.2 Virial Theorem and Stability

The virial theorem (VT) is a global energy constraint for a mechanical system that comes directly from the conservation equations for mass, momentum, and energy density. It relates $\mathscr{W}$, the gravitational potential energy, and $\mathscr{T}$, the kinetic energy. Magnetic fields can also be included since they exert stresses that enter the dynamical equations. The usual approach, which we used in Chap. 4 to discuss virial mass estimates, is to start with the equations of motion (including the gravitational acceleration)

$$\rho \frac{d\mathbf{v}}{dt} = -\nabla P - \rho \Phi + \frac{1}{4\pi}(\nabla \times \mathbf{B}) \times \mathbf{B}, \tag{11.1}$$

take the scalar product with $\mathbf{r}$ and integrate over the mass of the body (assuming it is finite). This gives the Lagrangian form of the virial theorem

$$\frac{1}{2}\frac{d^2 I}{dt^2} = 2\mathscr{T} + \mathscr{W} + \mathscr{M}. \tag{11.2}$$

Here $I = \int r^2 dm$ is the moment of inertia. The total energy is $E = \mathscr{T} + \mathscr{W}$. Thus, a self-gravitating system in strict virial equilibrium ($\ddot{I} = 0$) is bound, $E = \frac{1}{2}\mathscr{W} < 0$. Note that surface terms have been ignored. This is only so simple a form if the mass is isolated, having neither neighboring masses nor merging into any surrounding medium; but in the diffuse ISM this is less obvious. The *surface* is defined by optical depth and by trace species that are used to find the cloud in the first place. The treatment in Lagrangian coordinates is the most familiar since the mass M and mass elements dM are clearly independent of the geometry (Chandrasekhar 1967).[1] In this form the virial theorem uses a fixed mass to obtain global quantities and treats the coordinate $\mathbf{x}$ as a function of time. A "virialized" mass estimate is obtained provided $\ddot{I} = 0$ as we discussed in Chap. 4. The total internal cloud stress, the combined thermal pressure, and macroscopic random motions that is treated as an additional pressure, become $\mathscr{T} = P_{tot}V$ and the gravitational energy is $\mathscr{W} = aGM^2/R$, where a is a geometric factor. If the total pressure is dominated by turbulence, and if that dynamical fields can be approximated by a single velocity dispersion $\sigma > c_s$ (like an isothermal gas but with the motions being supersonic with respect to the cloud's sound speed c_s),

$$M_{VT} = \frac{5\sigma^2 R}{K_{VT} G} \tag{11.3}$$

[1]The first application of the full VT to the ISM, including magnetic fields and rotation, appears to have been Chandrasekhar and Fermi (1953) but it was already standard for treating stellar structure for decades before that.

Here K_{VT} is a scaling constant that depends on the internal structure of the cloud whose value in the literature has ranged from $1 \leq K_{VT} \leq 2$ (see e.g. MacLaren et al. 1988). If the mass is stable against collapse, M_{VT} should be of the same order as the mass obtained from spectroscopic analysis. The mass is obtained from integrating the column density over the cloud surface area. Any single transition comes from only a minority tracer of the total mass, but if the structure behaves according to this prescription any optically thin line profile of sufficient strength can provide the dynamical information that is necessary to obtain M_{VT}.

A different approach that one better suited to discussions of interstellar diffuse structures uses the Eulerian (space) frame of reference (McKee and Zweibel 1992). The results should be invariant with respect to the representation, but there are several insights provided by the Eulerian equation that are quite relevant to clouds. By choosing to work within a fixed volume, the continuity equation enters explicitly since the densities can vary in space and time. In the Lagrangian treatment, since the mass is held constant, compressibility issues are sidestepped. The Eulerian version considers volumes dV instead of mass elements dm. In this, the density of the medium is different between the cloud and diffuse gas (called "intercloud" in the earlier papers). Writing the fluid equations in conservative form, since this corresponds to the relevant quantities of mass and momentum densities, we have (including the gravitational potential)

$$\frac{\partial}{\partial t}\rho + \nabla \cdot (\rho \mathbf{v}) = 0, \tag{11.4}$$

$$\frac{\partial}{\partial t}\rho \mathbf{v} + \nabla \cdot [\rho \mathbf{v}\mathbf{v} + S] = \rho \nabla \Phi. \tag{11.5}$$

The most important quantity here is the stress,

$$T_{ij} = \rho v_i v_j + S\delta_{ij} \tag{11.6}$$

where S is a symmetric diagonal uniform function that we will now identify with the gas pressure but it could also include random motions on a scale larger than the kinetic. In other words, we might have some turbulent motion. Defining the moment of inertia for a substructure in the medium, confined within some volume V, to be

$$I = \int dV r^2 \rho, \tag{11.7}$$

and taking the first derivative and inserting the continuity equation gives:

$$\dot{I} = \int dV \frac{\partial \rho}{\partial t} r^2 = -\int dV \nabla \cdot (\rho \mathbf{v} r^2) + 2\int dV r \rho \mathbf{v}\mathbf{v}. \tag{11.8}$$

Taking the next derivative (guided by the Lagrangian form of the virial theorem and the usual definitions of the energies as before), equation 11.8 becomes

$$\ddot{I} = -\int d\mathbf{S}\frac{\partial}{\partial t}\rho\mathbf{v}r^2 + 2\int dV\frac{\partial}{\partial t}(r\rho\mathbf{v}\mathbf{v}). \tag{11.9}$$

The first term is resolved with substitution of the momentum equation, introducing the surface term for the stress tensor. The gravitational energy is $\mathscr{W} = -\int dV\rho\mathbf{r}\nabla\Phi$ and, without the surface terms (the case if the mass is isolated) the result would be the standard VT with an additional term $\frac{1}{2}\ddot{I}$ that is well known from studies of self-gravitating masses in equilibrium (e.g. Chandrasekhar 1967). The end result is:

$$\frac{1}{2}\ddot{I} = 2\left[\mathscr{T} - \frac{1}{2}\int r_i T_{ij} dS_j\right] + \mathscr{W} - \frac{1}{2}R^2\ddot{M} \tag{11.10}$$

where the first bracket becomes:

$$\frac{1}{2}\int dV[3 < P_{th} + < \rho v^2 >] - \frac{1}{2}\int r_i[< P_{th}\delta_{ij} + < \rho v_i v_j >]dS_j. \tag{11.11}$$

Let us examine the stress tensor a bit more closely. For pressure, S is diagonal so the term $\int dV\mathbf{r}\cdot\nabla P = 3\int dVP$. But if the velocity is decomposed into mean, $\mathbf{u}$, and random, $\delta\mathbf{u}$ contributions, then the average of the fluctuations over volume necessarily vanishes and the stress tensor becomes

$$T_{ij} = \rho(< u_i >< u_j > + < \delta u_i \delta u_j >) + P\delta_{ij}, \tag{11.12}$$

thus, isotropic, dynamical fluctuations produce a so-called *turbulent pressure*. There are issues related to the separation of the fluid motions into average and fluctuating components, to which we will return below (Sect. 11.5.2), but for the moment they will not be necessary. It suffices to assume that any motions in this treatment are random. This allows the temperature and density to differ significantly across a boundary and still maintain equilibrium since the stochastic macroscopic velocities may also be different. In the Lagrangian form, where the thermal properties are averaged over the mass (distinct separate self-gravitating structures), the condition for stability is that the total energy is negative. Radiative losses result in further contraction and may lead to collapse. In the Eulerian approach, the surface terms are explicitly kept. Any discontinuities in pressure between the cloud and its surroundings are counterbalanced by shears and anisotropies in the turbulence (and, possibly, magnetic fields that we have ignored).[2]

[2]A simple way to see what happens if we include magnetism is to again think of the virial condition and set $B^2V \sim GM^2/R$, where B is a mean value of the field regardless of its detailed configuration and the proportion allows for internal (structural) differences among arrangements of magnetic

11.3 Reprise: What Is a Cloud?

Throughout this book we have been alternating between discussing the diffuse medium, as a sort of global environment, and clouds. Intuitively, this is like the Earth's atmosphere where it is easy to distinguish between the white, fluffy things and the more transparent, blue sky. The difference becomes much less clear on a completely cloudy day, when there seems to be no clean separation either by opacity or composition. Which analogy is appropriate for the ISM is not obvious. We talk about clouds as though they have physical surfaces instead of compositional or opacity differences. This is necessary to apply the VT. Either for a self-gravitating mass described in Lagrangian coordinates or a more diffuse structure for which an Eulerian description imposes a fixed volume, the notion of some sort of boundary is implicit. The conundrum repeats in our description of turbulence and its effects in the mechanics of the ISM. Referring to a structure as "virialized" implies it is in mechanical equilibrium and static but the virial theorem in its full form requires neither condition. It states the relation, averaged over either the mass or volume of the cloud, between the total energies of all constituents that determine its dynamical structure. As in the atmosphere, every "piece of a cloud" is transient and the surface is defined by the photosphere. Even if several tracers agree on the location of the boundary, in this case R, matter can pass from one regime to another. Take, for instance, MBM 40 (see, e.g., Shore et al. 2003). The outer neutral hydrogen envelope, mapped in the 21 cm line, has a turbulent velocity that appears to be the same as the internal velocities traced by ^{13}CO although the temperatures are much different. The boundary seems to be only a location where the gas passes between atomic and molecular phases. There is no indication of a net outflow or inflow, so if this picture of the cloud as a coherent structure whose constituent gas is constantly being exchanged is correct, the boundary is not a border but an open frontier.

11.3.1 Isothermal Cloud Models

In discussing the structure of the Galactic disk we have already applied one of the static self-gravitating models commonly used in astrophysics. Taking a slab, the equation for mechanical balance for the stars is the same form as that for the gas, $\rho \nabla \Phi = \nabla P$ where P is some form of isotropic distribution in random velocities and ρ is the density of the principal constituent. If we are talking about a collection of stars, or even dark matter, the "gas" is collisionless and what we are

---

field and density. Then for the virial condition to hold (equilibrium with only the magnetic field opposing self-gravitation), so a very low β configuration [in a plasma, the ratio of the plasma pressure ($P = nkT$) to the magnetic pressure ($P_{mag} = B^2/2\mu_o$)], $\Phi_m/M \approx$ constant. where Φ_m is the magnetic flux through a surface of radius R. Although such considerations are important for star forming clouds, this magnetized state is very far from what is observed in the diffuse medium.

calling a pressure is not the same as an ideal gas. You can imagine an ensemble of interacting (but not colliding) point particles, each of identical mass (for simplicity) that have random components of their momentum. In galactic dynamics, the stars move around the center of mass with some mean orbital velocity that depends on their distance from the center, and with a range of eccentricities because of from their individual kinetic energies and angular momenta. If they also have random orbital inclinations they behave as an aerosol of harmonic oscillators relative to the midplane of the disk (the isothermal slab model we used in Chap. 10). Those with greater energies have higher turning points but since they are rarer, the "gas of stars" is more tenuous at greater height. The calculation in Chap. 10 assumed a single dispersion in this quantity, thereby mimicking an ideal isothermal gas, so in the equation for the structure of the disk the local approximation of the combined Poisson equation and mechanical balance yields a hydrostatic equation in which the motions obey the stationary form of the virial theorem. The transformation of this pair of equations into a single *nonlinear* second order differential equation is possible because the equation for the pressure depends only on the density, $P = \rho\sigma^2$ so

$$\nabla \cdot \nabla \Phi = \sigma^2 \nabla \cdot \frac{1}{\rho} \nabla \rho = 4\pi G \rho. \tag{11.13}$$

This is a form that will also be familiar from stellar structure, for a sphere it is the isothermal Lane-Emden equation. But a more important feature is that it contains a *characteristic length*:

$$L^{-2} = 4\pi G \rho_0 / \sigma^2 \tag{11.14}$$

where the density has been scaled to its central value ρ_0. Note that this length depends on the density and velocity dispersion and those are really the zeroth and second moment of the distribution function for the constituents. So we have a macroscopic structural model that depends only on the "microphysics".

This does not mean an isothermal sphere is infinitely stable (see, e.g., Harrison et al. 1965). One such example is when the sphere is pressure bounded. Bonor (1956, 1958) and, independently, Ebert (1957) treated the condition that an isothermal sphere is bounded by an external medium. Calling $P_{int} = \rho\sigma^2$ the gas pressure within the cloud (whether thermal or turbulent pressure is the cause is not important for the moment), and P_{ext} that of the external medium, including the surface term in the virial equation gives:

$$3(P_{int} - P_{ext})V = 3(M\sigma^2 - P_{ext}V) = \lambda \frac{GM^2}{r} \tag{11.15}$$

where λ is a geometric factor. Minimizing this for r gives

$$r_{crit} = \left(\frac{GM^2}{P_{ext}} \right)^{1/4} \tag{11.16}$$

and substituting this back into equation 11.14, the mass of such a virialized sphere bound only by external pressure is

$$M_{BE} \sim \sigma^4 P_{ext}^{-1/2}. \tag{11.17}$$

This, with geometric factors included, is the Bonnor-Ebert mass. Notice that as the external pressure increases, the mass of a stable sphere decreases. This is because the gravitational potential energy decreases as the mass is compressed. The result is not too different than the Jeans estimate (next section) if we say that the internal and external densities are the same and therefore the pressure becomes $\rho_{ext}\sigma^2$. But the consequence of the difference is important for translucent clouds because the clouds may be underpressured relative to the external gas.

11.3.2 Gravothermal Instability

For the isothermal slab, as we used to represent the Galactic disk, the structure equation alone says nothing about the stability or time development of the structure. We transformed this to one dimension by assuming a thin disk. In effect, that is the same as assuming the length L is small relative to the distance from the center of the system so the disk is flat—at least locally. Then if σ is a constant, or we assume a single constituent species with a unique mass, we arrive at the representation described for the distribution in space of the disk gas. To know if this layer is stable, we have to extend the treatment to include time. The layer is virialized. There are no net accelerations in the system although there can be structural fluctuations whose time averages vanish. If, instead, we can locally compress or expand the medium, variable densities and density gradients produce local changes in the gravitational acceleration. If shear or simple dispersion is large enough, these perturbations will not grow. But if these local accelerations (which means the gradient of the gravitational potential around the density fluctuations) are large enough, matter will stream toward the local potential minima and increase the perturbation amplitude by accumulation. This runaway is the signature of a mechanical instability and was first found by Jeans at the start of the 20th century.

The new feature of this treatment compared to a sound wave is the gravitational feedback. For simplicity, start with a slab that is much larger than any scale on which we impose a fluctuation so we can ignore the boundaries. Any form can be chosen for the fluctuations but since this is analogous to an acoustic wave it suffices to choose a simple wave with spatial frequency k and temporal frequency ω and, for any quantity,

$$\delta Q(x,t) = \delta Q_{k,\omega} \exp i(\omega t - kx). \tag{11.18}$$

In what follows we will drop the subscripts as understood. The equations of motion and continuity transform as

$$i\omega\delta\rho - ik\rho v = 0; \quad i\omega v = ik\frac{\delta\rho\sigma^2}{\rho(0)} - ik\delta\Phi \tag{11.19}$$

where now $\rho(0)$ is the mean unperturbed density with any spatial variation ignored. The gravitational potential depends only on the density through the Poisson equation

$$\nabla^2\delta\Phi = 4\pi G\delta\rho \rightarrow \delta\Phi = \frac{-4\pi G\rho(0)}{k^2}\frac{\delta\rho}{\rho(0)}. \tag{11.20}$$

The Poisson equation is independent of time but the density variations are not, so the gravitational perturbation is phase locked to $\delta\rho$. Equations 11.18–11.20 yield a single dispersion relation:

$$\omega^2 = \sigma^2(k^2 - k_J^2) \tag{11.21}$$

where

$$k_J^2 = 4\pi G\rho(0)/\sigma^2 \tag{11.22}$$

is called the Jeans wavenumber. This is the *same* length scale as before, but now it represents a stability limit. If the perturbation grows to this size the layer is unstable while the self-gravitating isothermal system was constructed to have the core far smaller and therefore the gravitational acceleration lower. Note that this critical length assumes only an isotropic random velocity distribution for the constituents. If only gas pressure supports the medium, σ is replaced by c_s, the sound speed, and the limiting mass is $M_J = K_J T^{3/2}\rho(0)^{-1/2}$, where K_J is a geometric scale factor. This is the same scaling law as the Bonnor-Eberts mass for equal external and internal pressures.

11.3.2.1 Isothermal Slabs and the Toomre Criterion

In Chap. 10, we discussed how to use an isothermal slab to model the vertical distribution of gas and stars in the solar neighborhood in the Galaxy choosing the stars as the main agent for the local gravity. A molecular cloud can also be thought of as a slab if it is elongated and sheetlike, for instance as seen in Taurus-Auriga. Self-gravitation may produce further sheetlike collapse if the *surface* density is high enough. Gas is internally supported against the catastrophe by pressure, no longer just a velocity dispersion but an isothermal equation of state $P = \rho c_s^2$, where c_s is the sound speed. As in the Jeans criterion, which is a slab in an unbounded medium, collapse occurs when the crossing time for an acoustic wave is longer than the collapse (freefall) time, if there are no other internal means of support. But there can also be ordered flows in the gas, a possibility we explicitly ignored in deriving k_J. Uniform motion is irrelevant. But if the structure shears, the situation changes. Now there is a new *timescale* in the problem that depends on neither the length scale nor density, the *shear* (or vorticity). For the Galaxy on the larger scale, differential motion depends on the global mass distribution. Locally, however, there may be a shear because of external forcing of the slab. The transverse velocity gradient is the same as an epicyclic displacement around the mean flow whose frequency is κ. For a differentially rotating disk this depends on the radial gradient of the angular momentum. For a cloud, in contrast, it can be any shear that is ordered on a scale larger than the cloud. The critical state is found when the shearing timescale is about the same as the gravitational collapse time on the Jeans length scale,

$$c_s k_J / \kappa \approx 1 \tag{11.23}$$

where $k_J c_s^2 = G\Sigma$. On substitution of k_J, the stability criterion becomes

$$\frac{\kappa c_s}{G\Sigma} \sim \mathcal{O}(1). \tag{11.24}$$

This is, essentially, the Toomre (1964) criterion for self-gravitationally induced instability under shear. The critical parameter, Q, quantifies the result of equation 11.24 relative to unity. If $Q \ll 1$ the system is unstable, otherwise it either remains marginally stable or is stretched and twisted without collapsing. The Toomre stability criterion has been invoked in the context of models for global star formation as a substitute (or proxy) for the Jeans criterion since it is based on more easily measurable quantities (Kennicutt 1983). It depends on the surface density, which is linearly proportional to the visual extinction (see Chap. 6). For global dynamics, κ is derived from the Galactic rotation frequency $\Omega(r)$ through

$$\kappa^2 = \frac{1}{r^3}\frac{d}{dr}(r^2\Omega)^2 \tag{11.25}$$

For molecular clouds, κ is the vorticity that quantifies the shear. Although it is more difficult to measure than a galactic rotation curve, it can be inferred in many clouds from large scale surveys. The velocity dispersion comes, as usual, from some measure of the line widths. Thus, the Toomre criterion can be imposed without resort to larger scale (Galactic) dynamics.

11.4 Thermal Instability

Any gas in the diffuse medium is heated by background starlight (through collisions and ionization), and cooled through collisional excitation of optically thin transitions. If we treat the gas as ideal, a change in the thermal energy under pressure balance leads to changes in the density and both feed back into the local dynamics even without self-gravity. Since the losses are radiative, and "on the spot", very fine structure can develop on thermal loss timescales. Writing the energy loss function as $\mathscr{L}(\rho, T) = \Lambda - \Gamma$, where the last two symbols are the loss and heating rates, respectively, thermodynamic considerations give

$$\frac{dE}{dt} - \frac{P}{\rho^2}\frac{d\rho}{dt} + \rho\mathscr{L} = \text{conductive losses} \rightarrow 0 \tag{11.26}$$

with E being the internal energy and $d/dt = \partial\partial t + \mathbf{v}\cdot\nabla$ as the total advective time derivative (the term that couples the thermal properties to the dynamical response of the medium).

Optically thick gas is thermally stable since the radiative diffusion is on a long enough timescale that a self-gravitating gas can reach hydrostatic balance. This is, in the simplest picture, the condition that slows and eventually halts the collapse of an isothermal medium. The energy of a gas is governed by the balance of heating and cooling which together form the loss function $\mathscr{L}$ that depends on density and temperature. First examined by Field (1965), the criterion for *local* instability is governed by the density and temperature dependences of the total loss function $\mathscr{L}(\rho, T)$. The criteria for instability depend on how the perturbation affects the medium. If the pressure remains constant (two phases in mechanical equilibrium, as in a first order phase transition), then $(\partial\mathscr{L}/\partial T)_P < 0$ is the condition for instability. Since $P \sim \rho T$ for an ideal gas, this is the same as $(\partial\mathscr{L}/\partial T)_\rho - (\rho_0/T_0)(\partial\mathscr{L}/\partial\rho)_T$. Note that emission depends on both T and ρ hrough collisional excitations and ionization. If, instead, the density remains constant, then the condition reads $(\partial\mathscr{L}/\partial T)_\rho < 0$. Field emphasized that the dynamics require the imposition of the isobaric (constant pressure) condition between phases (see also Hunter 1966). The composition of the gas is important here. For atomic gas, the main losses are from collisional excitation of resonant state fine structure transitions such as [O I] 63 μm, [C I] 609 μm, and [C II] 158 μm. In the warm diffuse medium, collisional excitation is favored for atoms whose ionization potentials are slightly less than hydrogen (e.g. carbon and nitrogen). Therefore, the medium is not completely neutral but hydrogen still is. In HII regions, the optical forbidden lines are excited, especially [N II] 5755, 6548, 6583Å, and [O III] 4363, 4959, 5007Å, since the gas is hot enough that the electrons can collisionally excite these forbidden transitions. The hydrogen Lyman series, especially Lyα, are coolants in recombining media but are not collisionally excited unless the temperatures are extremely high, as in shocks. In molecular gas, the cooling is produced by collisions with H_2 instead of electrons so, being a neutral interaction, has a low cross section. At the highest temperatures, coronal lines (especially from high ionization states of C, N, and O) are effective coolants above a few million K but this is mainly important in shocks generated by supernovae.

Bremsstrahlung is always a coolant in ionized gas but line emission is the most efficient. An example that illustrates their effect is the extreme electron temperature obtained for compact HII regions (e.g. Churchwell 2002). If the densities are high enough, above 10^6 cm^{-3} or so, the forbidden lines are suppressed by collisional de-excitation and their efficiencies are substantially reduced. The electrons, consequently, cool mainly through thermal continuum emission. Further, optically thick media are more thermally stable since the cooling shifts to surface losses instead of from the bulk and will have a higher internal temperature. Finally, conductive transport of heat, essentially the same as photon trapping in dense regions but also kinetic transport by electrons even in nearly collisionless media, suppresses the instability. These combine to produce regions where, for a strict interval of temperature and density, different phases are in pressure equilibrium. The transition between these states of the gas is treated, as in thermodynamics, as a mean state. Heating processes are extremely important, recalling that $\mathscr{L}$ is the difference between the heating and cooling rates, and because these are external to the volume in question, they can be quite diverse. Cosmic ray ionization, shocks, photoionization and dissociation, and even recombination continua, are among the possible heating sources. Any volume of the WIM is pictured as having zones of different temperature and density that are in mechanical equilibrium. If the medium is unstable, it transitions from the higher to lower temperature if compressed and, on the contrary, to higher temperature if expanded. The original version of the *two phase interstellar medium* (Field, Goldsmith, and Habing 1969) was based on simplified cooling functions and is now only a schematic representation.

11.5 Turbulence

11.5.1 Introduction

The need to include turbulence in any model of the interstellar gas was realized through two different, but related problems. The first was the need to explain the ubiquity of superthermal widths of absorption lines seen against individual stars through the diffuse gas. The other indicator of turbulence is the requirement to maintain large clouds against collapse induced by self-gravity and to extend their lifetimes without excess heating. This requires some superthermal, pressure-like agent that acts on a large scale and extends throughout the structure. The requirements can be satisfied by a random dynamical component, that is turbulence. The widths are independent of atomic mass and would require equivalent temperatures far higher than permitted by either ionization or excitation. Often exceeding a few km s^{-1}, the velocities correlate among individual lines of sight. In addition, absorption lines yield inconsistent abundances if not broadened and desaturated. Although, as we discussed in Chap. 2, high spectral resolution data shows that at least a part of the contribution is from individual substructures distributed along the line of sight, the narrowest features still are broader than their thermal widths. The comparison was rapidly drawn with the same conundrum well known from stellar atmospheres, especially the Sun, where line broadening by some large macroscopic but spatially unresolved gaussian component was added in quadrature to the thermal widths to reproduce the line cores and saturation levels. Assuming that any fluctuations in optical depth are mainly from velocity broadening (hence the velocity and not the density), the convolving functions both being gaussian implies that the dispersions add quadratically so the total broadening depends on a velocity that is $v_D = (v_{th}^2 + v_t^2)^{1/2}$ and the analog of *microturbulence* that is also used in modeling stellar spectra, $b \equiv v_D/v_{th}$, is then applied to all lines regardless of their ionization state or atomic mass.

The assumption of gaussianity is quite strong. It was adopted as the simplest hypothesis but such a picture has much deeper implications. For stars, the idea that the motions are random in time in a relatively small volume—like a container in the laboratory—makes a some sense. This analogy depends, however, on the details of those mechanisms responsible for generating the presumably random velocity fields. Remember that the definition of a gaussian process requires identically distributed variations that are independent. In the diffuse interstellar medium, this hypothesis is more difficult to swallow since the distance over which a line is formed can be enormous—many times the size of a single substructure, and the resulting line profiles are always integrations along an unknown, complex line of sight toward a source. The coherence scale can only be gauged using neighboring sight lines and stars whose distances are very different to probe the thee dimensional structure. Nature does not afford many such opportunities.

Why has the physics of interstellar turbulence remained such a difficult nut to crack? A partial answer is that it is not just chaotic motion. To properly treat it, one must address how it is driven. In a broad sense, at some scale the motions damps and thermal motions and something like viscosity take over, so turbulent motion must be maintained. The temptation is to treat the observed motion as a blur of uncorrelated random fluctuations that, because they are macroscopic, add to the gas pressure in the dynamical equations, but this is misleading. The cascade is produced by density and velocity fluctuations extending across scales. It is not a dynamical analog of temperature, although the fluctuations look like a "statistical mechanical continuous field". The velocity dispersion depends on scale. The fluctuations also correlate not only longitudinally, among themselves, but transversely. Large scale driving produces transient, statistically steady state, scattering of the fluctuations in an inherently nonlinear way.[3]

[3]To see this last point, note that the derivative in the Eulerian version of the fluid equations contributes $ik_m\delta_{mj}$ to the interactions. Since the order of k and k' is immaterial, and can be reversed, this is the same as saying that $\mathbf{k} + \mathbf{k}' = \mathbf{k}''$ and $\omega + \omega' = \omega''$ for the frequencies. This three wave coupling is when two colliding waves combine to produce a third at a combination frequency in space and time (Whitham 1974; Craink 1988). In turbulence, this condition is the basis of the nonlinear couplings (Sagdeev and Galeev 1969; McComb 1990; Pope 2002). But these conservation conditions may be inconsistent. The frequencies are scalar, the wavenumbers are vectors, and the frequency of the wave may be linked (for the weak turbulence case) directly to the wavenumbers. For instance, for Alfven waves, $\omega = v_A \mathbf{k} \cdot \hat{\mathbf{B}}_0$ where the Alfven speed depends on the strength of the external field B_0. If two of these waves, moving at the same velocity, are oriented in the same direction because of a constant field strength they never interact if from different times. On the other hand, if oppositely directed along the field, they collide. What happens depends on the same condition as before. The frequencies add, the wavenumbers cancel, and if k

Fig. 11.1 An everyday example of the filamentary nature of laboratory turbulence. A laser was used to illuminate a smoke cloud, showing the individual vortex knots and filaments within the more diffuse cloud

11.5.2 Dynamical Separations

Laboratory turbulence studies separate the dynamics for the mean flow $\mathbf{v}$ and the fluctuations $\delta\mathbf{v}$ (see Fig. 11.1). Called the *Reynolds decomposition*, it was introduced for cases in which the experimental setup can be controlled and for which the three dimensional velocity field is measurable. The mean velocity, V, and the scale over which it is coherent, L, are the quantities used in the Reynolds number (see below). Note that δv is *not* a perturbation; it can be the dominant component and, indeed, is

---

was not initially present it cannot appear. The implication of this, realized in the earliest work on MHD turbulence by Iroshnikov (1963), Kraichnan (1966), and supported by Shebalin et al. (1983), Montgomery and Matthaeus (1995), and Ng and Bhattacharjee (1996) is that no parallel cascade is possible. The condition required for the K41 spectrum (Kolmogorov 1941*a*,) cannot occur. The waves interact through the distortion they produce in the field lines, something not available for normal fluid turbulence, and this coupling introduces an intrinsic anisotropy in the cascade. This is, nonetheless, only on a sort of microscale. It happens within the distance of a single wavelength for the scattering waves. Note that this is much smaller than the injection scale if we are within the range of the cascade. So likely it will not be seen, directly, in any imaging or dynamical probes. Its effect is, instead, to change the predicted spectrum and introduce a new scale for the lower end of the turbulence. The difference between weak and strong forms of turbulence depends, then, on the number of fluctuations it takes to change the velocity amplitude. The theories to date have been for incompressible turbulence. This works well for Alvfenic modes since they are strictly transverse (except when magnetosonic, fast, modes are included). The change in the amplitude is either a random walk of small increments, so varies like $(t_A/t_{dyn})^2$, or strong so it varies linearly with the ratio. The difference in the isotropy leads to a length scale, $l_\perp$, that becomes a free parameter of the theory. To limit it, Goldreich and Sridhar (1995) introduced the concept of balance, that the rate of transfer between collisions on the parallel scale l is balanced by the rate of transfer among modes on the orthogonal scale.

for isotropic turbulence. The fluctuation is a stochastic quantity such that

$$< v_i >= V_i, \qquad < \delta v_i >= 0, \qquad < (\delta v_i \delta v_j)^n > \neq 0 \quad (\text{when } i = j). \tag{11.27}$$

What is meant by "taking the mean" is not as clear in the astronomical context as it is in the laboratory. An experiment runs in real time and the experimenter chooses whether to take means over length or time. In fact, both can be performed and compared. The equivalence of these two averages is the *Taylor hypothesis* (or, better, *ansatz*). In the interstellar medium we view the world at only a moment as if the motions were frozen in time, so any averages must be over space. Thus, we are forced to assume stationarity even if at some time in the past or future a particular location may have been different. This is a more serious limitation than it may seem. We don't know the initial conditions and only imperfectly know the flow boundary conditions at any time, so we are forced to make some drastic simplifications if we are going to proceed. The quantities accessible to measurement are few in the laboratory and fewer in the universe since we observe only radial velocities and neither the density or mass within a sampled volume (like a map) is uniquely defined. All quantities are either projections toward the observer or surface quantities integrated along the same direction. To make matters more difficult, the optical depth changes depend on the internal state of each element of the medium along the line of sight and there is no guarantee that some part is not so opaque that it obscures a view behind it. This applies equally to a cloud in a planetary atmosphere as to an interstellar cloud. Density fluctuations are linked to the velocities in a compressible medium through the continuity equation, $\delta\rho/\rho = -\delta v/v$ where now v is the magnitude of the velocity fluctuation. Since the velocities and densities have power law distributions in a cascade, we expect a power law in optical depths that is biased toward larger scales (in the non-selfgravitating case) as $\delta\tau \sim r^{4/3}$. Although one might think that the smallest fluctuation will be denser, that is not the case for the inertial subrange of a turbulent spectrum since the number of elements increases as the scale decreases and each is progressively smaller in amplitude. There is a catch here. Depending on what happens at the dissipation length scale, there can be very large excursions in both density and velocity. Although rare, these will alter the line formation since they introduce large spatial variations in the optical depth at a specific velocity and change the line profile relative to that which is contributed by the rest of the line of sight.

11.5.3 The Source Scale and the Cascade

Turbulence originates with a large scale shear that creates vortices over abroad range of sizes. These collide, disintegrate, merge, redistribute kinetic energy, and dissipate in a process that takes place throughout the flow at increasingly finer scale. This process is irreversible on the smallest scale at which diffusive redistribution occurs through viscosity. The competition is between the driving and dissipation. The inertial acceleration, $\mathbf{v} \cdot \nabla \mathbf{v}$ varies dimensionally as V^2/L while the viscous term, $\nu \nabla^2 \mathbf{v}$ is dimensionally $\nu V/L^2$. Here V and L are the typical velocity and length, respectively, and ν is the viscosity (which is assumed to be kinetic in origin and constant within the flow). The ratio of these is the *Reynolds number*, $Re \equiv VL/\nu$; when $Re \ll 1$ viscosity dominates and the flow is laminar. The precise critical value of Re for which the flow turns turbid is still not known. The basic hypothesis is that, in steady state, colliding fluctuations establish a scale-invariant transfer of energy at a rate ϵ from any large scale L to any smaller, internal scale l such that

$$\epsilon = < v_l^3 > /l = < v_l'^3 > /l' \tag{11.28}$$

between two scales l and l'. The brackets indicate that a spatial average is taken over scales of order $L \gg l$. The process terminates when l is the scale on which $Re_l \approx 1$. Since $< v_l >= 0$ if there is no shear, only the second and even moments should remain after taking ensemble averages. There should be no correlated motions and the velocity dispersion substitutes for the second moment so that $< v_l^3 > \rightarrow \sigma^3$. This was assumed in the discussion of Jeans stability and the Bonnor-Ebert mass. Turbulent motions, however, are correlated.[4] They also are not fully homogeneous even if isotropic. One feature known from shear flows is the generation of *coherent structures* that retain their dynamical signatures even while being advected by the mean flow. In fully developed turbulence, the presumed constancy of ϵ and its independence of ν on the large scale introduce an essentially non-gaussian third moment, a *skew*. Different energy spectral distributions spectra have been proposed, depending on the physical hypotheses, and we briefly summarize them here.

11.5.3.1 Hydrodynamic Picture

The critical step is recognizing that the energy dissipation rate ϵ is independent of scale and also independent of the mechanism at the base of the cascade responsible for the final losses to the flow. Thus, by equation 11.28, we have

$$< (\delta v)^2 > \sim \epsilon^{2/3} l^{2/3} \tag{11.29}$$

[4]It is important to note that a confusion often arises in discussions of turbulent motions. The cascade derives its energy from driving on a scale that is much larger than the dissipation range. Numerical simulations actually have difficulty achieving this, especially shear flows, so instead they are often forced with gaussian stirring (but this is in time, not space). The spatial correlations are a distinctly non-gaussian feature. The randomness comes from events being independent in time but, wherever the fluctuation occurs, the coupling to different lengths establishes the cascade spectrum.

for every scale l so the third structure function,

$$S_3(r) =< (v(x+r) - v(x))^3 >\sim \epsilon r \tag{11.30}$$

This is a fundamental result of the theory. A related prediction is that for any order p, the structure function $S_p(r)$ should scale as $r^{3/p}$, which is *not* the case when there is intermittency in the dissipation and coherence in the cascade [see McComb (2014)]. If the density is assumed to be constant in time and space, so the medium behaves as if it were incompressible, then $\nabla \cdot \delta \mathbf{v} = 0$ and the only velocities are vortical and the only waves are transverse. Then dimensional analysis alone suffices to derive the energy spectrum (not the power) in wavenumber

$$E_{K41}(k) = C_K \epsilon^{2/3} k^{-5/3} \tag{11.31}$$

where C_K is the *Kolmogorov constant.* This formal law holds in three dimensions for scalar k assuming isotropy, homogeneity, and stationarity. Individual fluctuations are assumed gaussian in time and only spatially correlated (see, e.g., Frisch (1995) for an extensive discussion). Notice that the spectrum is bounded at both ends in wavenumber. The cascade exists between the lowest upper bound to the length scale (smallest k) on which the driving is applied (the source scale), and on the upper bound of the smallest scales for whatever dissipation scale may be taking place. The "flow of energy" is unidirectional in wavenumber, from small to large k. These formal boundaries in k-space must be handled separately. There is no specific recipe for how the energy passes between scales, simply that the amplitudes decrease as k increases. In the first picture (called K41 after the series of papers by Kolmogorov in 1941 that presented the essential results), the "bottom" of the cascade is reached when the timescale for a fluctuation (thought of as an eddy turn-over time) is comparable to the viscous dissipation rate. This minimum length scale, l_K, depends only on ν and ϵ as $l_K = (\epsilon/\nu^3)^{1/4}$. Further refinements (Kolmogorov 1962 (K62); Obukhov 1962; see Frisch 1995 and Lesiur 2008) incorporated intermittent bursts of dissipation around the upper boundary in k-space. This is more easily understood in a compressible medium since the density fluctuations accompanying those of velocity naturally produce dissipation that is biased toward random larger excursions of the continuous fields. It is unlikely that dissipation actually occurs in such a simple way in the interstellar medium. The viscosity is too low and kinetic effects set in long before a diffusive redistribution is possible. The theory also ignores radiative losses that must occur in the optically thin gas. But since the cascade initiates on a scale far larger than any on which such losses would be expected, it seems reasonable to take this power law as applicable to the bulk dynamics. Compressible turbulence is harder to treat because acoustic fluctuations are not suppressed and these couple various portions of the gas and also enhance the dissipation. The spectrum might be expected to deviate from the K41 case but recent numerical simulations (Norman et al. 2009; Kritsuk et al. 2013) show that if $(\rho\delta v)^{1/3}$ is used instead of the velocity as a tracer of the turbulence, the K41 spectrum is recovered. How an observer would distinguish these cases is more problematic since any observation of the velocity distribution is weighted by the intensity and, therefore, by the density.

11.5.3.2 Magnetohydrodynamic Picture

While hydrodynamic turbulence can be isotropic, the essential effect of including magnetic fields is that it cannot remain so. Even if chaotic, at any scale there is a preferred field direction for transmission of Alfven waves. If these are the means for coupling the material fluctuations, the resulting turbulence cannot remain isotropic and the spectrum of the cascade *must* be different from K41 (and K62) because Alfven waves are transverse oscillations. The difference was first realized by Iroshnikov (1963) and Kraichnan (1965).[5] Since the Alfven speed, V_A, depends on the field strength, for an incompressible medium one can form a single pseudo-velocity whose direction of propagation is aligned or opposite to the local field,

$$\mathbf{z}^{\pm} \equiv \mathbf{v} \mp \mathbf{b} \tag{11.32}$$

called Elsässer variables (Elsässer 1950). These are the sum of the velocity and magnetic perturbations, the difference and sum of the fluid and Alfven velocities.

The interactions are assumed to be between Alfven wave packets that propagate in a background magnetic field with speed V_A and change in amplitude due to randomly phased weak collisions. The assumption of random phase means the collisions between these Alfven waves are uncorrelated and essentially a random walk. Thus, the number of collisions varies as $n \sim t_A^2$. If we have $t_A \sim l/V_A$ then the spectrum becomes $E(k) \sim k^{-3/2}$ The exponent is not all that different from K41. But the mechanism is completely different. This is why the power law behavior of any correlation function, or power spectrum, is ambiguously determined if a very large scale is not sampled. Note also that, like K41, the waves are only transverse. The typical timescales, based on the Alfven velocities in the ISM, are of order 10 km s^{-1} for μG fields and densities of about 100 cm^{-3}. So this is one of the important implications of the MHD picture: The motions can be supersonic. A refinement and extension of the IK construction, this theory introduces the concept of critical balance, a marginally stationary state (see Goldreich and Sridhar (1995, 1997 and references therein; Sridhar 1995). The fundamental innovation for treating scattering of the waves is the decomposition of the turbulence into two components relative to the local magnetic field. A state of critical balance is imposed,

$$\frac{v_l}{l_\perp} = \frac{v_A}{l_\parallel} \tag{11.33}$$

based on the relative rates. If the Kolmogorov condition otherwise holds for the v_l components, $v_l \sim l_\perp^{1/3}$ so $l_\parallel \sim l_\perp^{2/3}$. This determines the anisotropy of the turbulence

[5]The ensuing set of ideas are known as IK theory.

and changes the spectrum from a $k^{-5/3}$ everywhere to one that depends on both the parallel and orthogonal directions.

11.5.3.3 Supersonic Turbulence

In contrast to low velocity hydrodynamic turbulence, the supersonic case is intrinsically compressible. Since shocks in three dimensions require jump conditions for the density, pressure, and velocity, all must change. The spectrum is expected to be very different for supersonic motions than either hydrodynamic or MHD turbulence. The argument, first outlined by Saffman (1971), is based on an ensemble of shocks that collide with some (unknown) probability distribution for their amplitudes. The picture assumes that any velocity fluctuations rapidly steepen, becoming discontinuities within the fluid.[6] These have a random and homogeneous spatial distribution so the number of encounters in one dimension depends on the length scale, L. The number should vary as k^{-1}. If the amplitudes are randomly distributed, hence if the individual discontinuities are uncorrelated in space and formally represented as δ-functions, the energy spectrum is $E(k) \sim k^{-2}$. Although this argument is very schematic, it is consistent with numerical simulations [starting with Porter et al. (1992*a*,); see also Mininni and Pouquet (2009)]. Again, notice that this is a cascade in the sense that individual interactions redistribute energy among the various scales but there is no loss.

In the absence of a fundamental theory for compressible supersonic turbulence, the best next thing is resorting to increasing sophisticated large scale numerical simulations. The laboratory provides no clues here since all supersonic analogs are finite size and suffer from boundary layer and finite domain effects. Boundary layers are especially difficult to treat because there must be some scale at which the flow becomes not only subsonic but dissipative and this is below the resolution of the simulations. In the engineering problem, a number of different prescriptions for treating the *subgrid scale* processes have been developed but none of them apply directly to the interstellar medium (see Elmegreen and Scalo 2004; Mac Low and Klessen 2004).

The two extrema of the spectrum, the source and dissipation scales, are explicitly left untreated when deriving the cascade, yet these are *the* signatures of turbulence. The first is imposed by the mechanisms responsible for the random motions in the first place, while the second is the reason the energy distributes only one way among scales. A number of mechanisms have been invoked over the decades for producing the large scale driving. So many, in fact, that the problem is not identifying which is the sole agent responsible for the motions but whether any signature remains in the turbulence of any single agent's predominance in any case. True universality

[6]On a technical point, this is because the picture presupposed the Burgers' equation, the one dimensional Euler equation with a viscous dissipation term (Burgers, J.M. 1948, "A mathematical model illustrating the theory of turbulence", Adv. Appl. Mech., 1, 171).

would imply that any such signs are washed out far enough from the *Taylor scale*, where the correlation function has its deflection point, but that requires that the motions be fully developed. The ISM displays such a wide range of processes and scales on which they act, that different environments may appear dynamically quite different. And also because of the drastically different environment in which it occurs, dissipation is a more complicated matter than in the laboratory. One key process missing in laboratory flows is radiative loss. Another is chemistry.

However, some features that make this a problem when discussing the translucent molecular clouds and the diffuse medium. Stirring is effected on a broad range of lengths and times and the amplitudes are also distributed over a wide range. The interaction of larger scales should inject vorticity, not merely one dimensional structures, into the flows in the clouds so such collisions are not one dimensional. The more serious issue, as we discussed for MHD turbulence, is the lack of a dissipation scale without which the fluctuations are just a set of random interactions. Heating should accompany shocks, along with compression, and this has been argued to leave a chemical signature in species synthesized in the post-shocked gas (e.g. Falgarone et al. 2006; Godard et al. 2009), especially CH^+, HCO^+, H_2O, and shock excited H_2 from the cold diffuse gas. Pan and Padoan (2009) found, from numerical simulations including a subgrid dissipation rate based on intermittency, that the heating from intersecting shocks can provide the necessary excitation to drive the CH^+ formation and account for at least some of the interstellar chemistry. An indication of this activity, whatever the spectrum (which is not sufficiently well resolved in most simulations to be sure) is the presence of extreme broadening in the molecular ion lines on the boundaries of clouds. The Pan and Padoan result requires an *rms* velocity of about 3 km s^{-1} over a scale of 1 pc. This is still larger than the shears inferred from translucent clouds, if we take those as the tracer of such flows. Consequently, the heating rate may be lower and the chemistry more sensitive to the detailed interplay of comic ray trapping in turbulence, phase transitions from cooling, and local ultraviolet irradiation.

11.5.4 Statistical Methods for Dynamical Analyses

The general theory of turbulence was developed for laboratory flows based on a very strong hypothesis: at some scale the flow loses all memory of its initial state and turns homogeneous. This works well, for example, in grid simulations in which a mesh is used to generate the fluctuations that interact at distance as they advect away from the grid.[7] The cascade resulting from collisions between the different scales imposed at the source find a dense set of resonances with which they interact and scatter. The resulting cascade is the one described by the K41 and related treatments,

[7]In many numerical simulations of non-decaying turbulence, the driving is effected in the model by gaussian stirring based on the power spectrum of the cascade.

as we have just seen. Is this at all relevant to interstellar flows? There are no grids in space, obviously, nor are there boundaries, and the dynamics are stirred from within by processes that are not altogether random. One issue is whether the velocity fields observed in these clouds qualify as turbulence in the sense of a self-similar cascade and, if so, what are the length (and time) scales on which the energy is injected and dissipated. Another is the observational signatures that can distinguish this from a simple random process. So let us pass to an examination of the observational tools.

The problems with observationally studying *interstellar flows*, compared to those more familiar from engineering or atmospheric physics, arise from the restriction that we only know the velocity distribution along the line of sight and that is weighted by the transparency of the gas.

We simply cannot see the "another side" or even know, precisely, at what depth we are sampling the flows, and a measurement of anisotropy is not possible.[8]

We can, however, adopt some of the basic techniques developed for the laboratory and see how far they can be unambiguously applied to this case. Consider, for instance, the measurement of the radial velocity, v at two points within the boundaries of the molecular emission, $\mathbf{x}$ and $\mathbf{x}' = \mathbf{x} + \Delta\mathbf{x}$. Here the displacement is in two dimensions. Were the distribution of the velocities simply Gaussian, the points separated by more than a resolution element would be uncorrelated. With increasing distance, this does not change. The dispersion in the velocities should be stable, independent of the distance, and the distribution function should become progressively more Gaussian as the sample size is increased. There should be no odd moments in the distribution, if it is isotropic, and there is a simple relation between the kurtosis and the dispersion, a relation between the dispersion and its square. If, however, we have a dissipative and driven process, even if the driving is perfectly random and on all scales, a signature will be evident in the correlation.

A spatial correlation function is often used in structural studies, in cosmology for instance. An example is the two-point correlation function (see Neyman and Scott 1952; Peebles and Groth 1976; Peebles 1993), that gives the probability of finding another object within some angular (or linear) distance of any other. Higher than second order correlations are also used, each providing information about a cumulant of the underlying statistical distribution. The higher moments are measures of the anisotropy of the spatial distribution. For example, the two point

[8] A side comment is in order here. If the flow shows significant variations in multiple line tracers in two dimensions, the flow is certainly non-isotropic on any larger scale than that over which such differences are measurable. Even poorly sampled mapping will show this. Nonetheless, density and kinetic variations conspire to produce the line of sight optical depth that is what is really measured in channel maps. Were it possible to change our viewing angle through a structure, this ambiguity could be removed. Many of the statistical methods developed for galactic structure and stellar dynamics are like those we have described in this chapter. But stars are not a continuous medium. Proper motion of a point source is an actual, not apparent, displacement in the plane of the sky. The motion can be assigned to a single mass. This is not true for clouds or diffuse gas.

spatial autocorrelation function is defined for any quantity Q by:

$$R(r) =< Q(x)Q(x+r) >= \int_{-\infty}^{\infty} Q(x)Q(x-r)dx. \tag{11.34}$$

This is written for a scalar field Q for which the spatial distribution is assumed to be isotropic, hence for scalar *lag*, r. Instead of the correlation function in space or time, it is often more convenient to treat the power in the turbulent component. This requires passing from space to wavenumber (as it would in going from time to frequency). This captures many of the essential features of the fluid interactions and has the formal advantage of changing convolutions into products through the Fourier transform. Noticing that $R(r)$ is such a convolution in space, its Fourier transform is the product of the amplitudes in wavenumber k (again a scalar)

$$P(k) = \frac{1}{2\pi} \int_{-\infty}^{\infty} R(r)e^{ikr}dk \tag{11.35}$$

so the Fourier transform of the two point velocity correlation function of an isotropic field is the energy spectrum. The same thing happens when considering the transformation of the conservation equations, where turbulence arises as interactions between waves within a broad spectrum. For an arbitrary scalar quantity without assumptions about isotropy, $P(k)$ is really an intensity measure, the relative prominence of any scale. The autocorrelation, if it does not behave like a δ-function as it should for Gaussian process, will decorrelate on some finite scale and, therefore, there should be a decrease of P with increasing spatial frequency. If, however, there is some characteristic scale, this will be a peak in the power spectrum.

Since turbulence is, by construction, a statistical property of a fluid, it is natural to ask what its probability distribution function (PDF) is. In Gaussian representations, the PDF is defined by a single parameter and all correlation functions are related to the same condition. The dispersion, the only non-vanishing independent moment under Galilean invariance, means that the amplitude of the fluctuations in velocity or energy at any scale are precisely known. There exists a wide range of functional forms that behave correctly, according to the standard axioms of probability, but that do not possess this property. One such class, called *Lévy processes* have the peculiar property that their dispersion is infinite. The simplest, a Lorentzian (also called a Cauchy distribution) has a finite integral over infinite range so the mean exists. In contrast, the second moment is *infinite* and depends on the interval over which it is computed so the dispersion increases without bound. What this means for a representation of turbulence is that if we take a particular velocity scale, the dispersion is fixed by that scale and as the range increases so does the amount of energy in the measured flow.

One way of measuring the turbulence PDF is provided by the line profiles themselves. If the line of sight samples a large enough number of fluctuations over the entire size range, at least up to the source scale, a homogeneous medium would show the same profile at all locations. Enlarging the solid angle should only change the intensity but not the profile so even low resolution observations should then give the energy in the medium since the dispersion is constant. Since the area (hence volume) changes the measured total energy increases but the energy density should remain constant. Even a large beam single antenna will then suffice. If, however, the PDF is not Gaussian, enlarging the sample increases the probability of seeing a large deviation and increases the dispersion so the energy density depends on the size of the sampled region. The same holds for the correlation function. If the fluctuations are uncorrelated on all scales, as would be the case for a Gaussian process, the zero should occur in one beamwidth whatever its angular size. But this is not turbulence because there is no spectrum of fluctuation sizes. In a cascade, the transfer of energy from one scale to another is necessarily correlated and there is a signature of that in the correlation function and its Fourier transform, the energy spectrum. Changes in T_b from one point to another do not contradict the uniformity of the dynamics, density fluctuations depend on the sample and the line of sight. But this can also affect the dispersion and mean of the sample if the number of elements is small.

In effect, *a PDF is the same as a line profile if the turbulent field is homogeneous.* That is, looking along any line of sight, the sample contains the integrated contributions of all radial velocity fluctuations. The frequency with which any deviation occurs with respect to the main flow is the same as the histogram of the fluctuations. In a gaussian process, for which the individual fluctuations are uncorrelated and have a maximum probability of zero and fixed dispersion, the *pdf* should be gaussian. In contrast, with both correlations and a power law distribution of the fluctuations, those at large value are more frequently seen than expected from a gaussian. hence, the profiles have long tails. The same holds for the *pdf* (see Fig. 11.2). The difference is that the line itself may also contain mean flows that distort the profile weighted along the line of sight by the line emissivity and integrated over the beam of the telescope.

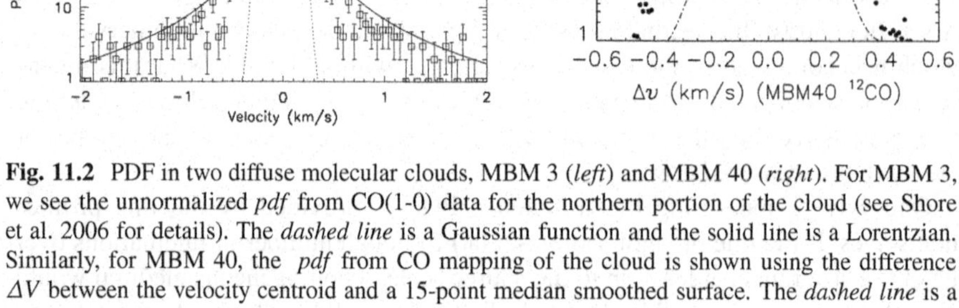

Fig. 11.2 PDF in two diffuse molecular clouds, MBM 3 (*left*) and MBM 40 (*right*). For MBM 3, we see the unnormalized *pdf* from CO(1-0) data for the northern portion of the cloud (see Shore et al. 2006 for details). The *dashed line* is a Gaussian function and the solid line is a Lorentzian. Similarly, for MBM 40, the *pdf* from CO mapping of the cloud is shown using the difference ΔV between the velocity centroid and a 15-point median smoothed surface. The *dashed line* is a Gaussian fit to the peak of the distribution (see Shore et al. 2003)

In any channel map, the intensity fluctuations per channel are insufficient to provide relevant density and dynamical diagnostics. This is because the profile may change with position inside a cloud boundary. Instead, integral measures—in this case moments of the profile—can be formed using the full intensity distribution at a given location. The limit imposed by the beam spacing determines the minimum scale in space on which the velocities can be resolved, and the assumption is that the emission does not vary in profile so rapidly that within a single beam (or half-beam) separation the change renders the moment useless. This last point is not trivial since the scale on which dissipation is presumed to occur in a turbulent cascade is almost certainly below the limit of the observational resolution. In forming the moments, it is best to use the intensity distribution without making an assumption of the form:

$$< v(\mathbf{x}) > = \int \Phi_v(\mathbf{x}) v dv \tag{11.36}$$

recognizing that this is the mean including any large scale flow that may be present. Any higher moment is then taken relative to this mean:

$$< (\delta v)^n >= \int \Phi_v(\mathbf{x})(v- < v >)^n dv \quad . \tag{11.37}$$

This is the deviation of the profile from the mean *within the beam* (e.g., Dickman and Kleiner 1985). For a Gaussian process with no correlation, all odd moments should vanish so within a single profile one can suspect that the presence of a skew is an indication that turbulence is responsible for the dynamics. However, there are other possibilities that cannot be excluded at this first pass observational stage. Systematic flows, since only the line of sight projected velocity is measured by the profile, can produce a deviation from a homogeneous structure. Only a change is angular sampling can rule this out. The same thing is true for a single channel in the data cube produced when mapping a cloud.

Although turbulence is not a discrete process, mapping is. To survey a cloud requires taking samples of a continuous velocity and density field in a grid with some fixed spacing and resolution, as determined by the point spread function of the telescope. To make matters worse, this is discrete in all three dimensions since the velocity is also sampled in channels of fixed bandwidth. The individual channels are, however, an *integral* along the line of sight, weighted implicitly by the individual emissivities of any structures. The emissivity, depending on density, abundance of the tracer, and temperature, hence the line profile averages over these. Thus, even if the medium is uniformly gaussian overall, individual deviations that are unrelated to turbulent motions may locally produce a deviation that appears to be the same as a departure from isotropic motions. These, if small enough, will decorrelate over some spatial scale. This should serve as a caution: discrete features always result from any turbulent flow when sampled discretely. That said, the signature of turbulence is that the moments should display a specific correlation function. As we discussed, this function is the real space statistic that becomes the wavenumber space energy spectrum. A power law autocorrelation function translates into a power law $E(k)$ in the cascade (inertial subrange). A useful aspect of the statistical approach is how it illuminates the underlying processes of the energy transfer. The ratio of the third moment to the second moment, $\left[< v^3 > / < v^2 > \right] \sim \epsilon l/E \sim l/t$, where E is the energy in the turbulent flow and here t is a transfer time between scales.

The velocity autocorrelation function (ACF) is conventionally obtained from the velocity centroids. It measures how the mean velocity of two points diverges with separation, either in space or time (Lumley 1967). In general, as we have been stressing, for cosmic sources, only one component of the velocity is available so this is not the same as the function for which the Fourier transform yields the power spectrum. The ACF is therefore defined in two dimensions and for one velocity component (here $\mathbf{x}$ refers to angular position with integration along the line of sight assumed)

$$C(\mathbf{x}) = \int d\mathbf{x} < v(\mathbf{x} + r) > \cdot < v(\mathbf{x}) > \tag{11.38}$$

This is a one beamwidth δ-function for a gaussian distribution. For more complex cases, the function can also turn negative. Take, for instance, an ensemble of vortices. At some distance there is a chance of finding, for the largest eddies, an anticorrelation so C can turn negative.

The structure function, introduced in Sect. 11.5.3.1, is similar to the moments of the PDF in that it is the mean value of different powers of the velocity fluctuations taken over the whole map. Although it is not a correlation function, its definition shows that it is related to $C(\mathbf{r})$,

$$S_p(\mathbf{x}) =< |\delta v(\mathbf{x} + r) - \delta v(\mathbf{x})|^p > \tag{11.39}$$

for an arbitrary moment p. This difference contains the mean velocity at each point in the sample so it is again essential that any systematic large scale motions be removed by some smoothing procedure. The correlation functions (CFs) suffer from finite domain effects, there are no correlations between the interior (mapped) and exterior (unmapped) domains, yet the integrals must be taken over the entire domain. Thus, the CF methods require either imbedding the mapped region in some smoothly connected background or limiting the range over which the integral is taken to avoid the boundaries. The structure function is, instead, a cumulant that gives the range of fluctuations in space so this does not require the separation of the flow into components or even dense sampling. Any correlated motions will fall out naturally. It is insensitive to the boundaries because it is defined for each point and uses only the observed values, without requiring embedding.

Returning to the cascade, it is important to remember that the rate of energy transfer is assumed to be independent of scale, in a steady state, and structure function was the representation chosen in the first treatments of the cascade (e.g. von Karmann and Howarth 1938; Kolmogorov 1941*a*,; Batchelor 1953). They can be formed for each component separately, as can the correlation functions, but for isotropic turbulence there is a general relation that allows for the recovery of the (unmeasurable) transverse component when the longitudinal (line of sight) function is available [see, e.g. Pope (2002)]. Other methods attempting to encapsulate the velocity field include Δ-variance (Stutzki et al. 1998; Ossenkopf et al. 2001), wavelets (Farge 1992), and Principal Component Analysis (also called Proper Orthogonal Decomposition—see §4.3.2) (Berkooz et al. 1993; Brunt et al. 2013). They have the common aim of producing robust *statistical* characterizations of the velocity and density variations.

In a continuous spectrum, the ACF varies as a power law in space, so its Fourier transform, the power spectrum, is a power law in wavenumber. Passing now to a component of the velocity, in three dimensions at each $\mathbf{x}$ we have three components of $\mathbf{v}$ so there both the auto- and cross-correlations can be formed between the different components, $R_{ij}(\mathbf{r})$. Similarly, cross-correlations exist between other primitive variables of the flow. But maps of the ISM are made in position and always within limited regions. Although wavelets are less sensitive than Fourier transforms to the finite domain effects, they too are not immune. There is the problem of knowing what v means in the astronomical case. In the laboratory, we can choose a specific point in the flow in three dimensions and know, to the accuracy of the measurement, the point components of the fluctuation in time or its snapshot variation in space. But in astronomical observations we are forced to measure the temporal snapshot line profile from which we can form channel maps that show the change in structure with radial velocity. The aggregate statistic is the mean velocity derived from the profile itself within some interval and the line width. Thus, we define the profile to have an intensity distribution $I_{ij}(\Delta v)$ at each point in a map (i, j) and the mean and variance are obtained from the intensity-weighted integrations over the velocity interval

$$< v_{ij}^n >= \int (v_{ij})^n I_{ij}(\Delta v) d\Delta v. \tag{11.40}$$

This mean depends on the line formation that depends on local conditions, such as excitation temperature and optical depth. Were we to use an optically thick flow tracer, for example CO(1-0) in regions of high column density, the mean would be biased toward the surface layers of the flow along the line of sight. If, instead, a weak line or an isotope that has a sufficiently small abundance (yet high signal) is chosen, the statistical characterization more closely approaches more the actual distribution of velocity and density along the line of sight.

These must be the *deviations* from any organized, large scale flows in the medium. Actually, this is the hardest part of correlation analyses in translucent clouds. It is tantamount to the application of the Reynolds decomposition to a one dimensional measure of a three dimensional velocity field. Since we can only determine the line of sight velocities, before attempting to obtain R it is necessary to remove the mean velocity, to change the form of v to δv such that $< \delta v >= 0$. How this is done depends on the extent of the survey. The confines of the cloud must be well delineated and, more important, it must be possible to obtain the large scale flows, if any, and remove them. In the laboratory this isn't a problem since that is not what is measured; you have a temporal spectrum of the fluctuations from the distribution of combs along the flow. Each provides a snapshot at different distances so the correlations can be taken in space at a fixed instant or in time at any one of the sites. In the interstellar case that is not possible so the greater the lag, the more of any large scale flow (hence skew) will be included in the correlation. Since the signature of turbulence is the non-vanishing third moment, the skew, deriving the energy injection in the cascade requires careful analysis.

11.5.5 Empirical Line Width-Size Relations

Much of the work on interstellar turbulence traces back to the classic study by Larson (1981) of the velocity dispersion in clouds. The work is one of the most cited papers in interstellar studies and the paradigm of many modeling and empirical analyses. The original data set was both heterogeneous—several different gas tracers were used—and based on a variety of interpretations of the *linewidth*. What emerged, though, was a simple power law of the velocity, v, as function of the linear scale of the cloud (L) spanning 0.1 to 100 pc

$$\sigma(km\ s^{-1}) = 1.10L(pc)^{0.38}. \tag{11.41}$$

Significantly, this law was asserted to hold both within *and* between cloud complex for a range of tracers. Larson used mainly ^{13}CO to avoid self-absorption problems and assumed isotropy when interpreting the velocity dispersions. These, in turn, were taken from the literature so there is a variety of meanings for the parameter, as we have discussed. If the profiles are gaussian then dispersion is arithmetically linked to the FWHM. The skew is not given for the individual regions and size scales are averaged, so this is not the same as a structure function. A turbulence origin for the relation is easily imagined from the tenets of K41. Taking Δv to be the velocity fluctuations on some scale l, then the constancy in the cascade of ϵ leads to $\Delta v \sim l^{1/3}$. There is no supersonic motion in the cascade, nor is there radiative dissipation. Systematic motions are also ignored, the medium is assumed to be homogeneous and confined. The conditions spanned by the clouds covered by the relation range from minimally bound (perhaps unbound) structures within individual translucent clouds to the scale of giant molecular complexes and self-gravitating cores. No account is taken of optical depth effects or of systematic motions. The possible role of virialization was discussed *if* the masses were measured independent of the size scale. That is, assuming the mass is known on a scale L, $\sigma^2 \sim M/L$ which, Larson noted, is an exact relation for polytropes with $\gamma = 3/2$.

Subsequent studies, both observational and through numerical simulations. have shown that the laws are neither fundamental nor universal but *are* indicative. From three dimensional supersonic direct numerical simulations, Kritsuk et al. (2013) found that the velocity-size relation is indeed expressing a cascade. Heyer and

Brunt (2004) performed structure function and principal component analyses (see Sect. 4.3.2) of an ensemble of clouds ranging in size from 0.1 to 50 pc. Their findings for a significant sample (27 clouds and cloud complexes) support the notion that, independent of self-gravity, there is a self-similar interval. They further proposed that the turbulent spectral energy distribution is invariant on these size scales ranging from a few hundredths of a pc to nearly 50 pc. This despite the degree to which the gas might be self-gravitating. In general, the turbulence in these clouds is driven and this is where the greatest divergence should occur in the various simulations. Magnetic fields appear to be the "significant other".

11.5.6 Some Open Questions

So ultimately what do the velocities observed in the translucent clouds indicate, turbulence or some other stirring? Without the essential indication of some scale on which the medium is dissipative, or some process that accelerates the instability of a cooling medium capable of becoming self-gravitating, we cannot definitively say. The spectrum would be important but, in general, is not accessible over a sufficient range of sizes to be certain. Any velocity width-size relation that spans physical environments is suggestive but again not definitive.

It is important to emphasize that there are no observation of the structure of any cloud yet so clean as to distinguish unambiguously between these different spectra. On the larger scale of observable structure, even on the scale of the cascade, the only observable features relate to the widths of the line profiles (that they are more consistent with the Alfven speed than hydrodynamic turbulence or thermal speed) and the spectrum. Since, in general, the sampled scales do not extend more than about one decade in wavenumber, the identification of the inertial range is possible but its spectrum is poorly determined.

References

Batchelor, G.K. 1953, *The theory of homogeneous turbulence*, (Cambridge: Cambridge University Press)
Berkooz, G., Holmes, P., and Lumley, J.L. 1993, ARFM, 25, 538
Bonor, W.B. 1956, MNRAS,116, 35
Bonor, W.B. 1958, MNRAS, 118, 523
Brunt, C.M. and Heyer, M.H. 2013, MNRAS, 433, 117
Chandrasekhar, S. and Fermi, E. 1953, ApJ, 118, 116
Chandrasekhar, S. 1967, *An introduction to the study of stellar structure*, (New York: Dover)
Churchwell, E. 2002, ARAA, 40, 27
Craink, A.D.D. 1988, *Wave interactions and fluid flows*, Cambridge University Press
Dickman, R.L. and Kleiner, S.C. 1985, ApJ, 295, 479
Ebert, R. 1957, ZA, 42, 263
Elmegreen, B.G. and Scalo, J. 2004, ARAA, 42, 211

Elsässer, W.M. 1950, Phys. Rev., 79, 183
Falgarone, E., Pineau Des Forîs, G., Hily-Blant, P., and Schilke, P. 2006, A&A, 452, 511
Farge, M. 1992, ARFM, 24, 395
Field, G.B. 1965, ApJ, 142, 531
Field, G.B., Goldsmith, D.W., and Habing, H.J. 1969, ApJ, 155, 149
Frisch, U. 1995, *Turbulence. The legacy of A.N. Kolmogorov*, (Cambridge: Cambridge University Press)
Godard, B., Falgarone, E., and Pineau Des Forêts, G. 2009. A&A, 495, 847
Goldreich, P. and Sridhar, S. 1995, ApJ, 438, 763
Goldreich, P. and Sridhar, S. 1997, ApJ, 485, 680
Harrison, B.K., Thorne, K.S., Wakano, M., and Wheeler, J.A. 1965, *Gravitational Theory and Gravitational Collapse*, (Chicago: U. of Chicago Press)
Heyer, M.H. and Brunt, C.M. 2004, ApJ, 615, L45
Hunter, J.H., Jr. 1966, MNRAS, 133, 239
Iroshnikov, P.S. 1963, Soviet Astron., 7, 566
Kennicutt, R.C., Jr. 1983, ApJ, 272, 54
Kolmogorov, A.N. 1941, Dokl. Akad. Nauk SSSR, 26, 115
Kolmogorov, A.N. 1941, Dokl. Akad. Nauk SSSR, 30, 301
Kolmogorov, A.N. 1962, J. Fluid Mech, 13, 82
Kraichnan, R.H. 1965, Phys. Fluids, 8, 1385
Kritsuk, A.G., Wagner, R., and Norman, M.L. 2013, J. Fluid Mech., 729, R1
Larson, R.B. 1981, MNRAS, 194, 809
Lesiur, M. 2008, *Turbulence in Fluids*, Kluwer Academic Publishers
Lumley, J.L. 1967, in *Atmospheric Turbulence and Radio Wave Propagation*, eds. A.M. Yaglom and V.I. Takarski, (Moscow: Nauka), 166
MacLaren, I., Richardson, K. M., and Wolfendale, A. W. 1988, ApJ, 333, 821
Mac Low,M.-M. and Klessen, R.S. 2004, RMP, 76, 125
McComb, W.D. 1990, *The Physics of Fluid Turbulence*, Clarendon Press
McComb, W.D. 2014, *Homogeneous, Isotropic Turbulence*, (Oxford: Oxford Science Publications)
McKee, C.F. and Zweibel, E.G. 1992, 399, 551
Mininni, P. D. and Pouquet, A. 2009, Phys. Rev. E, 80b, 5401
Montgomery, D. and Matthaeus, W.H. 1995.ApJ, 447, 706
Neyman, J. and Scott, E.L. 1952, ApJ, 116, 144
Ng, C.S. and Bhattacharjee, A. 1996, ApJ, 465, 845
Norman, M.L., Reynolds, D.R., and So, G.C. 2009, AIP Conference Proc., 1171, 260
Obukhov, A.M. 1962, JGR, 67, 3011
Ossenkopf, V., Klessen, R.S., and Heitsch, F. 2001, A&A, 379, 1005
Pan, L. and Padoan, P. 2009, ApJ, 692, 594
Peebles, P.J.E. and Groth, E.J. 1976, A&A, 53, 131
Peebles, P.J.E. 1993, *Principles of Physical Cosmology*, (Princeton: Princeton University Press)
Pope, S.B. 2002, *Turbulent Flows*, Cambride University Press
Porter, D., Pouquet, A., and Woodward, P. 1992*a*, *Theoretical Computational Fluid Dynamics*, 4, 13
Porter, D., Pouquet, A., and Woodward, P. 1992*b*, Phys. Rev. Lett., 68, 3156
Saffman, P.G. 1971, Studies Appl. Math., 50, 93
Sagdeev, R.Z. and Galeev, A.A. 1969, *Nonlinear Plasma Theory*, (New York: Benjamin)
Shebalin, J.V., Matthaeus, W.H., and Montgomery, D. 1983, JPlPh, 29, 525
Shore, S.N., Magnani, L, LaRosa, T.N., and McCarthy, M.N. 2003, ApJ, 593, 413
Shore, S.N., LaRosa, T.N., Chastain, R.J., and Magnani, L. 2006, A&A, 457, 197
Sridhar, S. 1995, NYASA, 773, 44
Stutzki, J., Bensch, F., Heithausen, H., Ossenkopf, V., and Zielinsky, M. 1998, A&A, 336, 697
Toomre, A. 1964, ApJ, 139, 1217
von Karmann, T. and Howarth, L. 1938, Proc. Roy. Soc. London, 164, 192
Whitham, G.B. 1974, *Linear and Nonlinear Waves*, (New York: John Wiley & Sons)

Chapter 12
Coda

All's well that ends well

Abstract Molecular clouds, especially diffuse and translucent ones are very difficult to define. We may not be able to define them precisely, but we know them when we see them.

12.1 What Is a Molecular Cloud?

The famous songwriter Joni Mitchell expressed it best in her song *Both Sides, Now*, first recorded by Judy Collins in 1967, and subsequently by Mitchell in her album "Clouds", which was released in 1969. The relevant portion of her song was quoted by Mark Heyer and Tom Dame at the beginning of their excellent 2015 review article on molecular clouds. Rather than repeating their quotation, we decided to have a little fun with the lyrics. And so, with apologies to Joni Mitchell and song writing in general, to the tune of *Both Sides Now*,

With CO, CH, and OH lines
Continuum emission of all kinds
and even with atomic lines
I've studied clouds that way

Turbulent or virialized
With models realistic or contrived
Their characteristics I've derived
But how well I cannot say.

I've studied clouds of all types now
Both near and far and still somehow
It's observations I recall
I really don't know clouds At all.

L. Magnani, S.N. Shore, *A Dirty Window*, Astrophysics and Space Science Library 442, DOI 10.1007/978-3-662-54350-4_12

Maybe we can't quite define exactly what molecular clouds are–especially the diffuse ones –but we know them when we see them.

Reference

Heyer, M.H. and Dame, T.M. 2015, ARAA, 53, 583

Appendix A
Regulatory institutions for Radio Frequency Interference (Chap. 4)

In the United States, the regulatory commisions involved in adjudicating the use of the electromagnetic spectrum are the FCC for non-federal-government uses and the National Telecommunications and Information Administration (NTIA) which regulates federal government use. At the international level, the Radiocommunications Sector of the International Telecommunications Union (ITU-R) formulates regulations through World Radiocommunication Conferences which are held every 3-5 years. The regulatory work relevant to radio astronomy takes place through a study group which is organized into working parties and task groups. Technical and procedural aspects of radio communications occurs under Study Group 7 which is responsible for the use of the electromagnetic spectrum for scientific research. In particular, radio astronomy is the concern of Working Party 7D.

Additional scientific input to Study Group 7 is provided by the International Council of Scientific Unions (which operates under a mandate from UNESCO) via the Scientific Committee on Frequency Allocations for Radio Astronomy and Space Science (IUCAF). IUCAF is tied to the astronomical community because it is sponsored by the IAU (and also URSI and COSPAR).

Additional information on the regulatory structure of the radio portion of the electromagnetic spectrum can be found in *Handbook of Frequency Allocations and Spectrum Protection for Scientific Uses* a publication of the National Research Council of the National Academies, The National Academies Press (2007).

L. Magnani, S.N. Shore, *A Dirty Window*, Astrophysics and Space Science Library 442, DOI 10.1007/978-3-662-54350-4

Appendix B
Radio Band Designations (Chap. 4)

The table below lists the radio frequency bands according to various standards (Table B.1).

L. Magnani, S.N. Shore, *A Dirty Window*, Astrophysics and Space Science Library 442, DOI 10.1007/978-3-662-54350-4

Table B.1 Radio band designations

IEEE band designation	*Frequency range*
HF	0.003 to 0.03 GHz
VHF	0.03 to 0.3 GHz
UHF	0.3 to 1 GHz
L	1 to 2 GHz
S	2 to 4 GHz
C	4 to 8 GHz
X	8 to 12 GHz
Ku	12 to 18 GHz
K	18 to 27 GHz
Ka	27 to 40 GHz
V	40 to 75 GHz
W	75 to 110 GHz
mm	100 to 300 GHz
NATO band designations	*Frequency range*
I	100 to 150 MHz
G	150 to 225 MHz
P	225 to 390 MHz
L	390 to 1550 MHz
S	1.55 to 3.9 GHz
C	3.9 to 6.2 GHz
X	6.2 to 10.9 GHz
K	10.9 to 36.0 GHz
Ku	10.9 to 20.0 GHz
Ka	20.0 to 36.0 GHz
Q	36.0 to 46.0 GHz
V	46.0 to 56.0 GHz
W	56.0 to 100.0 GHz
Waveguide band designations	*Frequency range*
R	1.70 to 2.60 GHz
D	2.20 to 3.30 GHz
S	2.60 to 3.95 GHz
E	3.30 to 4.90 GHz
G	3.95 to 5.85 GHz
F	4.90 to 7.05 GHz
C	5.85 to 8.20 GHz
H	7.05 to 10.10 GHz
X	8.2 to 12.4 GHz
Ku	12.4 to 18.0 GHz
K	15.0 to 26.5 GHz
Ka	26.5 to 40.0 GHz
Q	33 to 50 GHz
U	40 to 60 GHz
V	40 to 75 GHz
W	75 to 110 GHz
F	90 to 140 GHz
D	110 to 170 GHz
Y	325 to 500 GHz

Appendix C
The Kramers-Kronig Relations and the Relation Between C_{abs} and C_{sca} Through Reciprocity (Chap. 6)

The reciprocity of the real and imaginary parts of the magnetic permittivity, dielectric constant, and actually any linear response function is the method by which optical constants are obtained for solids. The derivation is based on only one broad assumption, that the reaction of the substance (in this case, the induced dipole of a solid) is linear in the forcing, an example of which is an incident electromagnetic field. Rotation of a dust grain in an ambient magnetic field with lagging permittivity is another. Since we are discussing the optical properties of the dust, we will focus on the polarization,

$$P(t, \mathbf{x}) = \int_{-\infty}^{\infty} dt' \int_{-\infty}^{\infty} d\mathbf{x}' \alpha(\mathbf{x} - \mathbf{x}', t - t') \mathbf{E}(\mathbf{x}', t') \quad \text{(C.1)}$$

Since time and frequency are conjugate variables, the frequency dependent polarizability is related to the susceptibility as a product with the electric field, so the emissivity (which is the square of the polarization) is directly related to the incident intensity. The Kramers-Kronig relations (KKR) connect the two parts of the (generally complex) refractive index (refraction/scattering and absorption), hence it suffices to have laboratory measurements of the absorption to compute the scattering terms. This applies only to bulk properties and single composition material (at least in basic modeling). Multiple (grain) contributors to the interstellar extinction law are each treated separately and their contributions add to give the optical depth weighted by the fractional abundance of each species. For molecules, however, one cannot talk about bulk characteristics and, for example, fluorescent processes are possible (as we discussed in Chap. 2 for the small grains). Linear theory, the heart of this approximation and the KKR, breaks down. But to return to the grains, taking the Fourier transform of Eq. (C.1)

$$\mathbf{P}_\omega = \hat{\alpha}_\omega \mathbf{E}_\omega \quad \text{(C.2)}$$

L. Magnani, S.N. Shore, *A Dirty Window*, Astrophysics and Space Science Library 442, DOI 10.1007/978-3-662-54350-4

assuming α is homogeneous (that we can ignore spatial variations in the internal properties). The real and complex parts are then related to each other in a general way. The scattering part can be expressed in terms of the grain albedo which is the ratio of the scattering to the extinction cross-section or, alternatively, Q_{scat}/Q_{ext}. With this formulation, purely absorbing grains would have an albedo of 0, whereas purely scattering grains would have an albedo of 1. Icy particles tend to be excellent scatterers so that $Q_{scat} \gg Q_{abs}$, but even for these particles the albedo is not zero. They absorb some radiation and eventually re-emit it in the infrared. For a dissipative system, the susceptibility and dielectric are both complex and to be as general as possible,

$$\mathbf{P}(t) = \int_{-\infty}^{\infty} \chi(t - t')\mathbf{E}(t')dt'. \tag{C.3}$$

Note that we now explicitly assume a homogeneous material so spatial dependences are neglected. The usual statement, that the system is required to obey causality, means that for χ, any reaction of the medium must come from a "positive time", that is for $t - t' > 0$. Using Fourier transform translates the relation into a set of linear relations in the frequency domain since the integral is a convolution in the time domain. Thus

$$\mathbf{P}(\omega) = \chi(\omega)\mathbf{E}(\omega) \tag{C.4}$$

By Cauchy's residue theorem

$$\chi(\omega) = \frac{1}{2\pi i}\int \frac{\chi(\omega')d\omega'}{\omega' - \omega} \tag{C.5}$$

Then changing from the general response to the dielectric function and noting that it is complex, $\chi \to \epsilon_r + i\epsilon_i$, we have

$$\epsilon_r(\omega)) = \frac{1}{2\pi}\int \frac{\epsilon_i(\omega')d\omega'}{\omega - \omega'} \tag{C.6}$$

with a corresponding integral for ϵ_i. Recall that $\epsilon = n^2 - \kappa^2 + 2in\kappa$ since the imaginary part of the refractive index is the absorption coefficient. The absorption is most easily measured in the laboratory using transmission spectra and by using the KKR can be used to obtain the frequency dependent scattering coefficient. Further, irradiation and transmission, if the bands are excited, can be used to measure $\kappa(\omega)$ since the emissivity from solids at temperature T is $j_\omega = \kappa_\omega B_\omega(T)$.

Index

A

Absorption coefficient, stimulated emission, 34
Anomalous microwave emission. *See* Dust, Anomalous microwave emission (AME)
Antenna temperature, 109–111, 119–121
Atomic coolants, 14
Atomic fine structure emission. *See* Fine structure atomic transitions
Atomic gas, abundances, 58
Autocorrelation function (ACF), 127. *See also* Correlation function

B

Boltzmann distribution, definition of LTE, 33
Bonor-Ebert sphere, 272–273
  mass of, 273
Brightness temperature, 34, 119

C

Carbon monoxide (CO)
  absorption studies, 240–241
  excitation by molecular hydrogen (X_{co}), 200, 206–214
  formation of, 78, 174–177, 239
  low emissivity gas, 214–223
  and *Planck* maps, 132–137
  rotational transitions, 84–85
  telluric, 114
CH^+
  and extinction, 212–214
  optical transitions, 97–102
  rotational transitions, 91
CH^+, optical transitions, 98
Cirrus. *See* Dust, cirrus
Classical damping rate, 35
Cloud, definition of, 42, 495
CMB. *See* Cosmic background radiation (CMB)
CN, 47–48, 50
  optical transitions, 98
CO. *See* Carbon monoxide (CO)
Cold neutral medium (CNM), 9, 14–16, 77
  thermal velocities, 14, 77
Collisional excitation and de-excitation, 33, 49–50
  CN excitation, 50–51
  critical density, 49
Color excess (R), 158. *See also* Dust, extinction
Column density
  from absorption line equivalent width, 40
  from emission line, 42–43, 45
Compact H II regions, 276
Coolants, atomic (C^+), 14
Correlation function, 287, 290
Cosmic background radiation (CMB)
  excitation of optical CN rotational lines, 47
  space observations of, 41–42, 132–138
  temperature, 50
Cosmic rays, 23–24
  composition, 23
  energy distribution, 23
  gamma-ray production, 151–152, 178–180, 210–211
  gas to dust ratio, 178–181
  ionization, 81–83

L. Magnani, S.N. Shore, *A Dirty Window*, Astrophysics and Space Science Library 442, DOI 10.1007/978-3-662-54350-4

Curve of growth method, 36
illustrated, 41
Cyanogen. *See* CN

D
Dark gas. *See* Carbon monoxide (CO), low emissivity gas
Depletion factor, definition, 58
DIBs. *See* Diffuse interstellar bands (DIBs)
Diffuse Galactic ultraviolet emission, 163
Diffuse interstellar bands (DIBs), 63, 99, 100
Dispersion measure, 65
Draco complex. *See* intermediate velocity clouds
Dust
anomalous microwave emission (AME), 100, 169, 170
atomic depletion on grains, 57–60
cirrus, 140, 235–239
extinction, 157–159
extinction and Wolf diagram, 255–258
extinction efficiency, 160
extinction per baryon relations, 162, 173
far infrared, 131–144
ice spectra, 177, 178
and light echos, 165–168
and neutral hydrogen, 46, 47
and polarization, 168, 169
ratio to gas, 156
size distribution, 165, 171, 172

E
Einstein transition probabilities, 33
Emission measure (EM), 12
Energy loss function, 9, 276
Equivalent width
definition, 39
and optical depth, 40
Extended red emission, 162
Extinction per baryon. *See* Dust

F
Faraday depth, rotation measure, 65
Fine structure atomic transitions, 14, 145
Fluorescence, 62–64
Flux, 31
units of, 109
Formaldehyde. *See* H_2CO, rotational transitions
Front-and back-end, radio telescope, 109

G
Galactic fountain, 11, 243
Galactic structure, local standard of rest. *See* LSR, radial velocity
Galactic surface density, stars, 254
Giant molecular clouds, 190–192
lifetime of, 191
Gravothermal instability, 273–275

H
H_2CO, rotational transitions, 94
Historical context, 4–7, 12, 20, 22, 70–72, 131–144, 146–151, 186–190, 227–246
Hot ionized medium (HIM), 10, 11
Hydroxyl. *See* OH

I
Intensity, 30
units of, 109, 120
Intermediate velocity clouds, 245, 263
Interstellar medium, states of, 2, 9–12
Interstellar radiation field, 19
Isothermal slab, 251–253
vertical acceleration, 252
Isothermal sphere, 271–273

J
Jeans length, 273–275

K
Kolmogorov length, 282
Kolmogorov spectrum, 279, 282

L
Larson law. *See* Line width-size relation
Line profile, 34–38
Cygni, P., 54
Gaussian, 35
Lorentzian, 36
Voigt profile, 36
Line width-Size relation, 292–293
Local thermodynamic equilibrium, 33
LSR, radial velocity, 250

M
Magnetic fields
Faraday depth, rotation measures, 65
line profiles, 117
measurement of in diffuse medium, 65

Mean intensity, 30
Methylidyne. *See* CH^+
Mie scattering, 161
Molecular clouds
classification, 17, 20–21, 192, 201
dark, 193–196
definition of, 24–25, 76, 271
diffuse, 199–201
distribution in isothermal slab model (*see* Isothermal slab)
Galactic distribution, 261–263
in Galactic halo (*see* Intermediate velocity clouds)
Galactic scale height, 254
high latitude, Galactic hemispheres, 242–244
H_2 self-shielding, 18, 19
infrared tracers, 246
intermediate velocity, 245
lifetimes, 17
masses, virial theorem estimates, 124–125
self-gravitation parameter, 17–18
surface density, 240–244
translucent, 192, 193, 196–198
Molecular hydrogen (H), 71, 146
dissociation mechanism, 61, 74
ortho-to-para ratio, 49, 74
self-shielding, 18, 19
UV absorption bands, 55, 148–149
Molecular reactions, gas phase, 76–81
Molecular ring, 187, 189
Molecules
ionization and dissociation energies, 73
reaction rates, 73–77
relative abundances, 103
typical values, 76

N
Neutral hydrogen (H I)
21 cm, 43–47
21 cm emission surveys, 46
column density from 21 cm line brightness temperature, 16, 146–147, 211–212
continuum absorption, 47, 180
Einstein A-value, 44
and extinction, 45, 47, 162, 180
narrow self-absorption lines (HINSA), 192
North Celestial Pole Loop, 238

O
OH
and CO dark gas, 221–223
rotational transitions, 85
tracer of molecular gas, 87–90
On-the-fly (OTF) mapping, 123
On the spot approximation, 53–54. *See also* Sobolev approximation
Optical depth, definition, 31
Oscillator strength, 35

P
PAH, 165, 236
and bond stretching, 62–64
Photodissociation region (PDR), 2, 18–19, 145, 217–218
Photometry, Johnson-Cousins filters, 158, 159
Photon dominated region. *See* Photodissociation region (PDR)
Planck distribution, 33
Plasma frequency, 65
bifringence, 64–66
index of refraction, 65
Polaris Flare, 240
Polarization. *See* Stokes parameters, definition
Polycyclic aromatic hydrocarbons. *See* PAH
Principal component analysis, 125–126
Probability distribution function, 128, 285–292

Q
Q-parameter, 159. *See also* Dust, extinction

R
Radiation temperature, 120–121
Radiative transfer
equation of, 30, 32, 51–53
escape probability methods, 51–54 (*see also* On the spot approximation)
photon pumping, 62–64
velocity desaturation, 54
Radio telescope
antenna temperature, 119–121
detection, 109
flux units, 109, 120
main beam efficiency, 118
point spread function, 31, 118
Rayleigh scattering, 161
Reddening. *See* Dust, extinction
Red luminescence, 100–102
Resonance lines, atomic, 39–42, 50–51
Reynolds layer, 13
Reynolds number, 281
Rotation measure, 65. *See also* Faraday depth, rotation measure

S

Serkowski distribution, polarization, 168–169
Sobolev approximation, 54. *See also* Radiative transfer, escape probability methods
Source function
  definition, 31
  LTE, 33
Spin temperature (21 cm), 44
Standard cloud (Spitzer), 199
Star formation, star-gas cycle, 4
Statistical equilibrium, 32–34
Stokes parameters, definition, 115
Stress tensor, 270
Structure function, 127, 282, 290
Surface brightness. *See* Intensity
Surface density
  gas, 12
  H_2, 3, 237, 240, 243, 244
  stellar, 3

T

Thermal instability, 9–10, 275–277
Three-wave interactions, 278–279
Toomre criterion, 274–275
Turbulence
  cascade, 281
  hydromagnetic, 278–279, 283–284
  source scale, 281
  supersonic, 284–285
Two phase ISM, 10, 277

U

Ursa Major clouds. *See* North Celestial Pole Loop

V

Virial theorem, 124–125, 268–270
  cloud masses, 208–210, 268, 292

W

Warm neutral medium (WNM), 9, 11, 13–14, 77
Wolf diagram, 225–258. *See also* Dust

X

X_{CO}, definition, 206. *See also* Carbon monoxide (CO), excitation by molecular hydrogen (X_{co})

CPSIA information can be obtained
at www.ICGtesting.com
Printed in the USA
LVOW06*1248230717
542326LV00001B/154/P

9 783662 543481